Detlef Egbert Ricken · Wolfgang Gessner (Eds.)

Advanced Microsystems for Automotive Applications 98

Springer-Verlag Berlin Heidelberg GmbH

Detlef E. Ricken • Wolfgang Gessner (Eds.)

Advanced Microsystems for Automotive Applications 98

With 253 Figures

Springer

Dr. Detlef Egbert Ricken
VDI/VDE-Technologiezentrum Informationstechnik GmbH
Rheinstr.10B
14513 Teltow
e-mail: ricken@vdivde-it.de

Wolfgang Gessner
VDI/VDE-Technologiezentrum Informationstechnik GmbH
Rheinstr.10B
14513 Teltow
e-mail: gessner@vdivde-it.de

Die Deutsche Bibliothek - CIP-Einheitsaufnahme

Advanced Microsystems for Automotive Applications 98 / Eds.: Detlef Egbert Ricken; Wolfgang Gessner. - Berlin; Heidelberg; New York; Barcelona; Budapest; Hongkong; London; Milano; Paris; Santa Clara; Singapore; Tokyo: Springer, 1998

ISBN 978-3-662-38795-5 ISBN 978-3-662-39696-4 (eBook)
DOI 10.1007/978-3-662-39696-4

Originally published by Springer-Verlag Berlin Heidelberg New York in 1998.
Softcover reprint of the hardcover 1st edition 1998

Typesetting: Camera-ready by authors
Cover-Design: de'blik, Berlin
SPIN 10666963 68/3020-5 4 3 2 1 0 - Printed on acid -free paper

Preface

The idea of a conference dedicated to microsystems for automotive applications was born at the Innovation Centre North Germany in early 1995 in order to improve communications in the supply chain..
A conference programme was drawn up a few months later with the support of H. Ehlbeck. The main topic of the first conference was sensors for safety applications. This event revealed the need for a discussion platform for microsystems in the automotive field and the conference 'Advanced Microsystems for Automotive Applications' was held subsequently in 1996 and 1998. The conference has grown from a event similar to a workshop to a large conference including contributions from Europe, Japan and the United States. Global suppliers and manufacturers as well as SME's and research institutes present their technology and solutions. This proceedings volume includes both oral and poster presentations from the 1998 conference. The technical topics extend from microsystems for safety applications and systems for driver support to intelligent systems for the power train. The main emphasis has shifted from the pure microsystem, such as the accelerometer, to the system as a whole. Thus the papers presented during the conference deal with all aspects of a smart system: sensors, communication, signal evaluation and evaluation strategies and packaging.
Future restraint systems will utilize a wide variety of microsensors, e.g. accelerometers, yaw-rate-sensors, subsystems for occupant detection (passenger-presence, -position -size and -mass), precrash-sensors and intelligent inflator technology for depowered airbags. Some smaller companies, as well as the well-known suppliers, present their developments within the system context. The inclusion of the presentation on pressure sensor mats is a good example for the system oriented approach of the conference; although the mat is not fabricated with state of the art microsystem technology it is an important component for intelligent occupant detection systems and the paper fits well into the sessions on passenger protection.
Driver support systems have always been a major topic of the AMAA conference. The AMAA '98 discusses systems for electronic imaging and vision enhancement. State of the art systems realised with CMOS or CCD technology improve driving safety (e.g. in night vision systems or in blind spot monitoring sensors) and offer an unparalleled degree of convenience for the driver (e.g. automatic park systems etc.) Initial applicatons are approaching the stage of preliminary production. The blind spot monitoring device which is described in this volume could, for example, be produced for evaluation in 1998 and large scale production could commence before 2001.
Although initial products will be available soon, electronic imaging for automotive applications is a very challenging task. Various technological questions have to be solved to allow for cost effective solutions even for non-luxury cars. The TFA or

„Thin Film on ASIC" technology discussed in one of the papers could be an interesting approach for meeting the target prices. Since the various vision systems are not as close to application as the passenger restraint systems the corresponding papers provide a discussion of more general aspects of the technology involved.
Another important field of application for microsystems is engine management and the power train. The use of micromachined sensors enables engineers to monitor a variety of important parameters. The intelligent power train session as well as some of the posters reflect important components like chemical sensors, innovative pressure sensors and flow sensors. Furthermore, control strategies for both direct injection engines and transmission systems are discussed.
We would like to thank the members of the Honorary Committee and the Steering Committee for their support. Their assessments provided the basis for the selection of papers for the conference and the proceedings. The authors provided excellent papers and we would also like to thank them for their co-operation. Finally, the editors wish to thank H. Johansson, B.. Kühl and P. Mulvanny for many fruitful discussions during the preparation of the conference. Without their support this book and the conference would not have been possible.

Teltow, January 1998
Detlef Egbert Ricken Wolfgang Geßner

Table of Contents

Oral presentations

A Sensor for Crash Severity and Side Impact Detection

D. Rich, W. Kosiak, G. Manlove, S. V. Potti, and D. Schwarz

Delphi Delco Electronics Systems, Kokomo, IN 46901, USA

Abstract. This paper describes a crash detection sensor developed for use in two difficult crash discrimination situations: side impact and crash severity sensing. By changing the calibration of the unit during manufacturing, the unit can function as either a side or frontal impact sensor. Furthermore, the sensor can characterize the severity of the crash event and pass the information to a central restraint system control module. The entire sensor, which is composed of a micromachined accelerometer, a custom integrated circuit, and a microprocessor, is contained in a small, rugged module that is designed to be mounted near the impact zone.

Keywords. accelerometer, crash sensing, micro-electromechanical systems, microsystem technology

1. Introduction

The increasing usage of safety belts and the addition of supplemental airbag restraints and seat belt pretensioners have brought about a significant decline in frontal impact injuries and fatalities [1,2]. Increased attention is now also being directed towards improving the level of occupant protection during side impacts, and adapting the restraint system response to the severity of the crash. For both frontal- and side-impact restraint systems, the crash sensor is a key component.

Frontal Impact Severity Sensing. During frontal impacts, the front structure, or crush zone, of the vehicle decelerates early in the crash event, intruding into the engine compartment and absorbing crash energy. As the front structure crushes, the occupant and vehicle interior continue to move at the same relative velocity. Thus, nearly the full distance between the occupant and the vehicle steering wheel or instrument panel is available for an airbag system to provide cushioning. In the case of a 50 km/h crash into a rigid wall, the algorithm in the sensor typically must make a deploy decision between 15 to 25 ms after the bumper of the automobile contacts the wall, in order to provide adequate protection for the occupant.

Most current production vehicles use single-point accelerometer-based crash sensors mounted in the passenger compartment to sense and discriminate frontal impacts. Typically, this location provides timely discrimination of impact events. All impacts are categorized as "deploy" (requiring restraint system action) or "non-deploy" events. The response of current restraint systems to a "deploy" event is the same regardless of the severity of the crash or the occupant weight or position.

To provide maximum protection to the occupant, the next generation of occupant restraint technology will tailor restraint system response to the conditions of that particular crash event. The decision of when to deploy pretensioners, airbags or both, will depend on several factors: occupant position, occupant weight, use of seat belts, and severity of the crash event. Furthermore, the airbag inflation rate and peak pressure will be changed using dual-stage or variable inflation. By combining information from a central passenger compartment sensor and a forward satellite sensor located in the crush zone, the airbag control system will have a very robust and timely determination of the severity of the crash event.

Side Impact Sensing. Vehicle manufacturers are investigating structural enhancements, energy-absorbing padding, and side airbags to improve occupant protection during a side impact collision. However, the requirements of side impact discrimination are substantially different than those for frontal impact sensing. In the case of side impacts, the space between the exterior and interior of the vehicle and the space between the interior and the occupant are much smaller than in the case of frontal impacts. As a result, intrusion into the passenger compartment begins very early in the event, and the allowed occupant displacement is much less. Therefore, the trigger times required to initiate the airbag inflation are often less than five milliseconds.

We have developed a crash detection system that reliably senses side impacts, frontal impacts, and frontal impact severity. By changing the calibration during manufacturing and using different package options, the system can be used for a wide variety of vehicles and mounting locations. In the following sections, we will discuss the overall sensing system design and the design of the key components: the custom IC, the microprocessor, the acceleration sensor, and the package.

2. Sensing System Design

Electronic crash sensors for restraint systems have been produced in high volume for many years and are continuing to become more and more sophisticated [3,4]. The performance requirements for crash detection systems in side impacts and for frontal crash severity sensing are very demanding. To reliably discriminate crashes in these situations, many vehicles require satellite sensors mounted in the crush zone in addition to the sensors located in the passenger compartment.

Figures 1, 2, and 3 illustrate the importance of placing sensors in the crush zone of the vehicle. *Figures 1* and *2* show accelerations and velocities recorded during standardized frontal impact tests of a lower mid-size car. The three standard tests represent situations where different deployment levels of the restraint system are required. The 50 km/h rigid barrier test and the 50 km/h 30° angled barrier test

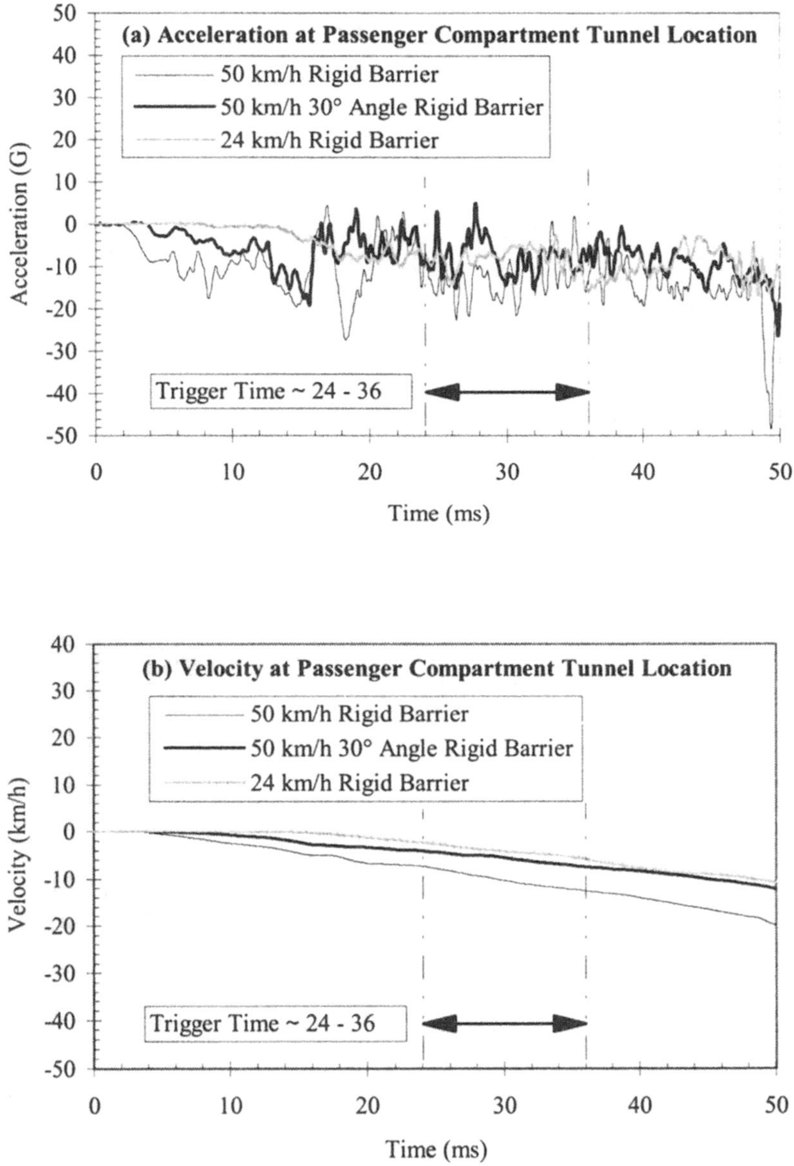

Figure 1. Acceleration (a) and velocity (b) at the passenger compartment tunnel location, measured during three standard crash tests of a lower mid-size car.

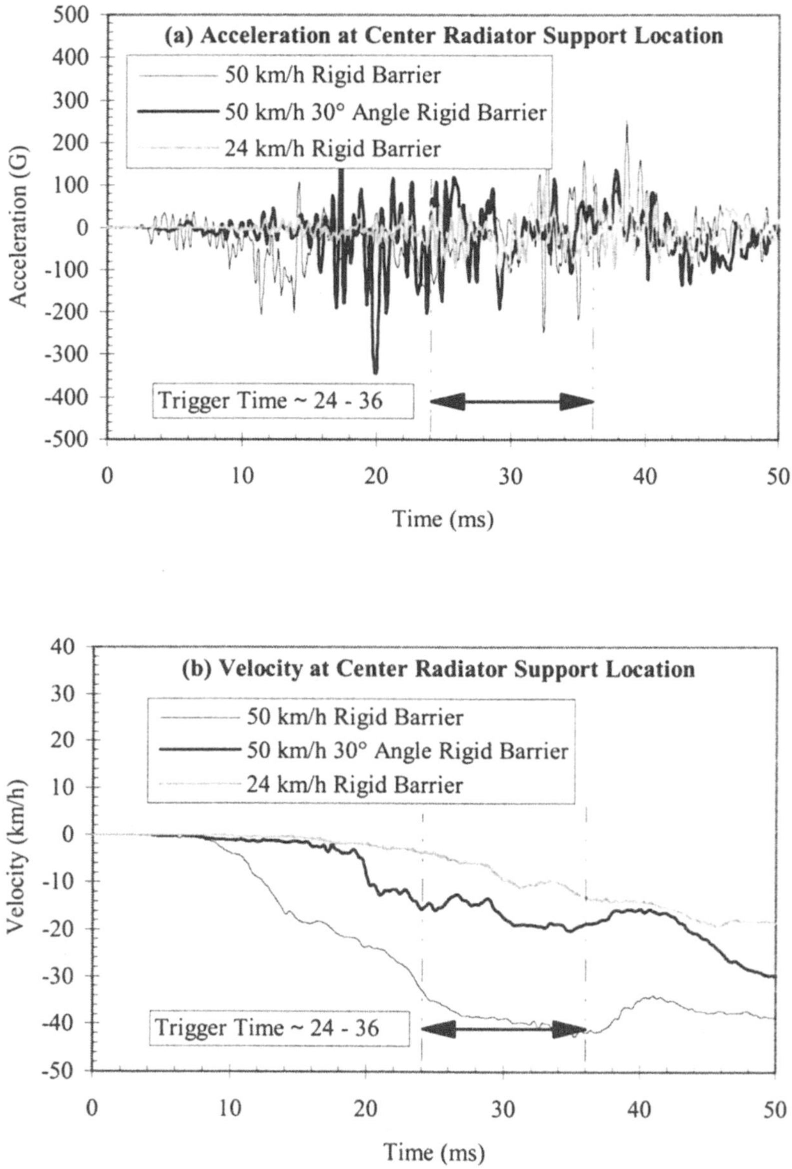

Figure 2. Acceleration (a) and velocity (b) at the central radiator support location, measured during three standard crash tests of a lower mid-size car.

both represent situations where the restraint system should provide the maximum response. The 24 km/h rigid barrier test represents a situation where the restraint system could provide a lower-level response, depending on occupant size, position, and belt condition.

The maximum accelerations at the central radiator support, shown in Figure 2, are at least an order of magnitude greater than those seen at the passenger compartment tunnel location, shown in Figure 1. (Note that the acceleration scale shown in Figure 2(a) is ten times greater than the scale shown in Figure 1(a).) Similarly, the relative velocity is greater by a factor of two or more. Table 1 summarizes a comparison between the lower-level 24 km/h rigid barrier and higher-level 50 km/h 30° angled rigid barrier tests for the two sensor locations. As shown in the table, the "discrimination factor", or the ratio between the results of the two tests, is much greater for the radiator location.

The results of side impact barrier tests for the same lower mid-size car are shown in Figure 3. The higher velocity 55 km/hr, 27° angle side impact test is a "deploy" event requiring restraint system response, while the lower velocity 15 km/hr, 90° angle side impact is a "non-deploy" event that requires no restraint system response. Plots (a) and (b) compare the effect of sensor location on the accumulated velocity, one parameter that would be used by a sensing algorithm. The velocity signal seen in the B-pillar location is much higher in magnitude than that seen in the tunnel location. At the approximate trigger time of 5 ms, the accumulated velocity difference between the "deploy" and "non-deploy" events is over 14 km/h for the B-pillar location, while it is less than 1 km/h as measured in the passenger compartment tunnel location.

Clearly, these data from both side and frontal collisions demonstrate that locating sensors in the crush zone can provide much higher signal amplitudes much earlier in the crash event. Discrimination, both between "deploy" and "non-deploy" events and among levels of crash severity, is more robust when the sensor is located in the crush zone.

Table 1. Comparison of maximum acceleration and accumulated velocity change at 20 ms for two frontal impact events: the 24 km/h rigid barrier and the 50 km/h 30° angled rigid barrier tests.

Sensor Location	Parameter	24 km/h R.B.	50 km/h 30° A.R.B.	Discrimination Ratio
Tunnel	Acceleration (G)	9	20	2.2 : 1
	Velocity (km/hr)	1.30	3.3	2.6 :1
Radiator	Acceleration (G)	35 G	340 G	9.7 : 1
	Velocity (km/hr)	2.0	8.7	4.4 : 1

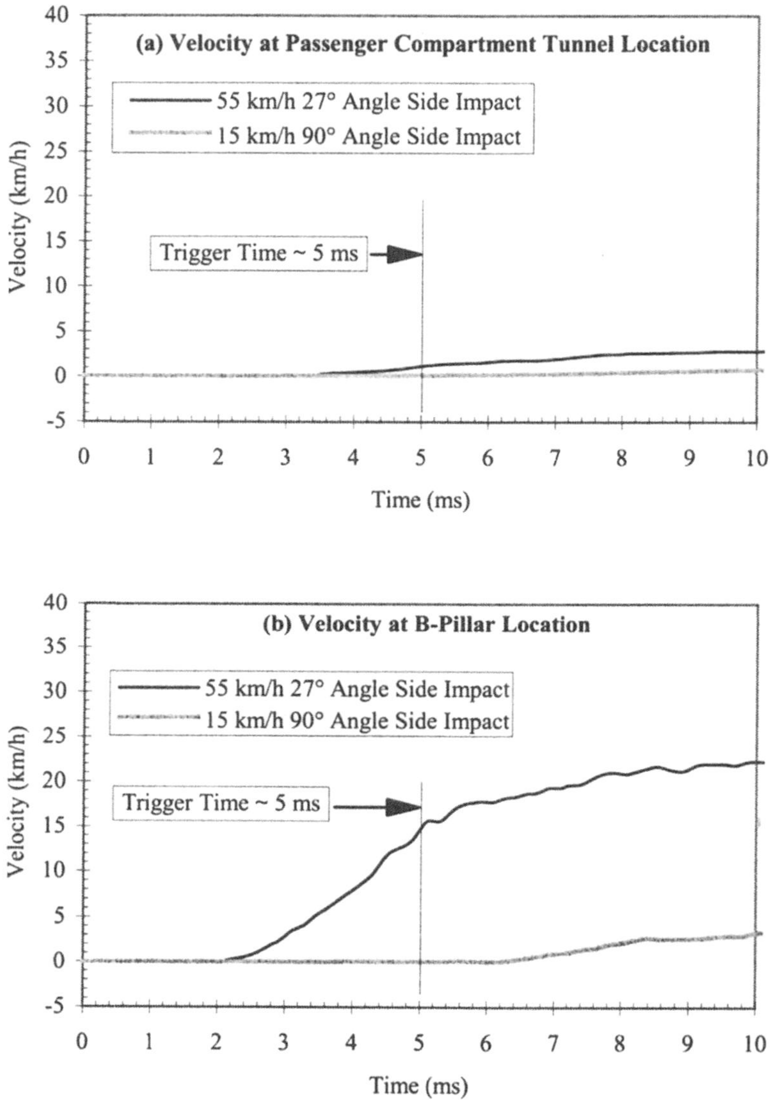

Figure 3. Comparison of the effect of sensor location on accumulated velocity during two standard side impact tests of a lower mid-size car. Data from the passenger compartment tunnel location is shown in (a), while data from the B-pillar location is shown (b).

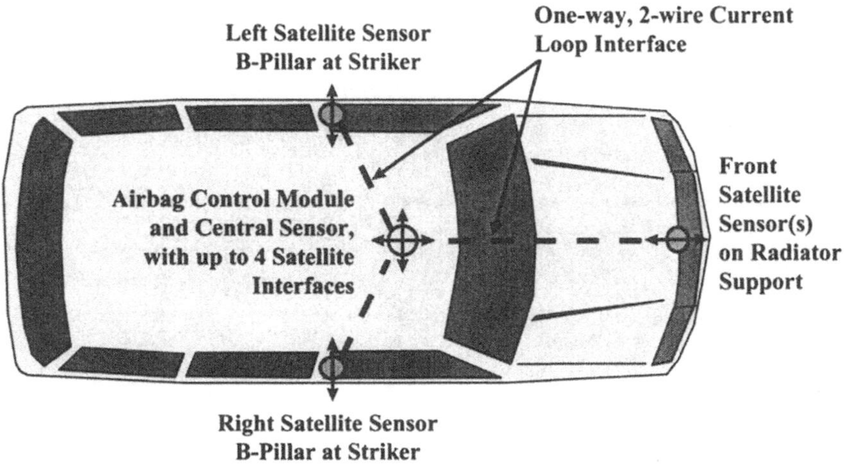

Figure 4. Vehicle configuration of the sensing system, with both front and side satellite sensors.

To provide reliable discrimination in these situations, our sensor is designed to be mounted in the crush zone. Figure 4 shows the vehicle configuration of a sensing system combining front and side satellite sensors with a central sensor located in the passenger compartment. These sensors interface with the central control module to communicate crash discrimination information as well as satellite state-of-health status. We chose to implement this interface as a unidirectional, 2-wire current loop that provides power to the satellite and information back to the central control module. In many instances, the sensor in the passenger compartment is used to improve system reliability against single-point system faults.

A block diagram of the satellite sensor is shown in Figure 5. The sensor-system is partitioned into a custom IC, an eight-bit microprocessor, and an acceleration sensor, plus a few discrete components. One key feature of this design is that it can function as either a side satellite sensor or as a front crash severity satellite sensor, by changing the calibration of the unit during manufacturing. Also, depending primarily on the communication requirements of the application, the functions of the microprocessor can be integrated into the custom IC [5].

3. Microprocessor

Algorithm. The algorithm uses acceleration and other parameters derived from the acceleration to determine the severity level in a crash. The algorithm logic to discriminate different severity levels is simple because of the timely information available at the location where the sensor is mounted. With appropriate placement of the sensor, it is possible to accurately and robustly determine the severity levels for different types of crashes, including offset rigid and deformable barrier tests.

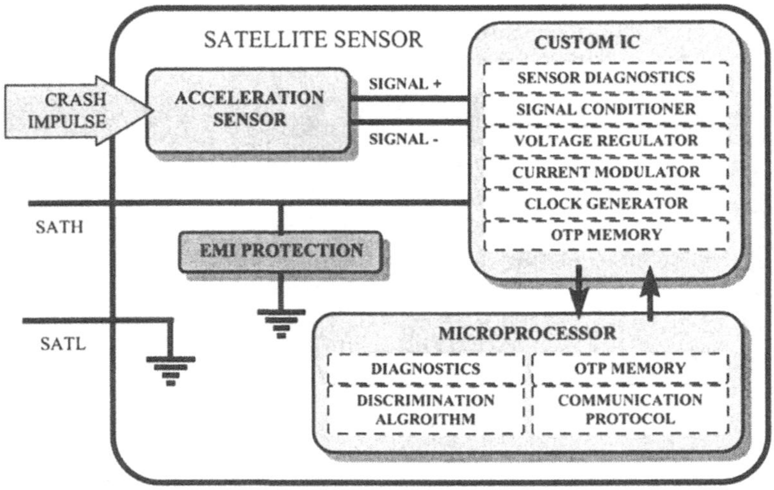

Figure 5. Block diagram of the satellite sensor.

The crash severity level is transmitted to the central controller where other inputs, such as status of the occupant's seat belts, are factored into the overall decision. Airbag inflation rates can be modified based on severity thresholds crossed and other occupant conditions registered. The sensor can assist achievement of higher non-deployment threshold velocities without loss of robustness, and the algorithm can be adapted in order to facilitate conformance to new legislation.

Communications Protocol. The microprocessor communicates the ID of the unit, the state-of-health, and the deploy or crash severity messages to the central controller. The controller is updated once every second for state-of-health and virtually immediately upon detection of a crash event. Because the microprocessor controls the transmission of messages, the message protocol can be designed to work with virtually any restraint system controller and virtually any number of crash severity levels.

Diagnostics. The diagnostics logic embedded in the microprocessor continuously monitor the sensor for accelerometer and circuit malfunction and will signal a fault condition if a malfunction is detected.

4. Custom IC Design

The custom integrated circuit provides amplification and signal conditioning of the acceleration sensor signal, and communicates with the central restraint system control module. This IC is implemented in a 1.2 µm CMOS process with twin-poly capacitors, high-voltage devices, and bipolar NPN transistors.

Some of the features of the custom integrated circuit include an accurate integrated oscillator and regulator, the acceleration sensor interface, two-wire serial communication, and non-volatile memory to calibrate for sensor and IC process variations.

Oscillator and Regulator. The regulator is an LDO with a p-channel pass transistor. It requires one external capacitor and regulates 5 volts to within +/- 5% from the battery input. The battery line can vary in amplitude from 6-40 V. The oscillator is an on-chip capacitive design using a temperature compensated current source [6]. The absolute value of the current source is modified with the non-volatile circuitry to adjust the center frequency to 4 MHz. The oscillator frequency is controlled within +/- 5% over the entire temperature range of -40 to 120 °C.

Sensor Interface. The acceleration sensor interface must amplify the input signal from approximately 1.8 mV-s^2/m (17.7 mV/G) to 408 mV-s^2/m (4 mV/G); this requires a nominal gain of 225. As the sensor has process variations which affect the sensitivity, the interface circuit can compensate the gain from 180-280.

The sensor has a sensitivity with a negative temperature coefficient of 0.22 %/K. A positive gain coefficient of the same magnitude has been incorporated into the interface circuit. The acceleration accuracy of the module over the entire temperature range is +/- 8%, including hysteresis, over the lifetime of the part.

A key feature to this sensor interface circuit is the output offset. Because the output acceleration from this circuit is directly applied to the crash detection algorithm, offsets directly translate to calculation errors. A minimized offset establishes the zero state acceleration response at mid supply, assuring a symmetric response to large input acceleration. Figure 6 shows a block diagram of the active offset cancellation circuitry and the input gain stage.

The sensor has a differential output which is converted to a single-ended output at Vout, and the gain is adjusted in this stage. The Vout signal is applied to the positive terminal of a comparator and compared to Vref. When Vout is above Vref, the up/down counter counts up. This increments the digital-to-analog converter (D/A) output, which is applied through a resistor to the negative terminal of the gain stage. The up/down counter counts up until the Vout signal drops below the Vref limit. At this point, the output is regulated to within one D/A count of Vref. The Vout signal will toggle one least significant bit (LSB) around this point. This is typically less than 10 mV of error, which is negligible in the operation of the algorithm.

Two-wire Serial Communication. The custom integrated circuit has a low nominal operating current, typically less than 5 mA. The communication with the central controller is achieved through current modulation of the module. The custom IC generates an additional current of approximately 38 mA, which the central controller interprets as a logical one, whereas the normal current of 5 mA is interpreted as a logical zero. Modulation of this current is controlled by the microprocessor.

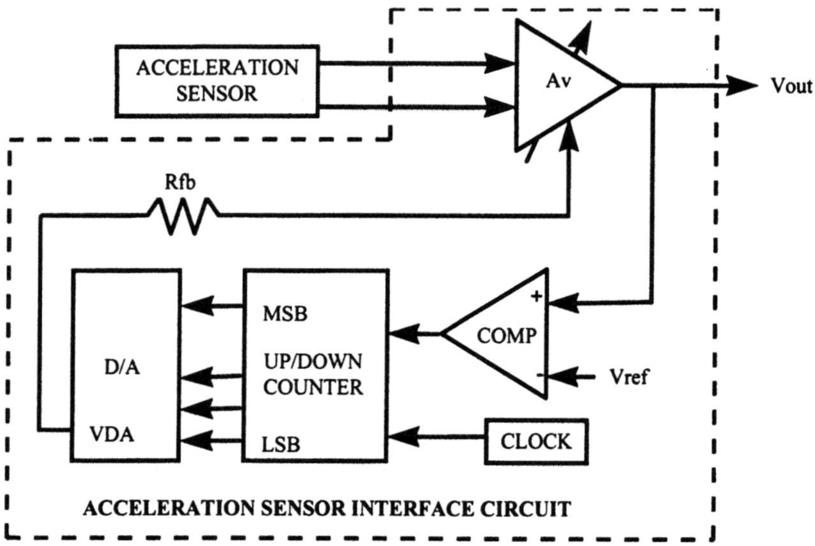

Figure 6. Block diagram of the acceleration sensor interface circuit and the input gain stage.

Non-Volatile Memory. The custom integrated circuit contains UV-EPROM to trim the sensor interface circuit and oscillator. UV-EPROM is incorporated in a latch which can be adjusted until the desired result is achieved; then the value is programmed. The non-volatile latch can only be programmed during the manufacturing calibration. Voltage margining of the programming depth and a parity chain are incorporated in the latch design to assure that valid data are maintained during the lifetime of the part.

5. Sensor Design and Fabrication

Piezoresistive accelerometers have been under development for many years [7,8]. Since 1993, millions of these sensors have been used in frontal impact detection systems [8,9].

This piezoresistive accelerometer is produced using a combination of bulk and surface micro-machining processes. It has a dynamic range of ±5000 m/s^2 (±500 G) and uses a patented "compensating beam" technique to reduce errors caused by internal and external stresses [10]. The cell, which is completely sealed at the die level, is mounted directly to the ceramic substrate.

Design. The sensing element of the device consists of a thin paddle connected to a frame in a cantilever fashion, as shown in Figure 7. The paddle acts as a proof mass and is free to respond to accelerations. As the beam flexes in response to acceleration, the resistance values of the piezoresistors implanted in the beam change.

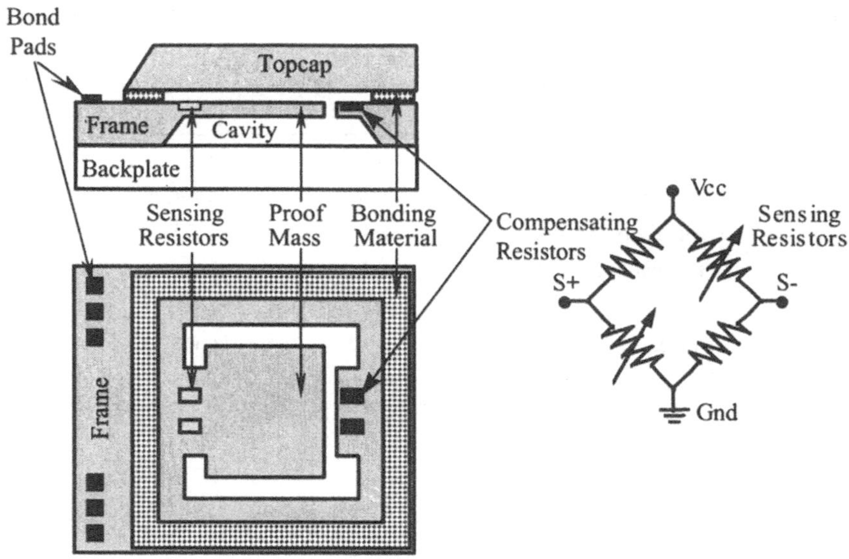

Figure 7. Cross-section and plan views of the accelerometer design, and the bridge network.

The resistors are arranged in a Wheatstone bridge circuit, which translates the small changes in resistance into a differential voltage. By arranging the compensating and sensing resistors on opposite sides of the Wheatstone bridge, the device rejects unwanted error strains and passes the desired strains due to acceleration. Performance of the device is summarized in Table 2.

Fabrication Process. Millions of high-reliability pressure sensors and accelerometers are produced each year using this process [9,11]. Summary diagrams of the fabrication process are shown in Figure 8. Starting with a p-type silicon substrate, an n-type epitaxial layer is grown to a well-controlled thickness on the top surface. Standard integrated circuit processes are then used to create piezoresistors and the sensor circuit (a). An electro-chemical etch process is used to etch a cavity into the back side of the wafer, forming a thin diaphragm of the epitaxial layer (b). At this point, the device is essentially a pressure sensor.

A silicon backplate wafer is bonded to the back side of the sensing wafer using a low-temperature fusion bonding process, forming a two-wafer stack (c). Then, a dry etch process is used to cut through the diaphragm from the front side, forming a cantilevered beam in a paddle shape (d). The sensing element is now free to move. A separate cap wafer is made using wet etch processes, and is bonded to the two-wafer stack using a screen printed glass frit process, (e) and (f). This combination of front- and back-side caps not only protects the accelerometer during the harsh wafer dicing and assembly operations, but also allows the device

Table 2. Performance of the accelerometer, including signal conditioning and amplification.

Parameter	Value	Units
Dynamic Range	±2450	m/s^2
Sensitivity	0.815	$mV \cdot s^2/m$
Sensitivity Accuracy	±8	% F.S.
Bandwidth (-3 db)	0.1-1500	Hz
Bias Error	±1.0	% F.S.
Noise, 0.1-1500 Hz	±0.5	% F.S.
Temperature Range	-40 to 120	°C

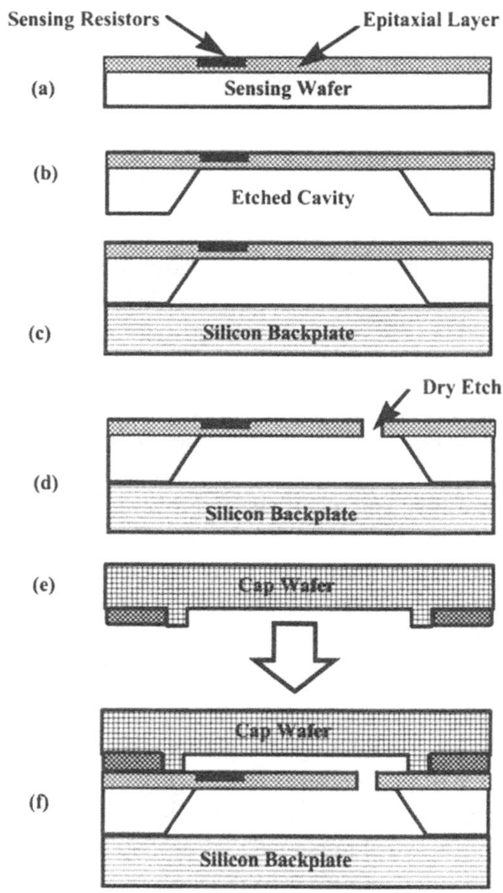

Figure 8. Diagram of the sensor fabrication process sequence.

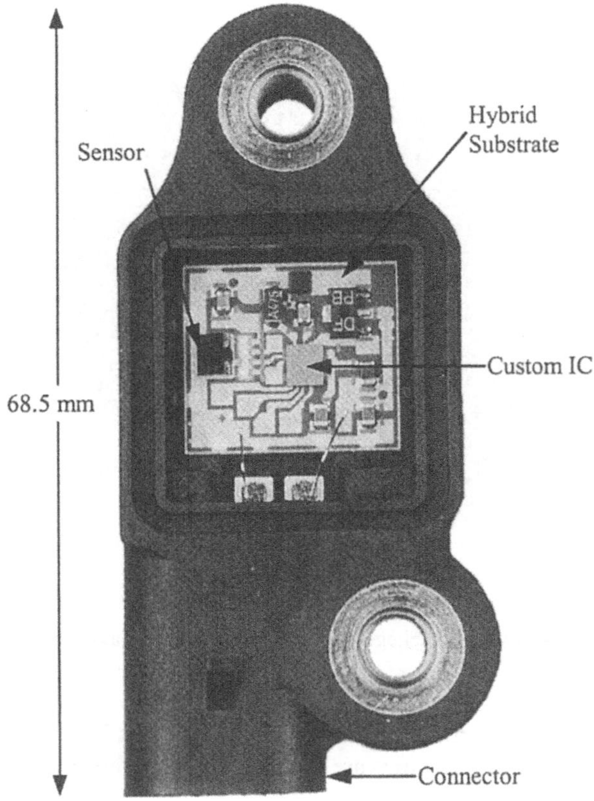

Figure 9. Photograph of the module with the cover removed. In this option, the sensing axis is perpendicular to the axis of the connector.

to be directly mounted to the ceramic substrate and covered with standard circuit passivation materials.

6. Package Designs

Two packages have been designed to meet the demands of both the passenger compartment and under-hood automotive environments. The packages differ in the orientation of the connector axis relative to the sensing axis. Particular attention was paid to balancing the requirements of size, rigidity, environmental protection, mounting flexibility, error proofing, and electro-magnetic interference (EMI) protection in a cost-effective way.

Several factors contribute to the compact size of the modules. Hybrid surface-mount technology, unpackaged integrated circuits and aluminum wirebond

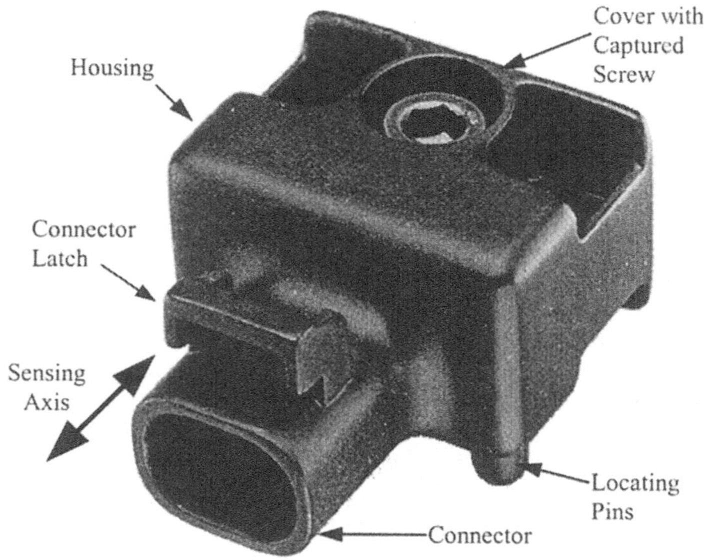

Figure 10. Photograph of the package option with the sensing axis parallel to the axis of the connector.

interconnects are used to produce a circuit area of less than 3.2 cm^2. In both cases, a connector and mounting bushings are integrated into the housing.

The package shown in Figure 9 is designed for mounting configurations where the axis of the connector is perpendicular to the sensing axis. In this case, the functions of the micro-processor have been integrated into the custom IC, and the microprocessor has been eliminated. The overall dimensions are 68.5 x 33 x 14 mm, excluding the connector latch and locating pin.

An alternate design, shown in Figure 10 is used for applications where the connector and sensing axes are parallel. A mounting bushing and captured screw are built into the housing cover.

In both package configurations, the rigidity of the module is important to faithfully transmit the crash signal to the acceleration sensor. Because the package size is small and the ceramic substrate is stiff, the first mode of resonance is typically greater than 3.2 kHz, as shown in Figure 11. When this signal is filtered by the two-pole filter within the conditioning IC, any unwanted high frequency response is further reduced (see Figure 12).

Environmental protection is achieved by passivating the circuitry and completely sealing the exterior of the module. A very compliant silicone gel passivates the circuit, while a cover attached with a silicone adhesive seals the exterior of the module. The connector is made weather tight through the use of a three-rib

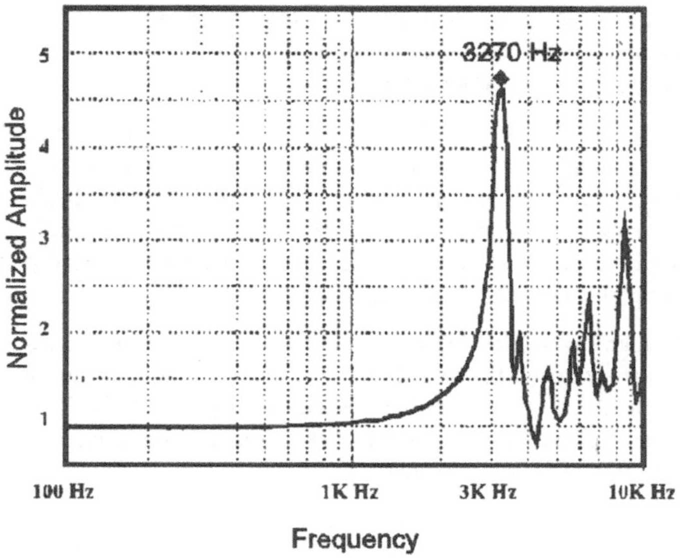

Figure 11. Frequency response of the package, measured using a reference accelerometer in the location of the sensing cell.

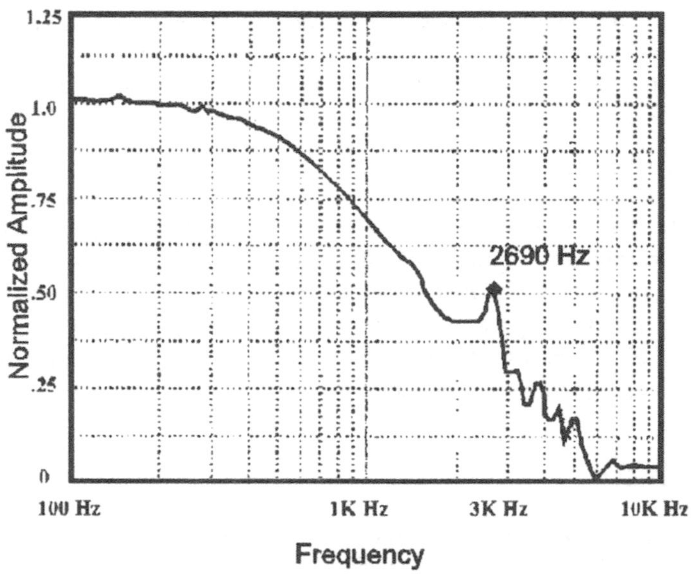

Figure 12. Composite frequency response of the package, sensor and filter circuitry.

compliant seal. Modules have successfully passed all applicable automotive testing, including powered temperature cycle, fluids compatibility, 96 hour salt fog, powered dunk, and water submersion testing.

To reduce assembly labor and part count at the vehicle assembly plant, an alternative mounting scheme was developed. In addition to a standard through-hole bushing design, an optional configuration using threaded bushings and pre-attached screws is available. Using this threaded bushing configuration, the module can be installed in a single operation by placing the head portion of the screws through keyhole slots and tightening the screws.

Due to the many possible algorithm calibrations, vehicle designs, and mounting locations, error proofing the installation is critical. In addition to the 63 possible electrical ID's, there are 17 possible mechanical keys that fit within the connector shroud. Each unique key insures that only a mating connector with a matching key configuration may be connected.

Many of these advantages were achieved through the use of a plastic housing. One disadvantage of using a plastic housing is the lack of EMI shielding when compared to a metal case. A combination of conditioning the input/output signals and the use of a ground plane was used to overcome this potential problem. Conditioning was achieved through the use of a pi network consisting of two 2200 pF capacitors and a 600 Ω ferrite chip impeder. This network effectively shunts high frequency energy coming in through the connector to ground. The second level of protection, the ground plane, is formed using a two-layer hybrid circuit technique. The bottom ground plane layer and the upper active circuit are separated by approximately 25 mm of dielectric material. This combination allows the module to withstand 150 V/m of radiated and conducted EMI.

7. Conclusion

A remotely mounted crash detection sensor has been developed that meets the sensing performance requirements of automotive restraints systems and the durability requirements for mounting near the impact zone. The sensor is well-suited for frontal impact, side impact, and crash severity discrimination applications on a wide variety of vehicles. One version of this design is currently being fitted on high-volume production vehicles.

8. References

1. C.J. Kahane, "Fatality Reduction by Air Bags: Analysis of Accident Data Through Early 1996." DOT HS808470. Washington, DC: U.S. Department of Transportation. 1996.
2. S.A. Ferguson, A.K. Lund, M.A. Greene, "Driver Fatalities in 1985-1994 Air Bag Cars", Insurance Institute for Highway Safety, Arlington, VA. 1995.

3. T.D. Hendrix, J.P. Kelley, W.L. Piper, "Mechanical Versus Accelerometer Based Sensing for Supplemental Inflatable Restraint Systems", SAE 901121; 1990.
4. R. Vogt, "Electronic System Design for Future Passenger Restraint Systems", SAE 960500; 1996.
5. D. Rich, W. Kosiak, G. Manlove, D. Schwarz, "A Remotely Mounted Crash Detection System", SAE 973240; 1997.
6. "Accurate Integrated Oscillator Circuit.," US Patent Number 5,699,024. December 16, 1997.
7. L. Roylance and J. Angell, "A Batch-Fabricated Silicon Accelerometer," *IEEE Trans. Elect. Dev.*, ED-26, (1979) 1911-1917.
8. W. Yun and R. Howe, "Recent Developments In Silicon Microaccelerometers," *Sensors*, vol. 9 (1992) 31-41.
9. D. Sparks and R. Brown, "Buying micromachined sensor in high volume," *Sensors*, vol. 12 (1995) 53-56.
10. "Self-compensating Accelerometer," US Patent Number 5,698,785. December 16, 1997.
11. D. Sparks, D. Rich, C. Gerhart, and J. Frazee, "A Bi-Directional Accelerometer and Flow Sensor Made Using a Piezoresistive Cantilever", ATA 6th European Congress on Lightweight and Small Cars: the Answer to Future Needs, Cernobbio, Italy July 2-3 1997, vol. 2, p 1119-1125. ATA-97A2IV40.

Peripheral Acceleration Sensor for Side Airbag

G. Bischopink, B. Maihöfer, D. Ullmann, M. Schöfthaler, J. Seibold, J. Marek

Robert Bosch GmbH, Department K8/STZ, Tübinger Str. 123, D-72703 Reutlingen

For a side airbag system it is necessary to measure the acceleration caused by a side crash within a time of less than 3 ms in order to inflate the side airbag in time.

The main function of the Bosch Peripheral Acceleration Sensor (PAS) is to measure the acceleration with a surface micromachined sensor element and to discriminate between fire or non-fire situation. The fire decision is made with an 8 bit microcontroller and is transmitted to the central airbag control unit via a two wire bidirectional interface (compare figure 1).

The electrical circuit is realized in hybrid technology. The hybrid is enclosed in a hermetical sealed metal housing. This results in an excellent EMI performance and a high insensitivity against mechanical stress and environmental conditions like water, humiditiy or dust.
The packaging of the PAS is a customer specific plastic housing with customer specific connector and bushings (compare figure 2).
The customer specific crash parameters as well as programmable sensor parameters like sensitivity or upper limit frequency are stored in the EEPROM .

Features:

- Supply voltage 7.0 - 16.5 V
- Current consumption < 50 mA
- Software controllable sensitivity range of +-50g / +-100g
- Software controllable upper limit frequency (two pole Bessel) 200 Hz/400Hz
- Sensitivity adjustment +- 3 %
- Nonlinearity of sensitivity< 1%
- Temperature shift of sensitivity < 5%

The interface from the airbag control unit to the PAS can be realized by using Bosch Pheripheral Integrated Circuit (PIC). The PIC can provide the power supply of the ECU and PAS and the communication with 2 PAS.

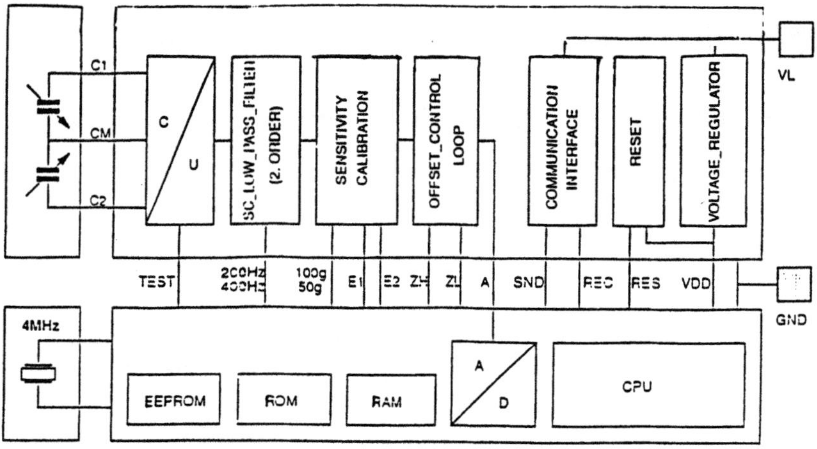

Fig. 1: Schematic block diagram

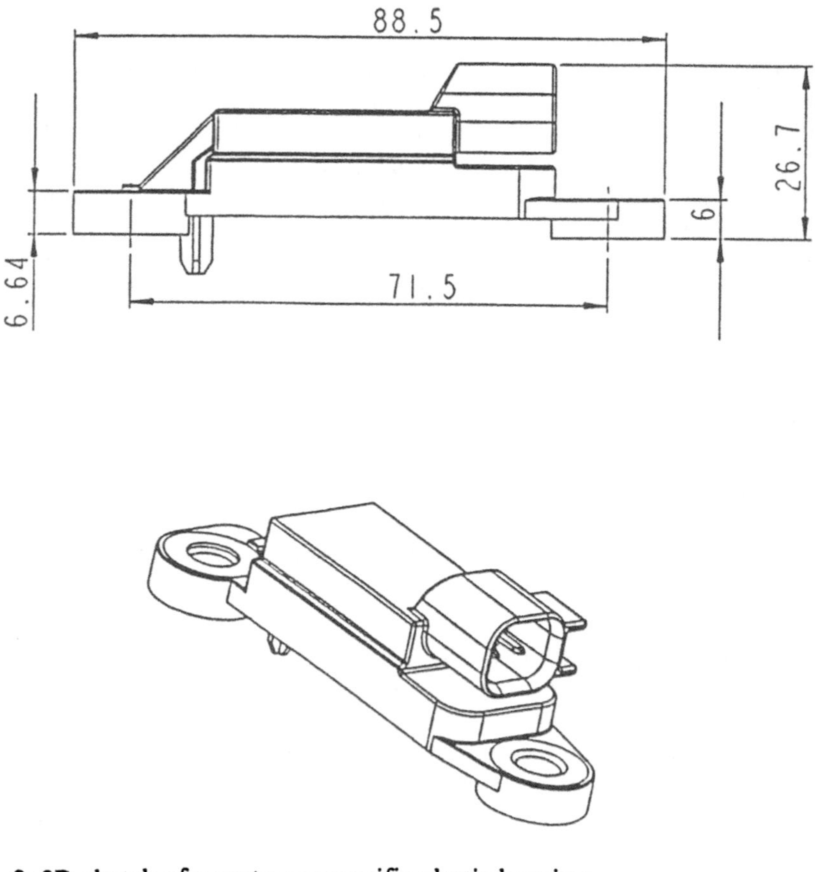

Fig. 2: 3D-sketch of a customer specific plastic housing

Future Applications of Microsystem Technologies in Automotive Safety Systems

Peter Steiner and Stefan M. Schwehr

TEMIC Telefunken microelectronic GmbH
Ringlerstr. 17, D-85057 Ingolstadt, Germany

Abstract. Since there is a strong demand for more intelligence in automotive safety systems, more information on additional parameters will be obtained by the electronics. This needs numerous electronic subsystems equipped with various kinds of sensors in order to obtain additional input on the crash environment. Microsystem technology will play an even more important role in the realization of these forthcoming tasks. The paper will present the application of advanced microsystems in future components of an entire airbag control system.

Keywords. Airbag, occupant detection, rollover, precrash, restraint systems, crash detection

1 Introduction

When TEMIC started to produce its first airbag deployment control units (ZAE) in 1992, these units contained two accelerometers, two microcontrollers, a safing sensor and some ASICs for single point detection of frontal and rear impact crashes. Four firing loops (two airbags and two belt pretensioners) could be applied to the unit.

Three years later in 1995 two satellite units each with its own accelerometer and microcontroller were added for additional side impact detection. The number of firing loops increased in this Multi-Restraint-System (MRS) by two sidebags squibs up to six loops. Simple monitoring of the passenger status (belted or not) was performed by reading out a so called buckle switch in the belt lock. Thus, the timing of airbag deployment can be adapted to the buckle status of the passenger. Additionally, belt pretensioning could be omitted in order to minimize repair costs. Nowadays, 12 and 16 loop units are in series production, which deploy additional headbags on the side and belt pretensioners also on the backseats of the cars. Weight sensitive mats in the seats transmit the occupancy status to the central unit, so that triggering of restraint devices only occurs, if there is any passenger present to be protected. State-of-the-art occupant detection systems (ODS) using rf-transponder technology in combination with car manufacturer specific child seats give also the information on child seat presence on the passenger side. Any injury to children caused by a deploying airbag can be avoided. More advanced child seat detection systems based on active infrared seat contour detection allow the use of numerous freely available child seats on the market. They will go into production at TEMIC in 1998.

Although the passenger is now well protected against frontal, rear and side impacts and the electronic control units contain some minimum adaptivity, there are a lot of tasks to do in the future.

Up to now, one special crash event has not been taken into account in the passive restraint systems: The rollover. Statistics show a decrease of the fraction of fatal injuries due to frontal or side impacts. The upcoming airbag systems do their jobs well. However, the fraction of fatalities due to rollovers increased because of the lack of any protection mechanism on the market. Because this task cannot be done by simply reading out the built-in accelerometers, a new kind of rotation sensors has to be applied to the control unit.

Furthermore, the strong increase of firing loops (up to 25 and more are feasible) and distributed sensors would lead to a dramatic increase of the complexity of the wiring harness of the car. Therefore, future systems will possess a restraint system bus connection which fulfills the special requirements of safety electronics in terms of reliability, safety and redundancy.

Although there is no doubt about the general positive effect of an airbag in lowering injury, some cases were reported where the rapidly deploying airbag injured persons or children more than it would have happened by the accident itself. These situations were caused by the persons being out-of-position (OOP). That means that those persons were not in the normal ideal position in their seat as assumed by designing the restraint system. They were too close to the inflator, so that they were heavily hit by the deploying airbag. This is valid for very small persons on the driver side or infants in rear faced child seats or unbelted persons on the passenger side, who are put forward in a precrash braking event. In order to avoid that negative effect, the OOP-situation has to be detected by suitable sensors which observe a keep-out-zone in a certain distance from the inflator.

Another approach to solve that problem is lowering the aggressivity of the airbag deployment itself. Such a "depowered" airbag has the disadvantage of a longer deployment time, which leads to the requirement for an earlier trigger time. That cannot be fulfilled by the current crash detection based on acceleration sensing only. Therefore, coming systems will include so called "precrash"-sensors, which deliver information on differential speed and distance between the car and its potential obstacle to the central release units. The forewarned central control unit can distinguish earlier and safer between crash and nocrash on lower acceleration levels. If the benefit in time is big enough, slow reversible restraint actuators can be considered which can also be used for several times without service and costs.

All these above mentioned features will be included in TEMIC ADONIS-System (Adaptive Omnidirectional Intelligent Safety). Complete market introduction is planned for the year 2003.

2 ADONIS Central Control Unit

The heart of the ADONIS-system is the ADONIS-ECU (electronic control unit). As can be seen in figure 1, it contains the well-known components:

- Two micromachined piezoresistive accelerometers (AS) for crash detection.
- A main microcontroller for information collection and triggering decision.
- A CAN (controller area network) controller for communication with further, non-safety, devices in the car.
- At least one ASIC for voltage regulation, energy reserve control and communication with the auxiliary satellite crash sensors.

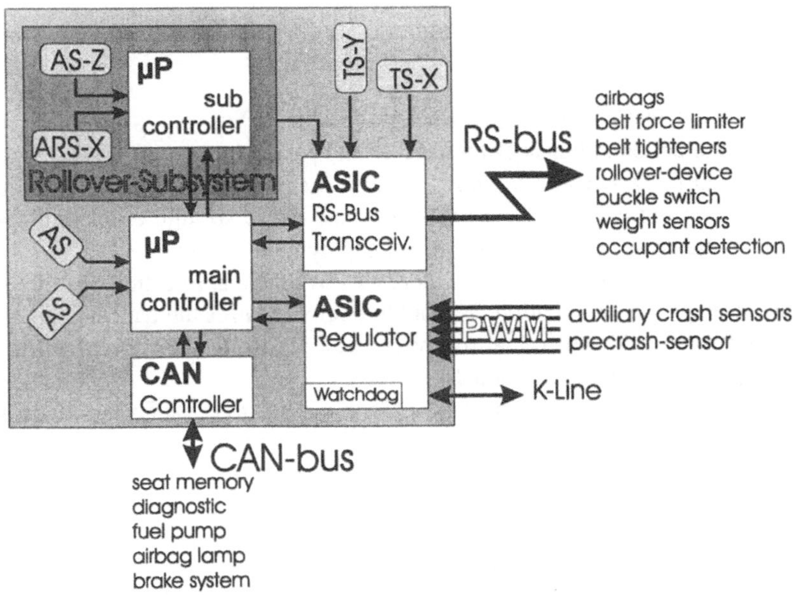

Figure 1. Block diagram of the ADONIS-ECU.

The new features are:

- A complete rollover subsystem including an angular rate sensor (ARS-X), a low-g-z-axis accelerometer (AS-Z) and an own microcontroller.
- A transceiver-ASIC for communication on the restraint system bus with various sensors and actuators.
- Two bi-directional trigger sensors (TS-X and TS-Y) for plausibility in all crash directions.

The microcontroller is programmed with a "smart" release algorithm, that takes the following parameters for calculation of the exact trigger timing into account:

- Crash severity

- Crash direction
- Rollover status
- Passenger presence
- Passenger weight
- Passenger position
- Passenger buckled
- Child seat presence
- Precrash information

In the following, the future components will be described in more detail.

2.1 Rollover Detection

In closed cars the following irreversible restraints should be actuated in the case of the dynamic and fast side rollover:

- belt pretensioners
- side bags
- headbags: window bags/inflatable curtains/ITS
- interaction airbags between driver and passenger

The more unlikely dynamic frontal rollover is assumed to be of minor importance, since any starting event for that rollover would cause so high acceleration levels that belt pretensioning followed by airbag deployment occurs.

Additionally for convertible cars, there should be the application of rollover bar release in any rollover situation. This additionally requires the bar release in rather slow quasistatic side and frontal rollovers. As a consequence, the sensor has to be more accurate and two dimensional.

When starting to work on rollover detection, one has to consider the fundamental physical characteristics of a rollover event:

1. High inclination angle of the car
   - with respect to road or environment
   - with respect to earth's gravity
2. Reduced gravity force F_g in direction of the car's floor.
   - A free flight of car is often involved in rollover event
3. Rotational movement in most cases round the longitudinal axis
   - lateral forces such as centrifugal forces F_r during fast rollovers
   - coriolis force F_C useful for detection

In the past, work at TEMIC was focused on rollover detection for triggering only an automatic rollover bar in convertible cars. The used sensors are based on the physical principles No. 1 and No. 2. The so called "spirit level" detector consists of a liquid filled tube with a bubble inside. If this tube is rotated, the bubble moves through a light path, when a certain inclination angle is exceeded. The operating principle is based on sensing the direction of earth's gravity force. If the car is free flying due to a rollover or a jump over a hill, this sensor is blind, because this force is missing. Therefore, a second sensor detects this reduced gravity situation by a

spring-mass system. The moving magnetic mass switches a hall sensor, when the gravity is below a certain limit. The rollover bars are released, if the time of reduced gravity reached several 100ms.
As already mentioned, these sensors are based on acceleration and thus they are easily distortable by lateral cross forces. They cannot distinguish between gravity force (heavy mass) and inert forces.
Although very tricky algorithms have minimized the principle problems of these sensors, some strange driving situations remained which lead to inadvertent triggering of the rollover bar. The restraint device is reversible and thus any mistriggering in very unusual driving situations is tolerable, especially since no better sensors were available on the market in the past.
Recently, the first cars on the market are equipped with a so called window bag, an inflatable curtain on the entire side line of the car. This restraint would be ideal for protecting passengers from being thrown out in rollover situations. However no suitable rollover sensor is available on the market. Misfiring cannot be tolerated with that kind of irreversible restraint. The reliability should be as high as known for conventional frontal airbag deployment.
Intensive studies were performed and it was found out that only sensors based on the physical principle No. 3 (rotational movement of the car) can only fulfill this requirement. We propose that a rotation sensor, for example an angular rate sensor, shall be the heart of the system.

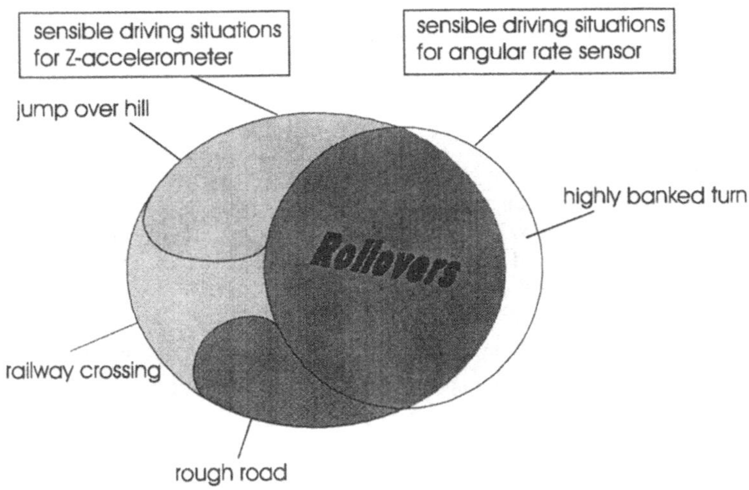

Figure 2. Group of events leading to triggering of a possible rollover sensor.
Light gray: Group of events, which trigger a rollover sensor based only on angular rate measurement. Medium gray: Group of events that trigger a rollover sensor based on z-axis acceleration measurement. Dark gray: Only rollovers trigger a combination of both sensors.

Intensive numerical simulations and practical tests with different driving situations, including nearly all kinds of rollovers, resulted in the figure 2.

When looking at the group of driving situations which can trigger a rollover sensor based on angular rate measurement, there is only one special event which is no rollover. This is when driving into or out of a highly banked curve. Here considerable angular rates appear although the car is in a stable driving situation. Fig. 3 shows the simulated signals from the angular rate sensor and the z-axis acceleration sensor. The actual inclination angle was calculated.

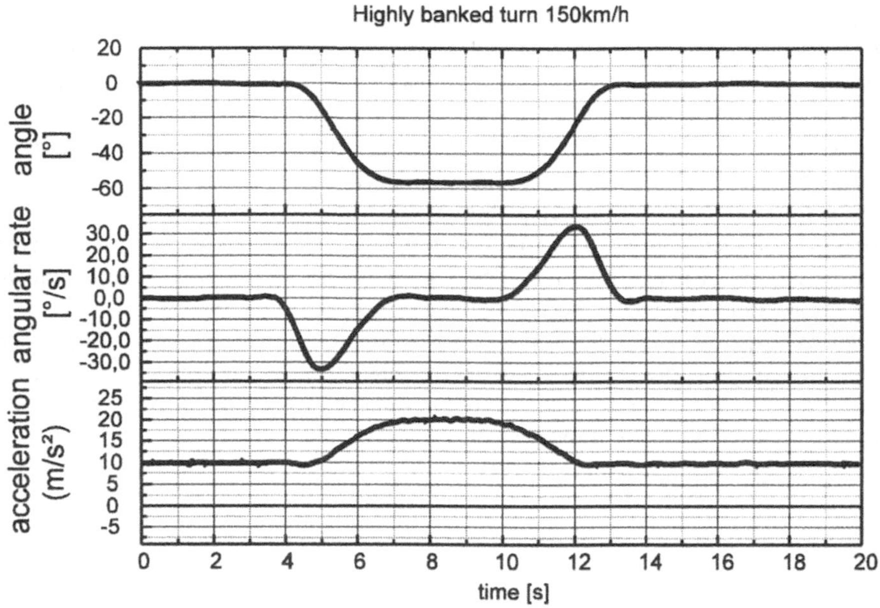

Figure 3. Simulated inclination angle, angular rate and z-axis acceleration of a car's driving on a plain road (0-4s), entering a highly banked curve (4-7s), being the curve (7-10s), leaving the curve (10-13s) and driving again on plain road (13-20s). The constant speed over ground was 150km/h.

The car reaches an inclination angle of 57° in the curve. By entering and leaving the curve the peak of rotational speed is 30°/s with a duration of about 2s. In terms of rollover detection this is a relative slow rollover. To solve this problem, either one can filter out such slow angular rates with reduction of performance in real rollovers, or one can take into account the z-axis acceleration.

The TEMIC concept is the latter case. The z-axis can be assumed as a measure for the ground contact of the car. Since there is an increasing gravity force (from 1g up to 2g) in direction to the car's floor, the car has pretty good contact to the road. This is in strong contradiction to a real rollover, where the car is under strong inclination (< 0.5g) or even free flying (0g).

Figure 4 shows the signal processing chain of the rollover microsystem. The analog values from both microsensors, the angular rate and the z-axis acceleration are evaluated in different parts of the algorithm. According to the understanding of figure 2, only if both parts see a rollover event, a trigger signal is messaged to the main airbag processor of ADONIS. This approach is the highest level in terms of reliability, since even different sensors based on even different physical characteristics of a rollover event are taken into account. Thus plausibility is not only guaranteed on electronic circuit level, operational safety is guaranteed from the physical point of view.

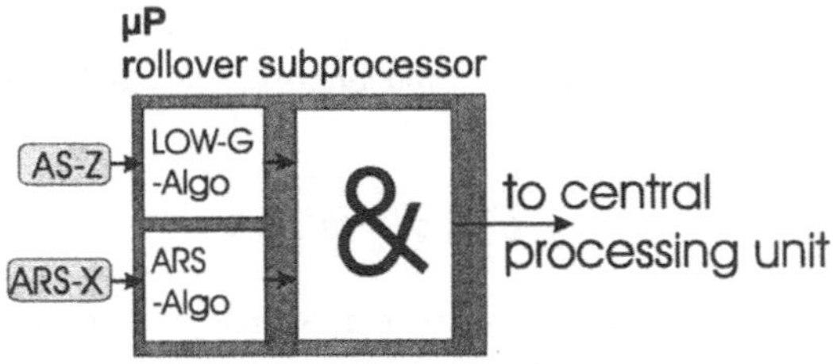

Figure 4. Block diagram of the rollover subsystem. The microcontroller evaluates the signals of the low-g-Z-axis accelerometer and the angular rate sensor. Only in the case when both paths of the algorithms agree on a rollover event, a trigger signal is generated.

In figure 5, the data resulting from a realistic rollover event are shown. The so called "delta µ" scenario happens when driving on an icy road with a friction coefficient of 0.1 and the driver looses control over the car.

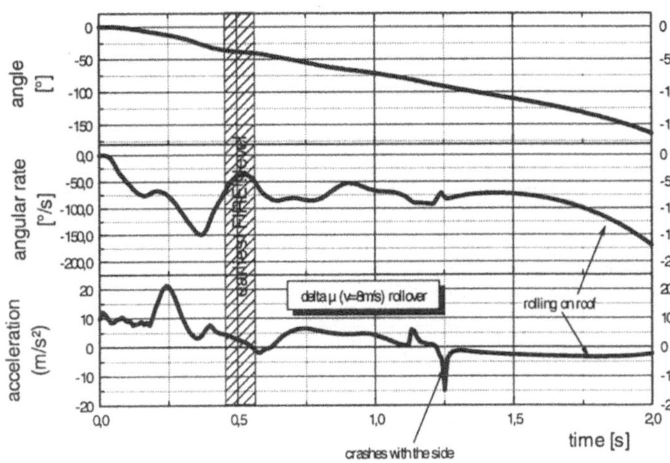

Figure 5. Simulated inclination angle, angular rate and z-axis acceleration of a car having a realistic rollover, when slipping from an icy road into the field with a rapid jump of the lateral friction coefficient (0.1 of ice → >1 of the field) at the tires.

The car leaves the road with a lateral speed of 8m/s into the field (friction coefficient >1). This rapid jump causes a momentum on the tires. The car gains rotational energy which has to work against the potential energy needed for lifting up the car before it is able to roll on the side. The values in this interacting system depend strongly on the geometrical properties of the car, such as position of COG, width and height.

From this energetic point of view, if there is enough rotational energy stored in the car to overcome the car specific potential energy, a forthcoming complete rollover of the car is unavoidable. This rotational energy can easily calculated from the angular rate ω with

$$E_{rot.} = \frac{I}{2}\omega^2,$$

where I is the moment of inertia of the car, which can be set as a car specific parameter in the release algorithm. That means, if the angular rate exceeds a certain limit, the angular rate part of the algorithm can deliver a trigger signal immediately.

In the example given in figure 5, the car rotates round itself with a peak angular rate of 150°/s. At the same time the gravity is reduced down to values well below 1g. The release algorithm will trigger after 500ms at an angle of 25°/s, when both conditions for a rollover, the high rotational speed and the low gravity are fulfilled.

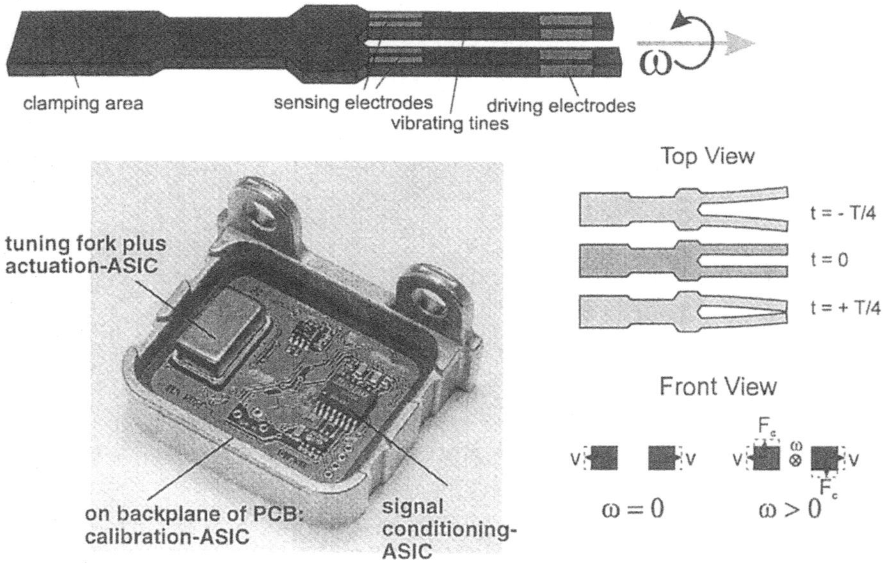

Figure 6. The TEMIC angular rate sensor. The drawing shows the function of the sensor when operated at its natural resonance frequency. The photo shows the arrangement of the tuning fork with its surrounding signal conditioning electronics in surface-mount technology (SMT).

The heart of the rollover microsystem is the angular rate sensor. It delivers analogue values proportional to rotational speed ω of the car round the longitudinal axis.
On the market several angular rate sensors are available for other applications, mainly for use in camcorders, for navigation purposes, or for cruise control applications in aircraft and cars. Most of them are not usable for the rollover application, either they are too sensitive against the harsh environment of the car (shock, temperature, etc.), or they are too expensive for mass production.
TEMIC has developed a special automotive angular rate sensor, which is available either for electronic stability programs (ESP) with a higher sensitivity or for rollover detection with a measurement range of about +/- 300°/s.
It consists of a quartz tuning fork in a metal housed multi-chip-module (MCM) with some peripheral driving and signal conditioning electronics (see figure 6). If the tuning fork is actuated electrostatically at its resonant frequency of 25 kHz by the driving electrodes, an in-plane vibration of both tines in an anti phase fashion appears. After rotation of the fork round its longitudinal axis the coriolis force F_c with

$$\vec{F}_{coriolis} = m\vec{v} \times \vec{\omega}$$

appears, being v the velocity during the periodic movement of the tines and m the moving mass of the tine. F_c is perpendicular to both the direction of ω and the forks plane given by the direction of v (see lower portion of fig. 6). The coriolis force strives to keep the original direction of v and, therefore, causes a momentum at the tines, which is proportional to the rotational speed ω. This momentum is read out at the sensing electrodes by the inherent piezoelectric effect of quartz.
The second sensor in the rollover ensemble is a bulk silicon micromachined low-g-accelerometer with a capacitive readout and a measurement range of +4g to –2g. Together with the signal conditioning ASIC, it is integrated into a hermetically sealed MCM in a metal can housing.
Further work at TEMIC is focused to integrate all inertial sensors - such as both crash sensors, the low-g-accelerometer and the angular rate sensor - onto one single inertial sensor MCM in order to lower the costs, when sufficient market penetration of this standard sensor setup for safety application is established on the market.

2.2 Trigger Arming Sensors

So called arming sensors are implemented in the ECU in order to have a second device for controlling the firing current. It has to be physically independent from the usual signal chain given by the accelerometer, microcontroller and ASIC. Any single failure appearing from this signal path should not lead to an inadvertent triggering of airbags.
In current ECU designs, this task is fulfilled by an electromechanical solution. It consists of a moving magnet, which is held by a spring in its normal off position. If acceleration of a certain amount is applied, the magnet moves against the force of the spring for several millimeters and the magnetic field causes closure of a reed

contact in the center of the tubular magnet. The firing current from the output stages is switched to the squibs. Usually, such sensors have a static sensitivity between 2g and 3g´s. They are designed to work on the safe side. In general, they have to be sensitive enough to close earlier than the airbag deployment from the algorithm is requested. This means that they will also close in some non crash situations, such as braking on a railway crossing or heavy rough road driving. In these cases, fine discrimination between crash and misuse is performed by the high sophisticated algorithm.

The function of the safing sensor has been proven since several years in the field. However, we made the experience that the reliability of the firing stages is so excellent that there is no further need for switching the firing current directly. Especially when we consider bus systems for distributing the firing information to the squibs, there is no more need for such a sensor.

The firing current has a value of several Amperes and thus makes the safing sensor rather expensive. Also from the manufacturing point of view, the electromechanical solution is macroscopic and cannot easily be mount automatically on the PCB.

Therefore, TEMIC proposes a micromachined solution, which has only triggering capabilities and is SMT compatible. The trigger sensor gives only a short pulse to the firing ASIC or bus-ASIC, when acceleration is applied. The trigger information is in the ASIC "AND" combined with the fire request from the microcontroller right before the last firing stage or the restraint system bus driver, respectively. Figure 7 shows the latter variant with a bi-directional trigger switch in a see-saw design.

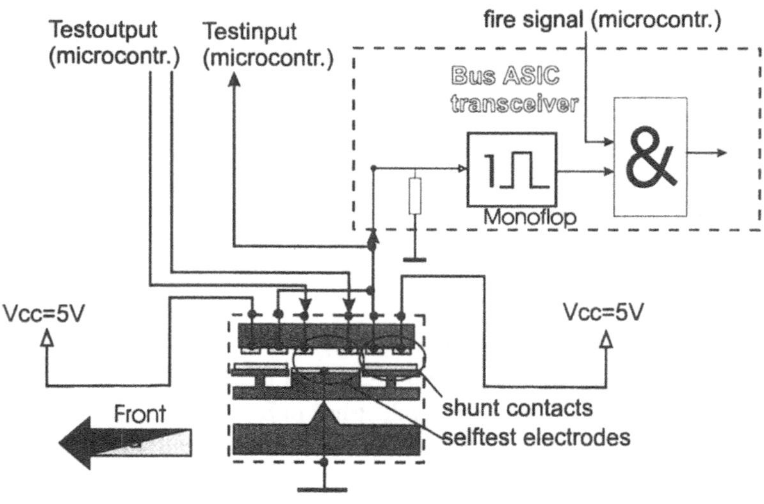

Figure 7. Block diagram of system design including a micromachined bidirectional trigger switch.

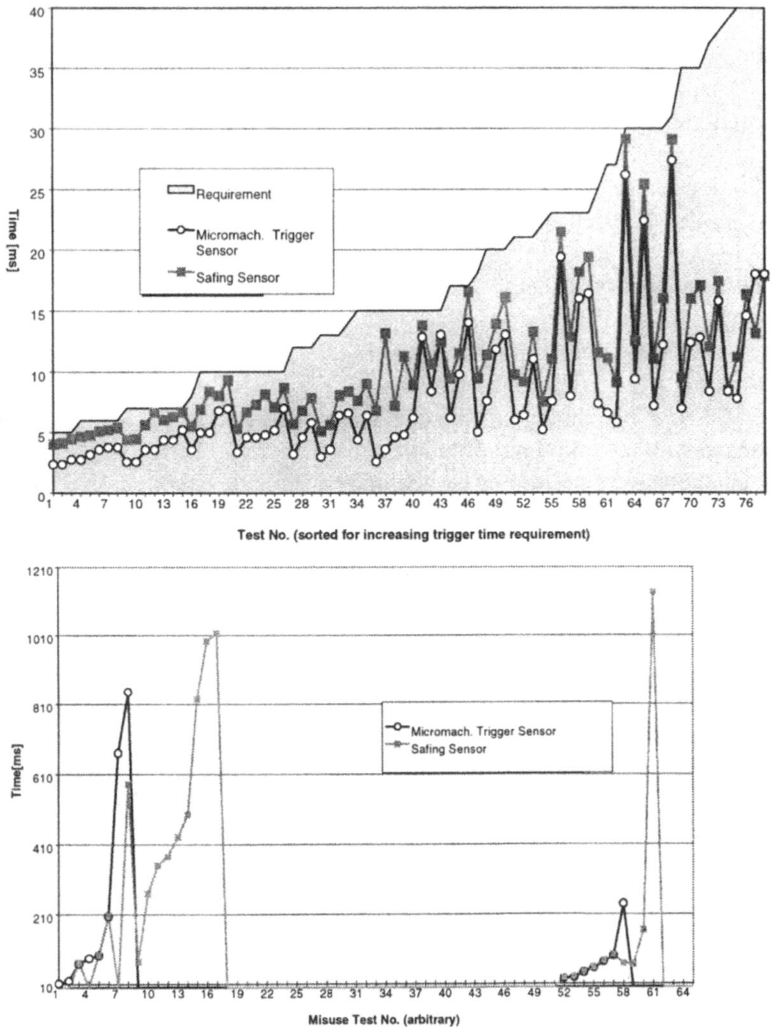

Figure 8a (top) and 8b (bottom). The figures show the closure time comparison between traditional safing sensor (squares) and a micromachined trigger sensor (circles).

The micromachining solution has some more advantages over the traditional safing sensor:

Faster trigger time due to the shorter switching distance from the millimeter range down to micrometer range according the law:

$$t = \sqrt{\frac{2s}{a}}$$, with s as the switching distance.

- It is usable also for the time critical side airbag application in stiff compact cars, where enough side impact acceleration is transmitted to the ECU location.

- It is ready for shorter trigger requirements in combination with depowered airbags and precrash detection.

Bi-directionality.

- Frontal crash and rear crash detection can be performed with a single device
- Side crash detection on both sides in ECU can be considered on modified sensitivity design.

Self test capability because low distances in μm range allow electrostatic actuation.

- Increase of reliability by repeated self test during power up of the system.

Figures 8a and 8b show the simulated performance of the trigger switch in comparison with the traditional solution.
The simulations were performed by assuming a static threshold of 2.2g and 3.5g of the safing sensor and the trigger sensor, respectively. Additionally, the trigger sensor was assumed to be critically damped in order to avoid effects appearing from resonances.
All solutions show a good performance within the timing requirement (top figure), whereas the trigger sensor behaves slightly faster, especially at the shorter trigger times. This can be explained with the shorter switching distance of the sensor.
In the bottom curve there can be seen the robustness of the sensor against misuse tests. As many as a dot appears in the graph, the sensor had an inadvertent closure. Here, the micromachined solution also shows the better performance due to an increased switching threshold (3.5g), which is a result from gaining time by the shorter switching distance.

3 Side Impact and Auxiliary Crash Detection

Since 1995, side impact satellite units are in series production at TEMIC. As smart microsystems on PCB level, they consist of an acceleration sensor, a microcontroller with a side impact discrimination algorithm and a voltage regulator. The fire signal generated by the satellite's algorithm is wired as pulse width modulated (PWM) information to the MRS-ECU's decoder-ASIC. In the central ECU a plausibility check on the external fire signal is made with the side acceleration signal measured in Y-direction.
Most of today's car designs use side crash detection either at the side member under the seat, or at the b-pillar near the door. Thus two different sensitivity axis' for the satellite units exist: Y-sensitive in direction of the PCB or Z-sensitive perpendicular to the PCB for the side member and the B-pillar mounting location, respectively.
Since cost reduction is an important issue, TEMIC decided to do a complete redesign concerning the fabrication technology and the system approach. The next generations of side impact satellite units will be technologically switched over from PCB technology towards MCM technology.

The first idea was to save the packaging costs of the silicon parts of the satellite by the assembly of bare chips. A second issue was to save the testing and calibration expenses of the system. As can be seen in figure 9, the module is fabricated in flex print technology, which is well known from the chip card application. The silicon dies of the micromachined piezoresistive sensing element, the signal conditioning ASIC and the microcontroller are attached at the flex print and contacted by wire bonding. The passive elements are conductively glued to the flex. The flex print itself is then assembled in a plastic housing, which includes the metal parts of the connector and the mounting holes. After complete assembly, the system is tested and calibrated while communicating with the internal microcontroller of the system. Thus all sensor related effects, such as sensitivity and offset (even over temperature), can be tested and calibrated simultaneously within a few steps.
By looking at the satellite from the (micro-)system point of view, perfect adjustment can be performed. Employing the resources of the microcontroller, such as on-chip voltage regulator, temperature sensor and EEPROM also for signal conditioning, the hardware costs (ASIC area) can be lowered. At the same time the accuracy increased significantly.

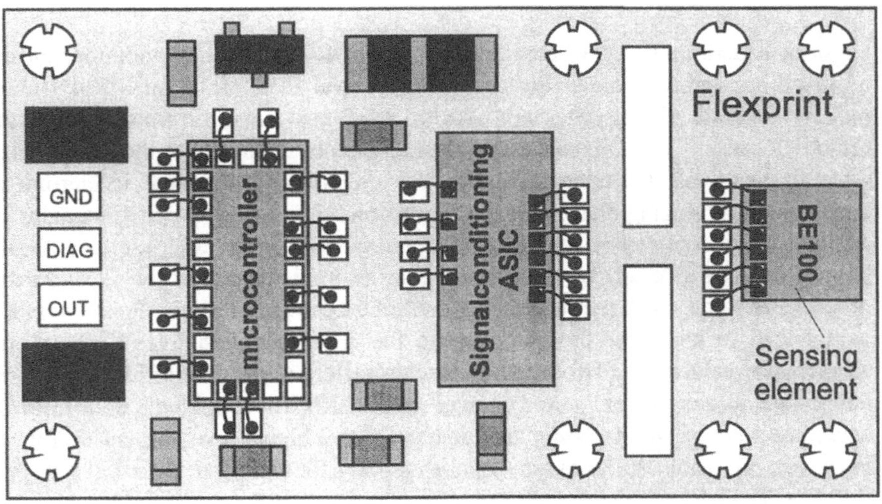

Figure 9. Principle design of the flex print based side impact satellite.

The reason for using the flex print instead of ceramics was the flexibility in the mounting direction. The sensing element on the flexible circuit board can be rotated to any customized angle given by tooling of the housing. This way, easy tailoring of the side member version or the b-pillar version can be performed with the same part as can be seen in figure 10. The only differences are in the housing and the software parameters.

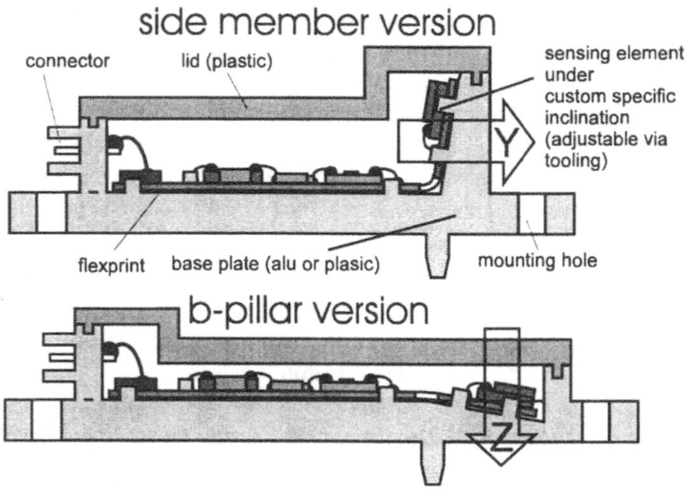

Figure 10. Different version of the flex print based side impact satellite. The same flex print can be used for different sensitivity directions by easy rotation of the sensing element.

Other considerations on the satellite sector at TEMIC are dealing with transmitting digitized acceleration values directly to the central ECU. Such an MRSD (multi restraint satellite digital) does not perform intelligent crash discrimination at the remote location. The advantage of this approach is to avoid the rather high material costs of the microcontroller chip. Only an enhanced accelerometer consisting of a sensing element and a signal conditioning ASIC which includes an analog to digital converter (ADC), a voltage regulator and a serial interface, would be needed. The side airbag algorithm would be implemented in the ECU's main processor and all crash parameter sets could be centralized at one location which would lead to an easier programming in the car production line. Furthermore, acceleration values from two build-in accelerometers in the central ECU and from the remote accelerometers would be available, which would open the door towards better and finer triggering going hand in hand with a higher level of plausibility.
However, to reach the goal, the challenge of safe *and* fast data transmission fulfilling all EMC-standards in the harsh environment of the car has to be solved.

4 Occupant Detection Systems

Occupant detection systems (ODS) have the task of gathering information about the status of the occupant. Hereby, the following parameters are of importance.
1st Occupancy of the seat at all. The repair costs according to a damage at the dashboard after airbag deployment are the reason for demanding this information. Insurance companies would give benefits if the airbag was not deployed when the seat is not occupied. Since the driver seat is assumed to be occupied in a running car, this is mostly an issue for the passenger seat and the fond seats. In the past, the detection was performed by weight sensitive mats, including an array of resistor that change their conductivity under the pressure of the passenger's body on the

seat. Problems arise on the long term stability and reproducibility of this measurement method.

2nd Child seat detection. Normally, rear facing infant seats are not allowed to be mounted on the passenger seat, because any airbag deployment would lead to a back shot of the seat when it is located too close to the airbag module. Some fatal injuries of babies, resulting from this effect, are already reported. Thus special, well readable warning labels must be installed in all newer cars with a passenger airbag. In the USA, cut off switches which allow the deactivation of the passenger airbag are in discussion. They should work as an intermediate solution until an automatic electronic solution is found for deactivation of the airbag when a child seat is installed or the seat is unoccupied.

Actually, there are some systems on the market that work with car manufacturer specific child seats which contain transponders in the body. An electronic in the seat detects the presence and arrangement of the transponders and thus recognizes the presence of a rear facing or front facing child seat. These systems work with high reliability of recognition, if the special transponder equipped child seat is used. However, they fail for all other child seats on the market.

Systems based on contour recognition employing ultrasonic or optical microsystems are currently developed in some laboratories. The contour of an empty seat is compared with the contour of a child seat or the contour of a passenger in the seat. The airbag is engaged only in the case of occupancy with a passenger. With that approach numerous child seats which are already on the market can be recognized and used by the customers.

3rd Passenger Status Recognition. In airbag design, the trigger time for airbag deployment is optimized in the sense of maximum protection for the passenger in the crash. Hereby, a compromise has to be found between a small female and a tall male person, since no adaptivity is performed concerning the distance to the dashboard or the weight of the individual passenger. In today's cars multi stage inflators, featuring variable inflation time and power, are going into series production. Beside the already implemented adaptation according the crash severity, they would allow adaptivity also concerning the passenger status, when suitable sensors will be available in the future.

Specially the measurement of distance to the airbag module is of major importance. Nowadays, the passengers are assumed to be in a well-fixed position in the seat at a typical distance to the airbag module. Problems arise, when they are "out-of-position" (OOP), i.e. when they are too close to the airbag in the time of deployment. The deploying airbag may hit the passenger and cause a more severe injury than by the same crash without airbag. In reality the situation happens, when very small persons are located too close to the steering wheel due to short arms. Recently some cases were also reported where persons were thrown forward during heavily braking prior to a crash and, therefore, been hit by the deploying airbag in the subsequent crash.

A so called "smart airbag" system should recognize such situations and then automatically depower or even deactivate the airbag. Fortunately, the problem can be reduced to the observance of a so called keep-out-zone round the airbag panel.

That means that a smart airbag system should contain a dedicated microsystem, which either watches an object entering the keep-off-zone or measures the distance of an object from the airbag panel.

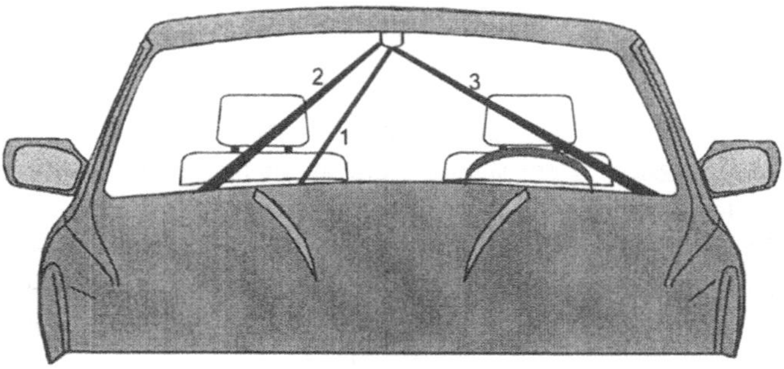

Figure 11. Various setups for occupant detection with three different arrangements of detection planes. Plane 1 is for child seat and occupancy detection on the passenger side. Planes 2 and 3 are for out-of-position detection on passenger side and driver side, respectively.

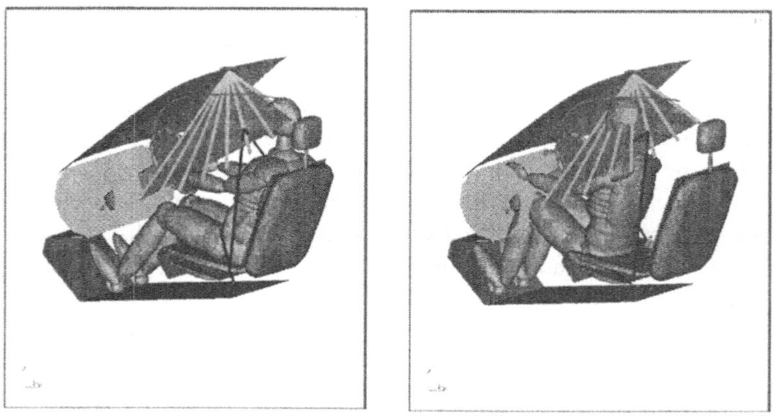

Figure 12. Side view on the OOP-detection for the driver side. A similar arrangement is made on the passenger side. When the head and thorax of the passenger move forward towards the steering wheel, they pass eight light beams spreading up a two dimensional triangular detection plane. The position of the passenger in his seat can be measured simultaneously. The location of the triangle is optimized to detect passenger of any size and in any seat position. Inflation adaptation according to the distance from the airbag module can be performed by this way.

TEMIC is now able to fulfill all three tasks with the same detection principle. The sensor is located in the mid header position of the car (near the doom light). It consists of one or more arrays of invisible infrared light beams that flash towards

the passengers in the seats. The detection planes can be arranged on the passenger side, either for occupancy and child seat detection (plane 1 in figure 11) or for out-of position detection of the passenger (plane 2). An additional plane can also be spread on the driver side for OOP-detection of the driver (plane 3).
Figure 12 shows the side view of out-of position detection on the driver side only, where the driver moves through the plane given by eight infrared beams. The forward movement may be caused by running onto a kerb or by simply tuning the radio. The occupant detection unit calculates the bodies actual position and speed while passing the light beams. Therefore, even prediction of the body's position at a later time can be made. If the body enters the keep-out-zone defined by the last beam nearest to the steering wheel, airbag deployment would be depowered or deactivated.
The microsystems block diagram is shown in figure 13. It contains an LED array as IR-emitter, a CCD-array as sensor and a microcontroller. The LEDs are actuated by the driver unit sequentially. After a single LED was fired, the reflected signal from the passenger's body can be read out from the CCD array.

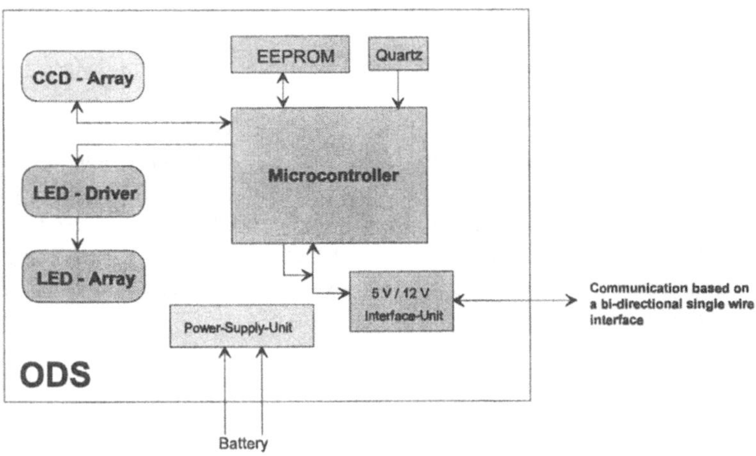

Figure 13. Block diagram of the passenger status recognition system. The microsystem employs strong interaction of the microcontroller with the light emitter circuitry and the detector. The microcontroller's software controls the LED timing and the subsequent read out of the CCD.

Resulting from the triangulation principle (explained in figure 14), the distance of the light spot is given by the signal distribution on the CCD-line. Thus any moving object through a light beam - such as the head of a passenger moving forward in a precrash situation - can be recognized as a drastic change in the distance. The path of the person passing the light beam on the way into the keep-out-zone can be monitored. The unwanted airbag deployment is omitted right in time.

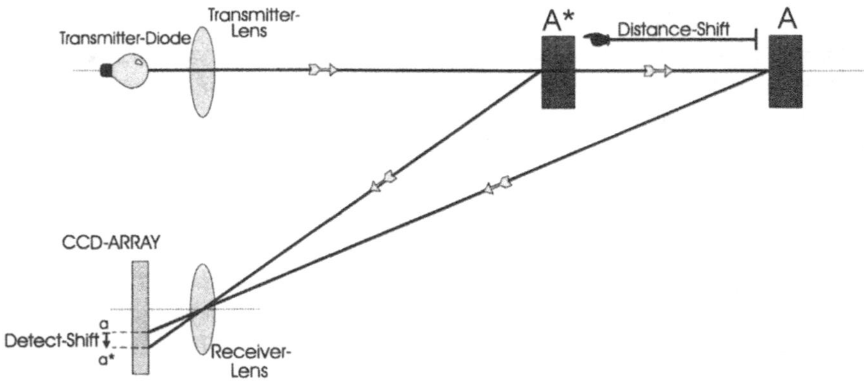

Figure 14. Triangulation principle. From the signal distribution of the reflected signal on the CCD, the actual distance of the light spot from the sensor location can be calculated. The emitted light is projected as an invisible light spot on the passenger's body. A distance shift of the spot results in a pixel shift on the CCD-array. With this method contour recognition for child seat detection on the passenger side (plane 1 in figure 11) and OOP-detection on both sides (planes 2 and 3 in figure 11) is performed. In the OOP detection part, any passengers passing the light beams are seen as rapid jump in the distance.

TEMIC has chosen this kind of detection principle, because it was possible to expand it from the simple single plane child seat and occupancy detection system of the past towards the multi plane occupant detection system (including out of position and keep-out-zone features) of the mid term future. For the long term future, we believe that three dimensional camera systems will monitor the entire interior of the car in order to watch all keep-out-zones round the numerous airbags (including side bags, window bags, etc.).

5 Precrash Detection

Precrash detection is defined as detecting the unavoidable crash prior to contact with the obstacle. The goal is, to gain additional reaction time for the restraint system.

5.1 Benefits from precrash detection for the restraint system.

The benefits of a longer reaction time are:

- **Earlier Triggering.** Especially for small compact cars the deformation space in the front and the side of the car is limited. Therefore, it is important to activate the restraint devices, i.e. belt pretensioners and air bags earlier for increased effectiveness. For instance, "ride down" effects can be minimized. The acceleration threshold for airbag triggering could be lowered proportional the detected differential speed relative to an obstacle. With this method, the deformation space can be extended electronically.
- **Pole crash.** Crashes with obstacles, that give only a weak acceleration signal to the accelerometers in the beginning, like the pole, are difficult to detect

right in time. This would be minimized by a recognition of these poles in the front or side area of the car.

- **Fire/Nofire distinction.** If the relative speed to an obstacle could be measured, discrimination between no fire low speed (<15-20km/h) crash and fire crash could be made more accurate. Thus, unnecessary repair costs could be avoided in some cases.
- **Use of depowered airbags.** Depowered airbags with a softer but slower inflation characteristic could be used. Problems with OOP (see previous section) could be minimized at its roots.
- **Higher reliability.** If there is an additional, physically totally independent signal of the crash available - like the differential speed to an obstacle - plausibility checks with the acceleration measurement could be carried out. Inadvertent triggering could be minimized.
- **Reversible restraints.** If the precrash detection gains enough time, reversible restraints - such as actuated bumpers or pneumatic airbags - could be considered in future restraint system designs.

5.2 Suitable sensors

Auxiliary acceleration sensitive satellite sensors. In the past, the major problem was the pole crash in very big and less stiff cars. In order to warn the restraint system right in time, a so called "pole catcher" sensor was used. This was an acceleration sensitive switch, which is located in the front area of the car under the hood. The sensor delivers a signal shortly after an obstacle touches the bumper. Then the acceleration threshold for triggering the airbag was lowered and the pole crash was detected right in time.

A more sophisticated solution of this idea of a "precrash" sensor is to put a satellite unit at this location. This approach is more robust against misuse scenarios, due to its intelligent algorithm capabilities. Problems arise on the requirement for temperature stability (-40 to +125°C) of such an electronic device. However, the time gain of such a solution (several ms) cannot be compared with the values expected from real precrash sensor, that looks out of the front of a running car. As shown in the previous section, there are several parameters which would give considerable benefits to the restraint system performance. The goal with the highest priority in precrash sensor design is to obtain the differential speed of the obstacle.

TEMIC actually considers the following sensor concepts for evaluation as precrash sensor:

A RADAR based closing velocity sensor for obtaining differential speed and direction of the obstacle (figure 15). A 5.8GHz frequency modulated continous wave (FMCW) RADAR detects the Doppler shift of the signal reflected by the obstacle. The differential speed of the obstacle can be calculated directly from this information. The microsystem consists of the antenna, a mixing unit and a microcontroller that evaluates the information. The differential speed information is transmitted to the central control unit in form of five speed classes. The speed range is 5-200km/h within a distance of 5m in front of the car. The 3dB angular characteristics are +/-45° elevation and +/-25° azimuth. A permanent self test is performed by read-out of the self speed of the car. The entire sensor subsystem including the antenna can easily be fabricated in traditional PCB-technology.

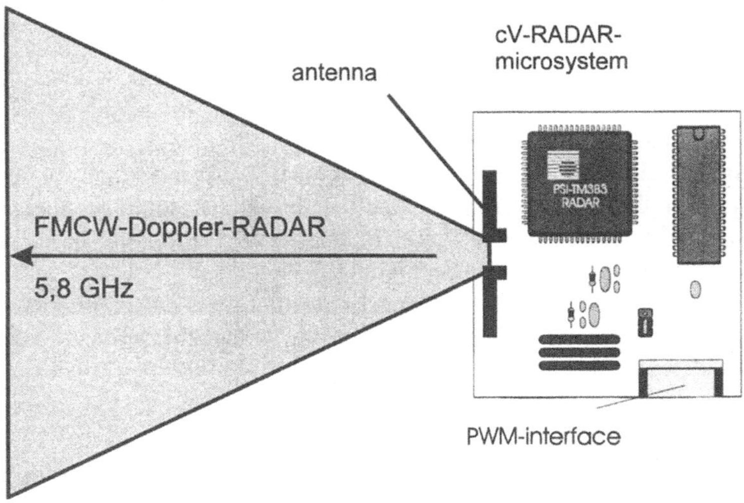

Figure 15. FMCW-Doppler-RADAR precrash sensor for closing velocity measurement in front of the car.

The basic application is the differential speed measurement. If it is necessary to obtain any information on the direction of the obstacle, two of such systems have to be cascaded.

A LASER based precrash sensor for the measurement of distance, differential speed, direction and size of the obstacle (figure 16). Time of flight measurement of short infrared laser pulses is employed for this second kind of sensor. It consists of 3-6 multiple laser beams, scanning the area in front of the car under certain angles. For each channel the distance from the obstacle is measured by measuring the time of flight between emission and detection of the reflected photons. From the distance information, also the differential speed can be calculated via differentiation. The direction and size of the obstacle can be recognized by comparison of the signals from the different channels.

This is the more sophisticated approach, since a higher level of information is available from this detection principle. However, work has to be done in future in order to size the macroscopic system down to a cost effective microsystem featuring micro-optics (e.g. micro assembly of lenses, laser diodes and photodiodes) and microelectronics (signal conditioning and driver ASICs, microcontroller).

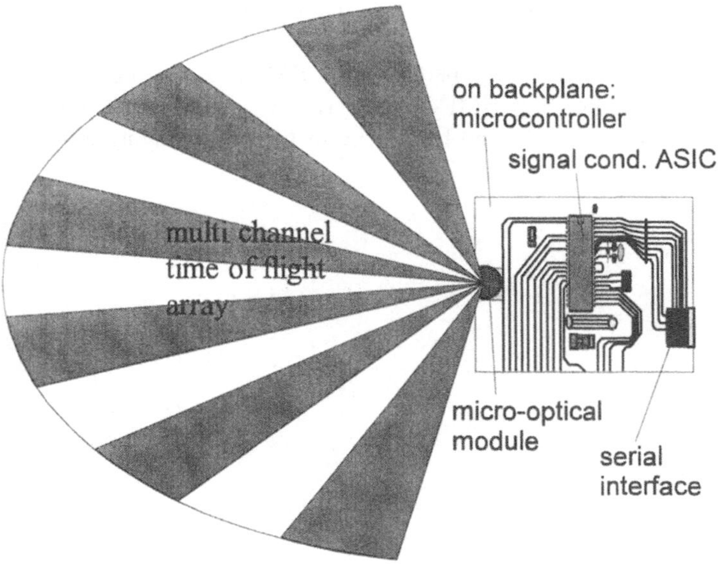

Figure 16. LASER based multi channel precrash sensor for time of flight measurement of distance, differential speed, direction and size of the obstacle.

6 Conclusion

Safety systems of the future will incorporate a wide area of different microsystems for automotive applications. In the sensor sector they will range from simple mechanical microsensors for bare acceleration measurement (in central control units and satellite units) up to complex laser or radar sensors for precrash detection. In all microsystems, a microprocessor is used with powerful algorithms for data evaluation and controlling of either the sensor itself (as shown in the case of the optical ODS and the precrash sensors), or the restraint actuators (in the case of the central ECU). Due to the high volumes of the systems, most of the rest of electronic needs will be integrated into applications specific ICs so that only a few discrete components will be needed further.

Not only this fact is the reason for a change in assembly technologies. They will switch over from macroscopic printed circuit design towards microtechnologies as shown for the satellite units. Other subsystems will follow featuring flip-chip and chip scale packaging technologies on MCM level. Micromachining, as for accelerometer fabrication already implemented will also be applied for microstructuring and mounting of lenses in optical microsystems.

7 Acknowledgments

The authors want to thank W. Kammerlander, M. Weinacht, G. Brinks, T. Ohgke, H. Seidel and V. Tiederle from the TEMIC Sensors Division. Special thanks also

to H. Spies and M. Spies at IBS for many fruitful discussions on ODS and precrash detection.
This work was supported in parts by the BMBF of the Federal Republic of Germany by funding the HMI²-project. Hereby, we want to acknowledge D. Ricken from VDI-VDE-IT and H. Laucht for their encouraging work in initiating this project.

Situation Appropriate Airbag Deployment: Child Seat Presence and Orientation Detection (CPOD)

Thierry Goniva
IEE s.à.r.l., Z.I. Findel, 2b, route de Trèves, L-2632 Luxembourg

Abstract: In the last years, the auto industry has grown increasingly aware of the potential dangers of passenger airbag deployments, especially for children placed in a child seat. As a solution to this problem, IEE has developped the CPOD system, based on its already hugely successful Passenger Presence Detection (PPD) sensor foil.

IEE's CPOD system is divided into two components: the transmitting and receiving antenna loops, integrated in the passenger seat, and a resonator circuit fitted in the socket of the child seat. The need for extremely low radiation and a cost-effective system has led to a very basic antenna structure, which consists of single-loop coils printed on the existing PPD sensor mat. Weak electromagnetic fields (emitted by the transmitting antenna) are used to establish a communication between the resonators and the processing electronics as well as to supply energy to the low-power resonator circuitry. Using both the information encoded in the received signals and their intensity, the processing electronics are able to take a decision on the presence, orientation and type of a child seat. Thanks to the redundancy and the multiple modulation of the transmitted information, the necessary reliability and interference immunity for a system which controls airbag deployment behaviour is obtained.

Future developments will integrate an Occupant Classification into IEE's sensor foil to offer even more parameters to modulate airbag deployment.

Keywords: smart airbags, child seats, occupant sensing

1. Introduction

1.1 Why Child Seat Detection?

While airbags have proven to be a very effective means for protecting car occupants during a crash, in some cases they lead to new dangers. One of these cases is the danger the passenger-side airbag represents for infants placed in a rear facing child seat mounted on the passenger seat. In this configuration, the child's head is too close to the airbag cover. In case of a deployment, the child's head is submitted to accelerations that can reach 100G.

Although the safest place for children in a car still is an appropriate child restraint system on the rear bench, some types of cars (e.g. roadsters, pickup trucks) do not offer this possibility. Often the argument that eye-contact with the driver relaxes a child and that the driver does not have to turn around to look after the child is also brought up in favor of placing infants on the front passenger seat.

In the USA, every year approximatively 2350 children seated in rear facing child seats mounted on the front passenger seat are involved in crashes with airbag deployment. As of November 1, 1997, 12 children have reportedly been killed by a deploying airbag in such crashes (source: NHTSA). In most of these crashes, the drivers have either not been injured at all or had only minor injuries. This has led to a growing public awareness of the problem, which in turn has created a demand in the automotive industry for a system which could provide a solution.

Such a child seat detection system should not only detect the presence of a child seat, but also its orientation (forward facing, rearward facing) and its type. The child seat type information (age category, orientation allowed by design) is important to check whether the child seat is mounted the way it has been designed to, and also because of the necessity to switch the airbag off for certain types of front facing child seats. To prevent misuse of a child seat, the system should also detect whether a child seat is incorrectly mounted (out-of-position).

1.2 The basis for IEE's CPOD system, the PPD sensor

In 1990, the german Airbag Task Force, constituted by five german auto manufacturers, defined the need for a passenger presence detection (PPD) sensor, that should assure detection of a person in the front passenger seat in order to prevent airbag deployment on an empty passenger seat. The purpose of this system is to reduce repair costs after a crash, a demand issued especially by insurance companies. In addition to the passenger airbag, the corresponding side airbag and seat belt pretensioner could also be deactivated.

Out of 30 different physical principles, the Airbag Task Force opted for a solution based on IEE's FSR technology.

The **F**orce **S**ensing **R**esistor (FSR) is based on a reproducible surface-effect. The typical construction of a FSR, as it is shown in figure 1, is based on a sandwich of two polymer films or sheets and a spacer. On one sheet a conducting pattern is

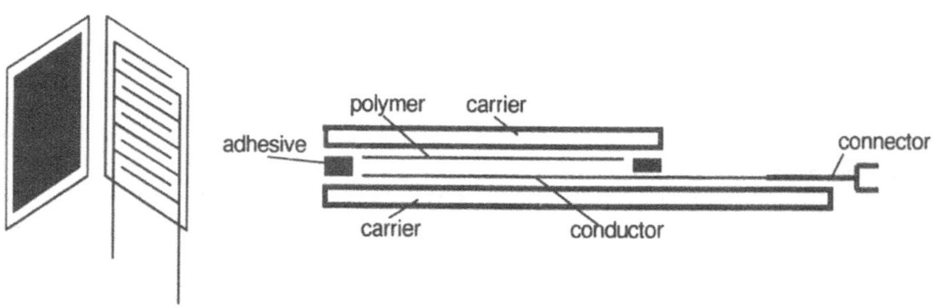

Figure 1: structure of an FSR sensing element

deposited (screen-printing) in the form of interdigiting electrodes. On the other sheet a proprietary polymer is deposited. Finally the two sheets are faced together so that the conducting fingers are shunted by the conducting polymer.

If a force is applied upon the FSR the electric resistance of the sensor changes according to the amount of force. Depending on the size of the used force actuator and the amount of the force, a more or less greater number of interdigiting electrodes are short-circuited.

In a PPD sensor, typically about 40 or more FSR sensing elements are distributed over the seat surface and connected in parallel, in such a way that a child with a weight of 12 kg is detected in any position in the seat. If the resistance is above a fixed threshold, the seat is "unoccupied". If on the other hand the measured resistance is below a fixed threshold, the system gives the message "seat occupied".

This requires every PPD sensor to be custom tailored to each seat design and seat assembly. Due to the different weight distribution, objects like briefcases or bags will generally not be detected as human beings.

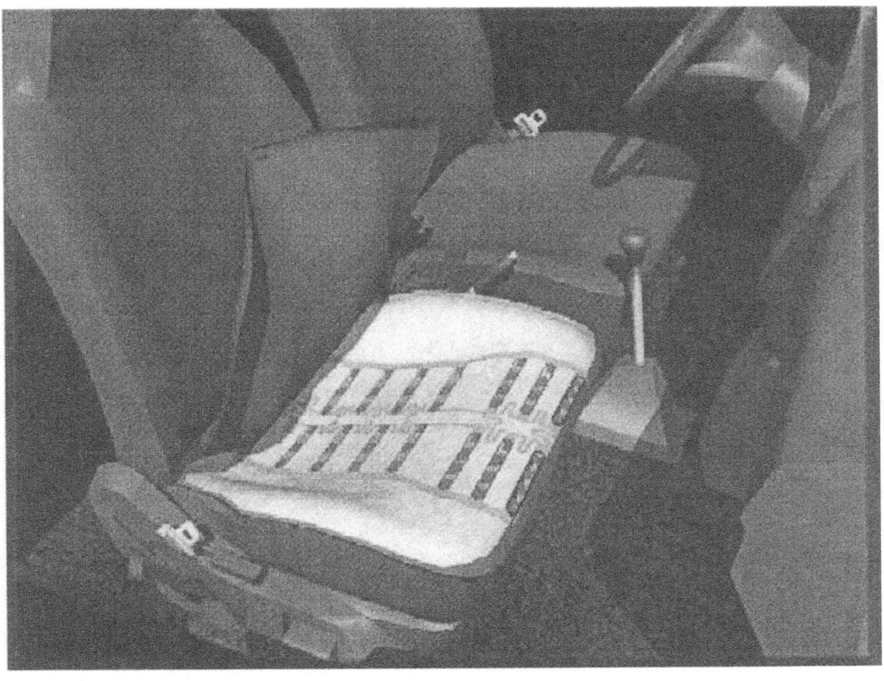

Figure 2: CPOD/PPD sensor mat in a passenger seat

The PPD sensor has been in high volume production since January 1994 and is thus widely accepted by the car and seat manufacturers. Together with the possibility to screen-print an additional antenna structure on the existing sensor mat, this has opened the way for developing the sensor into an electromagnetic CPOD at reasonable extra cost.

2 Functional Principles of the CPOD

2.1 System Overview

The CPOD system is basically a resonator tag system, consisting of the following components:

- 2 LC-resonator circuits with low power modulators, fitted in the socket of the child seat
- the antenna structure, printed on the PPD sensor
- the electronic interface, which generates and processes the signals for the CPOD/PPD system, decides on the occupancy status and transmits this status via a serial protocol to the airbag ECU

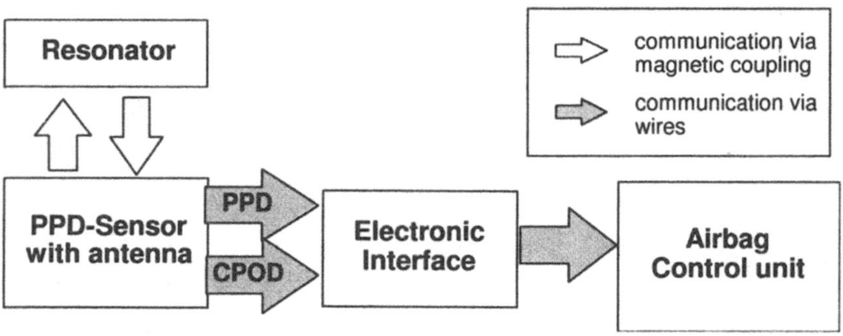

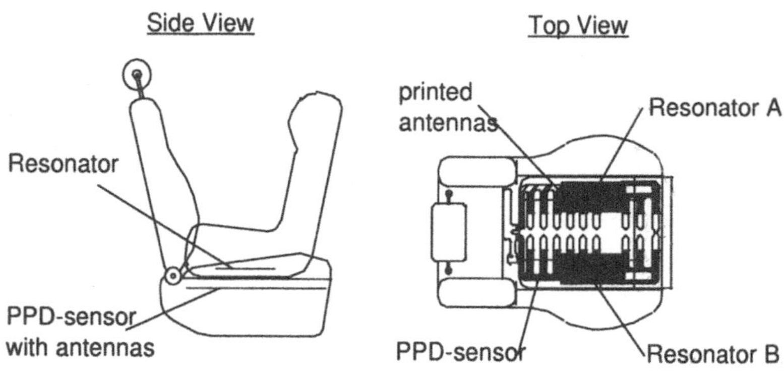

Figure 4: Position of the CPOD system in the passenger seat

The electronic interface is directly connected to the sensor mat, both parts are integrated into the seat cushion.

The antenna structure is made up of the transmitting antenna and two different receiving antennas (left and right side). The two resonators transmit different signals, so that together with the two receiving antennas, an orientation detection is made possible (left resonator on left antenna and right resonator on right antenna ⇒ child seat forward facing, and vice-versa). The fact that there are two resonators and two receiving antennas creates a redundancy, which is important for the safety of the system.

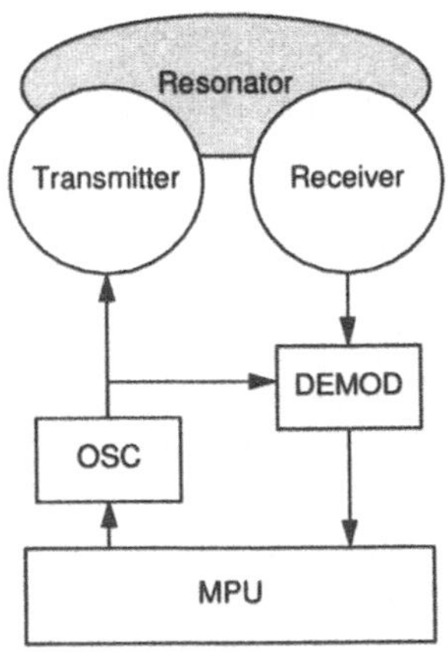

Figure 5

The functionality of the system is as follows (fig. 5): An oscillator generates the sine-wave current, that, after amplification, is sent through the transmitting antenna coil. The resulting electromagnetic field is exciting the resonators, supplying them with power. The resonators modulate the signal, transmitting information on their type (left (A) or right (B) resonator) and a digital protocol containing the child seat specific information. This information is extracted by the demodulator from the voltage induced in the receiving antennas. A comparison of the four received resonator signal amplitudes (left (A) resonator in left antenna, B resonator in left antenna, A resonator in right antenna, and B resonator in right antenna) enables the processing electronics to find out which resonator is closer to what antenna and thus to decide whether a child seat is present, forward facing, rearward facing or placed in an incorrect position.

2.2 The antenna structure

As shown in fig. 6, the antenna structure printed on the PPD sensor consists of a transmitter loop surrounding the two interconnected receiving antenna loops. To avoid a dead zone in the middle, the two receiving loops slightly overlap.

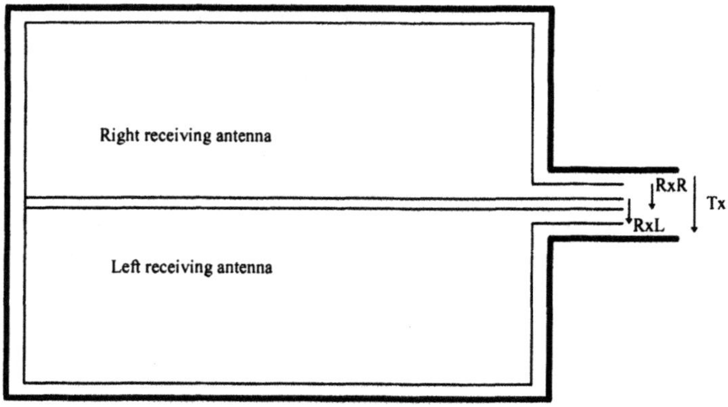

Figure 6: Antenna structure

Figure 7 shows a simplified representation of the magnetic coupling signal path, with one resonator and one receiving antenna.
A sinusoidal current I is sent through the transmitter loop. This current generates a magnetic field, which in turn induces a voltage in the resonator coil. The resonator circuits are powered by the energy provided by the magnetic field. The resonator information message is modulated onto the current in the resonator coil (function f''). Thus, the resulting magnetic field B is the superposition of the original transmitter field and the modulated field generated by the resonator. The voltage U

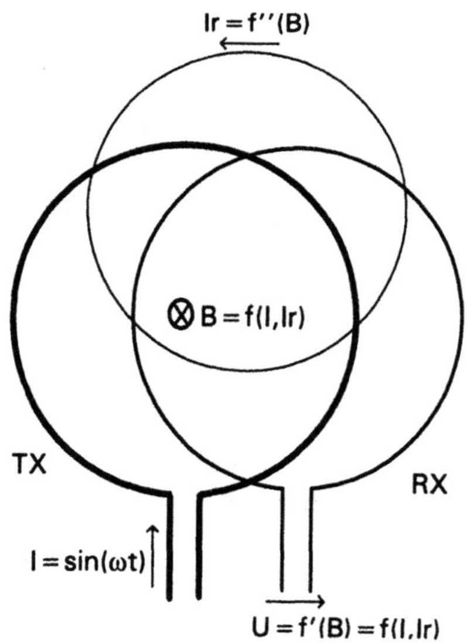

Figure 7: Schematic representation of the magnetic coupling of the antennas and one resoantor

induced in the receiving loop Rx is a function of the resulting field. The resonator information is extracted by demodulation of U. The closer a resonator is to a certain receiving antenna, the stronger the influence of Ir on U, or on the demodulated signal, will be.

A resonator placed over a certain receiving antenna causes a strong signal in this antenna and a much weaker signal in the other receiving antenna, thus enabling the processing electronics to decide which resonator is placed on which side. A resonator is considered detected if its signal in one of the receiving antennas exceeds a certain minimum and if its binary message was correctly received. If both resonators are detected and their position is well defined, the message "forward facing child seat" resp. "rearward facing" will be sent to the airbag ECU. On the other hand, if both resonators are correctly detected but their position is not clear, the message will indicate "out-of-position child seat".

2.3 Resonator signal modulation

As a system that controls airbag deployment, the CPOD must have a high degree of immunity against electromagnetic interference. There must be a clear differentiation between resonator signals and noise sources as portable TV sets, cellular phones, etc.. The signal and field strength (down to 0.1μT) being very low, the required immunity is obtained by a sophisticated modulation and the transmission of a binary message, which assures an unmistakable detection of the resonators.

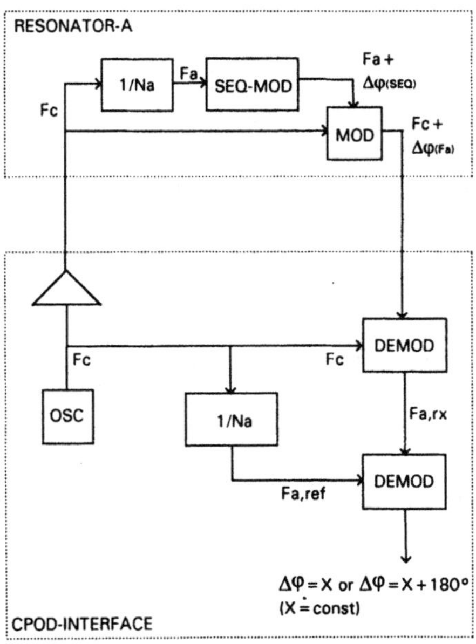

Figure 8: CPOD signal path

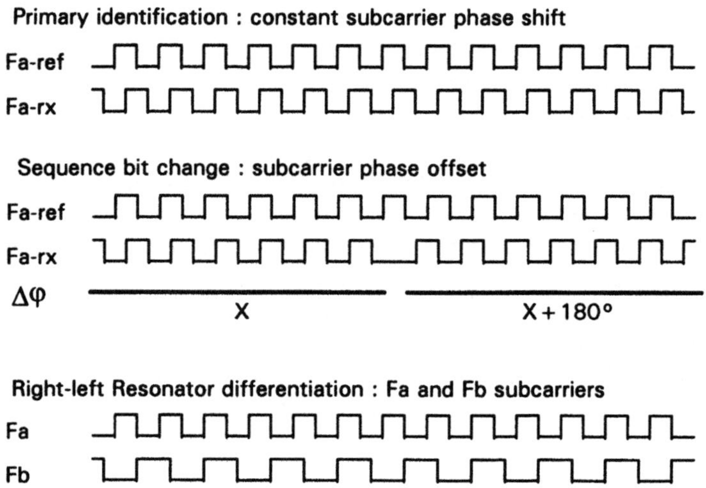

Figure 9:Subcarrier signals and phase modulation

Figure 8 shows the signal path from the transmitter oscillator to the demodulated binary resonator message. Figure 9 represents some of the signals involved. The oscillator genarates a sinusoidal wave with the frequency Fc (carrier frequency, 130kHz band). Each resonator generates a subcarrier Fa = Fc / Na or Fb = Fc / Nb, which is the primary resonator identification. In order to facilitate the separation of the two resonator signals in the receiver, Na and Nb are chosen in such a way that they contain different prime factors. In addition to the resonator type distinction obtained by the different subcarrier frequencies, a bit sequence containing a synchronisation header, information like child seat type, resonator type (redundant, for confirmation), etc., and a parity bit, is transmitted by the resonators. By phase shift (180°), both subcarriers are modulated with the bit sequence (SEQ-MOD). As a second modulation step, the carrier Fc is modulated with the modulated subcarrier (MOD). Physically, the phase shift is obtained by switching the resonance frequency of the LC-circuit.

The voltage on the receiving antenna terminals is demodulated using synchronous demodulation. A first demodulator stage extracts the subcarrier from the modulated carrier, the second stage extracts the bit sequence from the subcarrier Fa,rx.

Following a similar scheme, all resonator / receiving antenna combinations are evaluated: resonator A on the right antenna, resonator B on the right antenna, resonator A on the left antenna, resonator B on the left antenna. The bit sequence itself is preceded by a period of constant subcarrier. This is used to determine the four resonator amplitudes, that in turn are necessary to decide on a child seat orientation.

To consider a resonator as detected, both of the following conditions must be met:

- a minimal subcarrier amplitude has to be measured in at least one of the receiving antennas
- The bit sequence of the corresponding resonator must be correctly detected in the same receiving antenna.

3 Realization of the interface electronics

Figure 9 shows a block diagram of the CPOD interface electronics. As in many automotive ECUs, most of the circuitry is integrated into an ASIC.

3.1 The power supply

The power supply must provide sufficient voltage to drive the transmitter current (up to 500 mA) through the transmitter loop (max. 15 Ω) across the automotive voltage range (9 to 16V). in addition, a stabilized 5V power source must be provided for the microcontroller and the logic circuitry. Disturbances of the on-board electrical system (Schaffner pulses) must be filtered out in order not to disturb the correct functioning of the interface electronics.

3.2 The serial interface

The serial interface assures the communication with the airbag ECU. The interface hardware configuration and protocol are defined by the car manufacturer. Mostly, the communication is only from the CPOD to the airbag ECU, but a special dianostic mode used for production tests enables a two-way communication and a faster transmission rate. Like the power supply lines, the serial line has to show a very high EMC immunity.

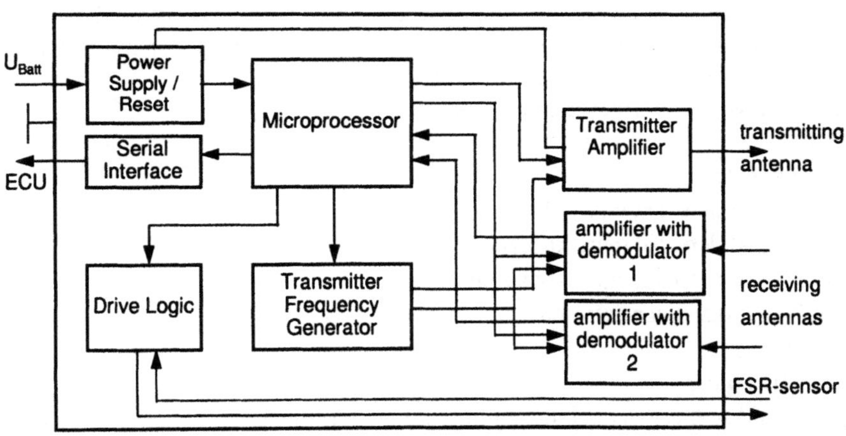

Figure 9: CPOD interface electronics block diagram

3.3 The PPD drive logic

The PPD drive logic provides the necessary signals for the PPD sensor, including resistance threshold detection and self-diagnostic (interruption or short of the PPD conductors). The PPD signals are fed to an AD input of the microcontroller.

3.4 The transmitter frequency generator

The transmitter frequency generator supplies the carrier signal (130kHz band) as well as the signals at subcarrier frequencies for the demodulators. To catch the exact resonating frequency of a resonator, the generator wobbles between 124..134 kHz.

3.5 The transmitter amplifier

The transmitter amplifier drives a sinusoidal current at carrier frequency through the transmitting loop. A current regulation adapts the transmitter current to the occupancy status, i.e. if a child seat is detected, the transmitter power is reduced. This scheme reduces heat dissipation and avoids saturation of the receiving circuitry.

3.6 The receiver circuitry

The receiver circuitry amplifies the signals from the receiving antennas and extracts the resonator information. Amplification is adjusted according to the incoming signal strength. The demodulation is done in a synchronous manner, using signals from the transmitter frequency generator. Very small phase shifts have to be detected (0.1°).

With the demodulator signals at its AD-inputs, the microcontroller determines the signal amplitudes and decodes the resonator bit sequences.

3.7 Microcontroller and software

The microcontroller directs and supervises all the operations of the CPOD. On hand of the signals at its AD inputs, it decides on the occupancy status and transmits it through the serial interface to the airbag ECU.

The following messages are possible:

CPOD:

- no child seat
- child seat detected forward facing
- child seat detected rearward facing
- child seat detected out-of-position
- only one resonator detected
- interference
- CPOD failure
- if a child seat is detected, the child seat type information is also transmitted.

PPD:

- not occupied
- occupied

- PPD failure

The strategy at what occupancy status to deactivate the airbag is left to the car manufacturer and implemented in the airbag ECU.

In order to be failsafe, the interface performs very extensive self-diagnostic functions covering the electronics itself and the sensor mat.

For production test purposes, a special diagnostic mode is implemented, allowing an external test device to call certain defined test and configuration procedures. These give insight into detailed measurement reports.

An EEPROM memory allows to configure certain seat-specific parameters during production,as well as to store some information for traceability.

4 Conclusion and Outlook

As it already happened with the PPD sensor, today IEE has the only child seat detection system in high volume production on the market.

In the future, smart airbags with multi-stage inflators will require an even more accurate sensing of the kind and position of car occupants. The next development of IEE's sensor mat will include an occupant classification system.

References:

[1] "The BMW Seat occupancy Monitoring System: A Step Towards "Situation Appropriate Airbag Deployment"", Klaus Kompaß, Michel Witte, SAE Paper

[2] "Automatic Passenger Presence Detection and Child Seat Orientation Detection", Andreas Hirl, Peter Popp, Joachim Uhde, Paul Schockmel, SAE Paper

[3] "Detection Of Passenger Presence And Child Seat Orientation", Bobby Serban, Paul Schockmel, Michel Witte, airbag 2000 Paper

Rollover Sensing (ROSE)

Dr. phil. nat. Gerhard Mehler[1], Dipl.-Ing. Bernhard Mattes[2],
Dipl.-Ing. Michael Henne[3], Dr.-Ing. Hans-Peter Lang[4],
Dipl.-Ing. (FH) Walter Wottreng[5]

[1..5] Robert Bosch GmbH, Automotive Equipment Division K8, Department K8/EES, P.O.Box 30 02 40, 70442 Stuttgart, Germany

Abstract

In order to prevent fatalities or serious injuries caused by rollover situations, Bosch has developed a system to detect these situations and initiate proper actions. To understand the need of this function, it is necessary to look at some statistical facts, which underline the importance of a rollover sensing device.

Taking into account different rollover types, the question appears: What physical unit describes the roll motion best ? After careful considerations, Bosch made the conclusion to prefer the angular rate concept. The advantages and disadvantages are shown and the physical principle of the algorithm is explained.

To compensate inaccuracies of angular rate sensors, acceleration sensors are implemented. They can take over, when the main sensors fail or do not show plausible results.

For the simulation of different rollover scenarios Bosch applies the multibody method in an overall environment of modeling, verifying, validating and visualizing tools.

1 Introduction

1.1 Definition Rollover

Rollover indicates vehicle overturn during the crash. It is defined as any vehicle rotation of 90 degrees or more, about any true longitudinal or lateral axis. Rollover can occur at any time during the unstabilized situation. Rollover does not apply to motorcycles. /1/

FARS (Fatality Analysis Reporting System) separates rollover into different situations:

- No Rollover
- First Event
- Subsequent Event
- Not Reported
- Unknown

1.2 Rollover Occurrence

How important is it to detect and handle this specific kind of crash ? To answer this question, it is necessary to see the distribution of all vehicle crashes:

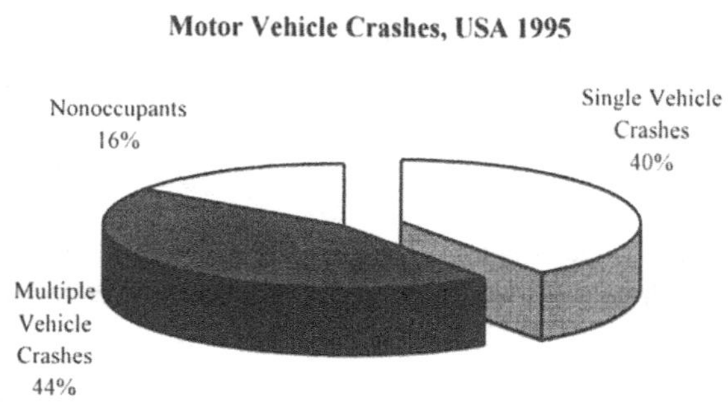

Fig. 1.1. Distribution of motor vehicle crashes in 1995 from a total of 41,823 crashes

Fig 1.1 divides motor vehicle crashes into 3 types: Single vehicle crashes, where only one car is involved, multiple vehicle crashes with more than one car, and crashes where pedestrians, pedalcyclists, and other nonoccupants are affected. Rollover only applies to single vehicle crashes. It plays an important role in this kind of crash (Fig. 1.2.).

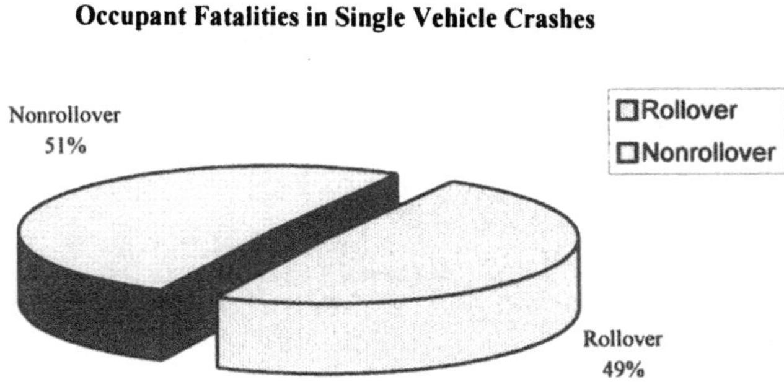

Fig. 1.2. Occupant fatalaties in single vehicle crashes, USA 1995, total: 16,725

Miscellaneous driving maneuvers can lead to a rollover. The most common ones are summarized in Fig. 1.3.:

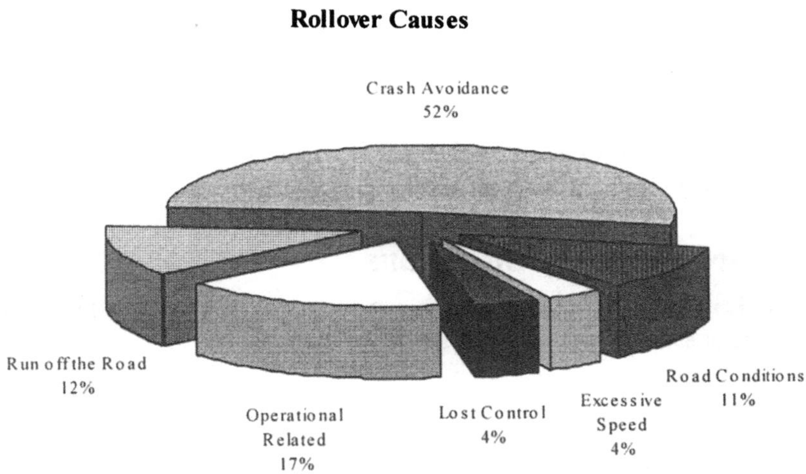

Fig. 1.3. First event rollovers, by crash type, USA 1992-1995, total 23,247

There is also a relationship between car type and rollover (Fig. 1.4.) . Location of center of gravity and track width are characteristics which can support roll propensity.

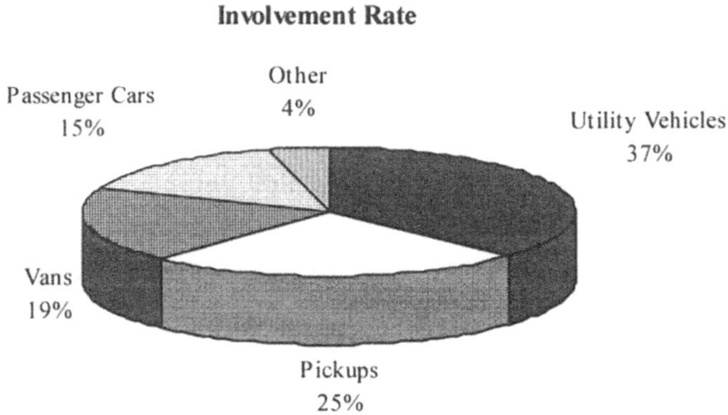

Fig. 1.4. Rollover involvement rate by vehicle type, USA 1995

The previous diagrams illustrate the importance, and need of a rollover protection device. They are based on NHTSA traffic safety facts. /2/

2 Typical Rollover Situations

In a typical rollover situation, the car gets out of control due to bad road conditions or steering maneuvers of the driver.

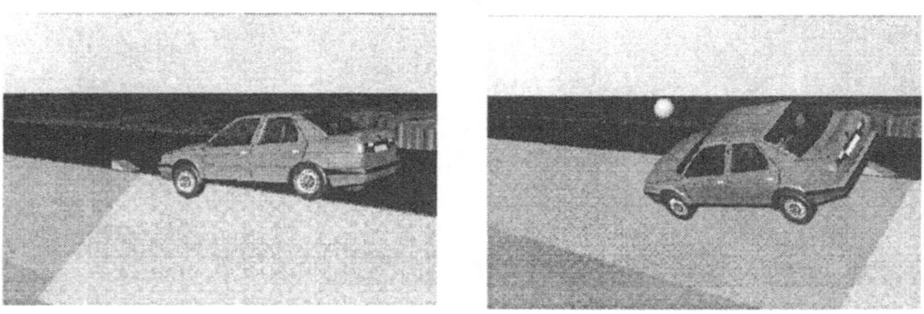

Fig. 2.1. Ditch Rollover, FASIM_C++ simulation

In the ditch scenario the vehicle loses ground contact (Fig. 2.1.). It is airborne in a free fall, where no acceleration can be measured. When the front tire contacts ground, it digs in the field and supports the started rollover process. Especially on unsurfaced roads or areas that commonly do not have guide rails, such as the US-Highways, the likelihood of this event is quite high.

The Curb Trip Rollover (Fig. 2.2) is also a realistic situation in real life. Because of a rapid decrease of the driving surface's friction coefficient, the car starts sliding off the road laterally.

The curb decelerates the car rapidly. This action causes a high moment of torque with respect to the center of gravity about the longitudinal axis. The MADYMO simulation shown below results in a short free fall.

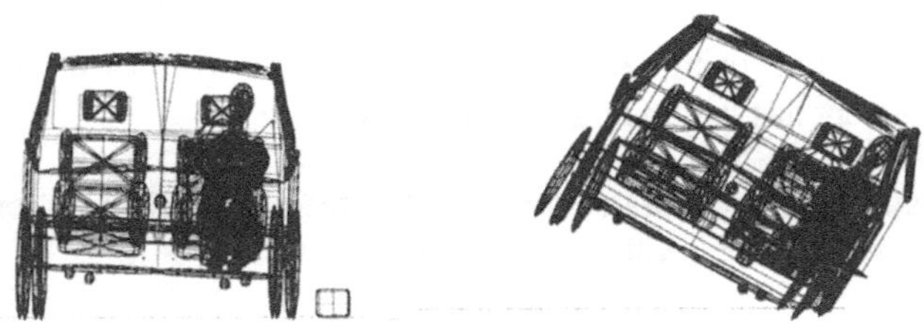

Fig. 2.2 Curb Trip Rollover, MADYMO3D simulation

The corresponding real world situation to the so called 'Ramp Test' is less likely, but it allows functional testing in crash facilities with fewer expenses.

The 'Ramp Test' is performed by driving the car with high speed (about 80kph) with one side onto a ramp (Fig.2.3.).

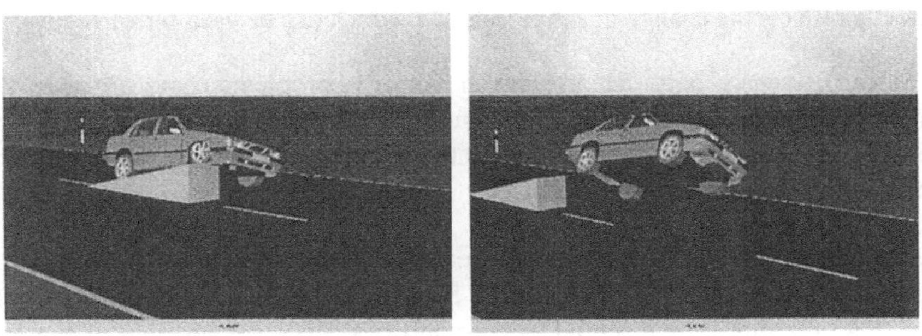

Fig: 2.3 Ramp Rollover, FASIM_C++ simulation

The car gains a high vertical acceleration component, which decreases until the vehicle is airborne. After leaving the ramp, it moves with an approximately constant angular rate in a free flight, till it hits ground.

3 Rollover Sensing at Bosch

3.1 Sensing Concept

In todays existing systems, reversible safety devices like rollover bars or extendablr 'headrests', are deployed to protect the passenger in case of a rollover. Irreversible devices, like side airbags and belt pretensioners are considered to be life savers for the future. To avoid unnecessary costs and to guarantee occupant safety, the system should be able to trigger precisely. This means, that only when a rollover occurs, the appropriate protection mechanisms are supposed to be fired exactly in the right moment.

First concepts to recognize these situations were made with accelerometers. The idea was to use the signal from the airbag sensors and add the rollover algorithm in the ECU, without using any additional sensors. It turned out that the acceleration concept has two main disadvantages:

1. Acceleration signals generated by simple steering maneuvers or different misuse tests can mimic a rollover situation.
2. When a freefall flight event occurs during rotation, the acceleration sensors can not measure the gravity components in this period of time.

A pendulum concept also failed because of unpredictable behavior (resonance).

The only concept, best suited for rollover sensing is measuring angular rates. An angular rate sensor provides a signal that is proportional to the rotational speed about its sensitivity axis.

The angular rate is the closest related signal to the vehicle orientation. It directly reflects the change of the orientation, which can be used for early firing.

Since the sensor only detects angular velocity, problems come up when the initial position of the vehicle is unknown. For a long period of time it can not be reliable, what the algorithm has integrated. Due to inaccuracies of sensor and software, the integrated value is flawed.

In addition, there exists a problem when the roll rate falls below the integration threshold of the algorithm. The effect : Slow rollovers could not be recognized. For these reasons, it is essential to constantly calculate the position of the car. This is accomplished by 'low g' accelerometers.

To cover general rollover maneuvers two angular rate sensors (about longitudinal and lateral axis) would be needed. Three accelerometers are necessary for calculating the initial condition and to perform plausibility checks (Fig. 3.1.).

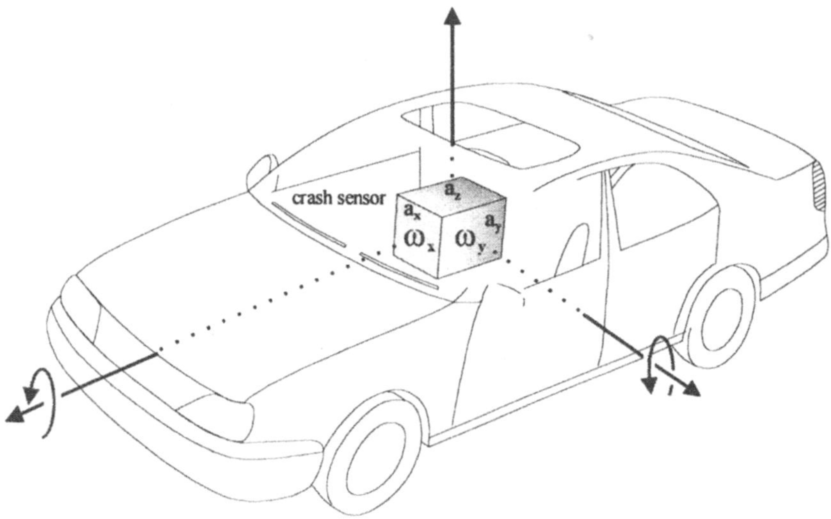

Fig. 3.1. Rollover function components integrated in the crash sensor

3.2 Main Algorithm Criteria

The main algorithm is based on angular rate signals. It contains two criteria:

1. Stability Criterion
2. Energy Criterion

While the Stability Criterion deals with the height of the center of gravity, the Energy Criterion evaluates the rotational and potential energy.

3.2.1 Stability Criterion

The correlation between a stable and unstable condition is shown in Fig. 3.2.

A vehicle (here presented by a box) can spin around three axes. In the case of a rollover we are only interested in motions about the x and y axis.

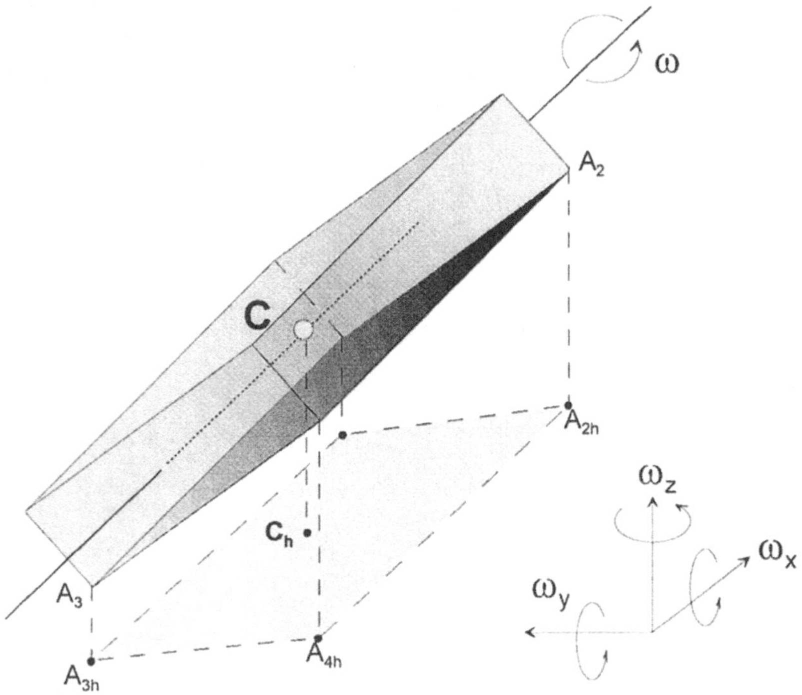

Fig. 3.2 Stability Criterion

It is open to examine when the system is going to be unstable. The projection of the four tire points ($A_{1..4}$) to the horizontal plane creates a parallelogram $A_{1h..4h}$. If the projected center of mass C_h is within the parallelogram the stability condition is defined to be fulfilled.

This criterion has been derived from statics. A point C_h out of the marked zone is defined as an unstable state.

The mathematical description of the orientation happens through the Kardan-Angles (/3/) and the projections are transformations of coordinates from the spun system into the horizontal system (road plane).

3.2.2 Energy Criterion

In fast rollovers, such as screw rollovers, the car reaches a high angular rate value (e.g. 300 deg/sec) in an early stage. It can be foreseen, that the gained rate will lead indubitably to a rollover.

The rollover can be predicted by comparing rotational and potential energy (Eq. 3.1. and Fig. 3.3.).

$$\Delta E_{rot} > \Delta E_{pot}$$

$$\frac{1}{2} \cdot \Theta \cdot \omega^2 > m \cdot g \cdot \Delta h$$

Eq. 3.1. Energy Criterion basic equation

The algorithm checks whether the rotational energy is sufficient to exceed the height Δh.

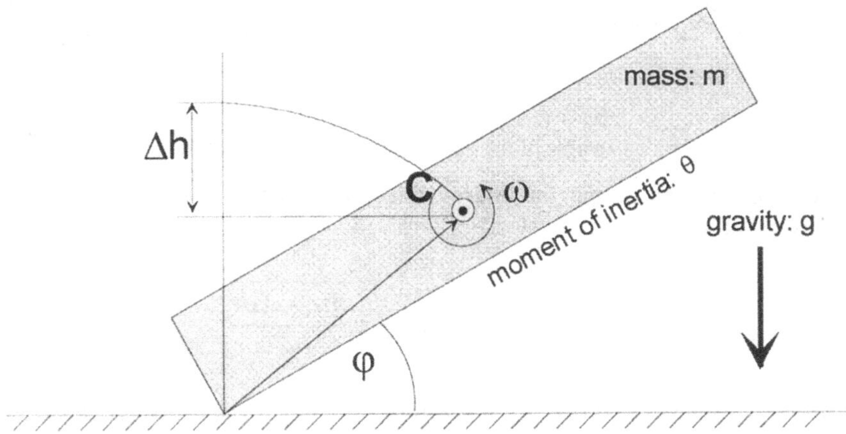

Fig. 3.3. Energy Criterion, two dimensional model for a rollover

Fig. 3.3 describes the mathematical terms used.

3.2.3 Simulation Example

Both criteria are simulated based on a Ramp Rollover Test (refer to chapter 2). The test conditions were set up in a way that the vehicle rolled onto its roof (180 degrees). Because there was no pitch component involved, it was not taken into account for this simulation. Therefore the orientation can be described with the roll angle only (Fig. 3.4.).

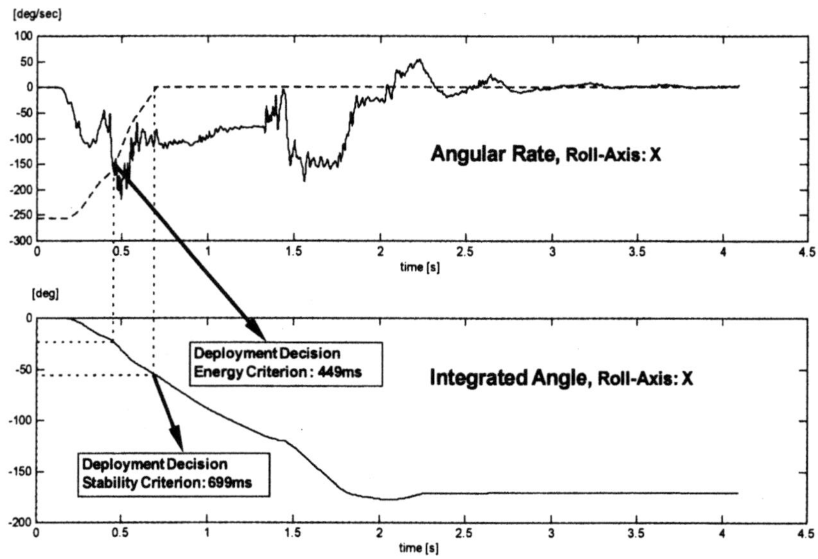

Fig. 3.4. Simulation example of the deployment effectiveness of the Energy Criterion

In the angular rate graph the energy threshold (dashed line) is included as a function f of ω:

$$f(\omega) = \omega_{crit} = \sqrt{2\frac{m \cdot g \cdot \Delta h}{\Theta}}$$

When ω exceeds the critical angular rate ω_{crit}, a rollover happens. The Energy Criterion in this case fires after 449ms and at 20 degrees, while the Stability Criterion recognizes a roll scenario after 699ms and at about 55 degrees. This strategy allows firing 250ms before the cars actually flips.

3.3 Plausibility Algorithm and Initial Condition

In order to guarantee a safe fire decision, acceleration signals are used. The evaluation of the signal of three orthogonal acceleration sensors lead to the initial orientation and provide, in a modified way, a plausibility check for the main algorithm.

The principle is shown in Fig. 3.5.:

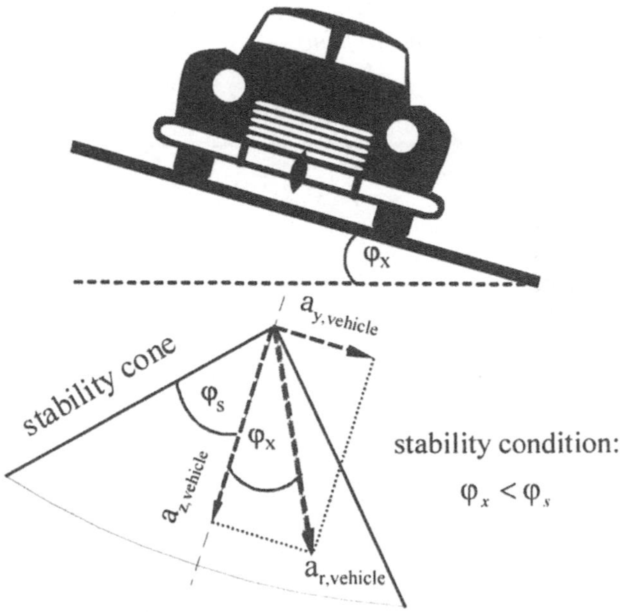

$$\varphi_x = \arctan\frac{a_y}{a_z}$$

φ_S : stability angle

Fig. 3.5. Initialization and Plausibility Algorithm for the Roll Angle

The result of the *arctan* function is the needed roll angle. It can be calculated as the initial orientation or as the plausibility check. To avoid interference and thus false angle values by driving conditions, it is necessary to implement the appropriate filter method. If φ_x exceeds the stability angle φ_S, then the plausibility algorithm will set the output of the main algorithm free and allows the system to fire.

For the pitch angle, a similar calculation is used. Replace acceleration a_y with acceleration a_x. The result is φ_y.

4 CAE - Tool Support

Together with the University of Duisburg, Bosch has set up an application and development environment (Fig. 4.1). /4/

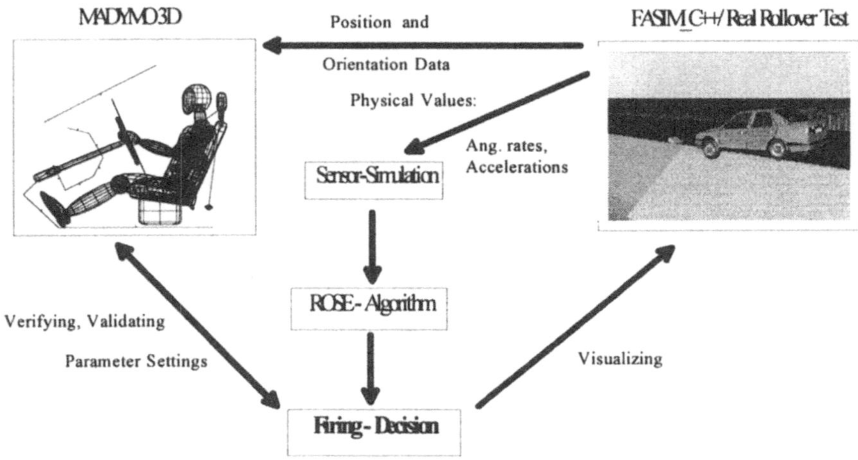

Fig. 4.1. Application and algorithm development concept

It consists of two multibody simulation programs: MADYMO 3D and FASIM_C++. While MADYMO 3D takes care of the occupant behavior, FASIM_C++ simulates the vehicle dynamics.

After implementing the car's multibody model in FASIM_C++, the specific car needs to be validated by real rollover scenarios.

The outcome data, either from a real test or generated by the simulation, can be used in two different ways. It serves as an input for:

1. the sensor model (e.g. angular rate sensor transfer function)
2. MADYMO 3D

In case no. 1, the sensor model feeds the ROSE-Algorithm with sensor specific data. The algorithm decides, if this data is sufficient for a deployment. The whole action can be visualized by FASIM_C++ by an animated video sequence. A ball, close to the front window, indicates a fire decision (refer to Fig. 2.1.).

In parallel to this task, MADYMO 3D calculates, based on position and orientation data, the passenger behavior. During the time the system deploys, the occupant reaction can be analyzed.

A closed application loop is realized. If the algorithm does not show suitable results, it can be modified by re-setting parameters.

There are some important benefits due to this concept:

- test expenses can be limited, because of the ability to simulate various scenarios
- appropriate sensor(s) can be analyzed, functional in every situation
- possibility of setting fire requirements
- algorithm weaknesses can be found, verified, validated and eliminated
- availability of sophisticated engineering tools and methods to realize satisfactory results

5 Conclusions

Rollover sensing is a new feature on the market. It represents a big enhancement in the restraint system area.

The concept to predict a rollover makes the function very strong and flexible. In contrast to the airbag function a future statement is made, based on the actual sensing values.

The idea is to include the rollover function into the actual airbag concept. So it would have synergy effects by using the present hardware.

The main challenge now is the development of a cost-optimized angular rate sensor. For this reason, Bosch is in an advanced state of developing a suitable and costwise acceptable silicon surface micromachined gyro for rollover application.. Deliberating the number of roll events in contrast to pitch events, the roll events are more likely.

In the first approach pitch rollover situations are neglected. The result is canceling one angular rate and one acceleration sensor. Also the algorithm can become less complex.

The statistic information (shown previously) helps increase customer safety awareness for rollover situations. Therefore the acceptance of this function should be high. It may possibly become a safety feature which nobody wants to miss anymore.

6 Literature

/1/ U.S. Department of Transportation FARS: *Roadsize Hazards 1995*, Washington D.C., 1996

/2/ U.S. Department of Transportation NHTSA: *Traffic Safety Facts 1995*, Washington DC, 1996

/3/ Prof. Dr.-Ing. habil. M. Hiller, : *Mechanische Systeme*, Springer Verlag 1983

/4/ Prof. Dr.-Ing. habil. M.Hiller, Dipl.-Ing. R. Bardini, Duisburg, Dr.-Ing. D. Schramm, Dipl. Ing. Th. Herrmann, Stuttgart : *Einsatz der Simulation zur Entwicklung von intelligenten Rückhaltesystemen*, VDI Berichte 1354: Innovativer Insassenschutz im PkW, Oktober 1997

Yaw Rate Sensor in Silicon Micromachining Technology for Automotive Applications

W. Golderer, M. Lutz*, J. Gerstenmeier, J. Marek*, B. Maihöfer*, S. Mahler*, H. Münzel* and U. Bischof*

R. Bosch GmbH, Automotive Equip. Div. K1, P. O. Box 30 02 40, D-70442 Stuttgart, Germany
*R. Bosch GmbH, Automotive Equip. Div. K8, P. O. Box 13 42, D-72703 Reutlingen, Germany

Abstract: A new generation of yaw rate sensor with a linear acceleration sensor, both based on silicon micromachining, is presented. The sensor is designed for mass production and high performance applications like the Vehicle Dynamic Control System VDC. For the yaw rate sensor a combination of surface and bulk micromachining is used . The linear acceleration sensor is build in surface micromachining. Both measuring elements and the ASIC are assembled on a hybrid. This leads to an advantage in design, signal evaluation and packaging. The designs of the measuring elements, the assembly, the interface and the characterization results of the current device are presented.

Keywords: Yaw Rate Sensor, Acceleration Sensor, Silicon Micromachining.

1. Introduction

For the Vehicle Dynamic Control System VDC BOSCH developed a yaw rate sensor with a vibrating cylinder which is in series production since 1995. To measure the lateral acceleration of the car a mechanical acceleration sensor is currently used. To decrease the dimensions and weight and to enhance the performance the development of a new generation of yaw rate sensor with linear acceleration sensor was started. The series production will start in the middle of 1998.

For the VDC the measurement of the yaw rate and of the linear acceleration is necessary. Therefore this new sensor measures the yaw rate with a measuring range of ±100°/s and the linear acceleration with a measuring range of ±18 m/s^2. The sensor includes two measuring elements, one for yaw rate and one for linear acceleration. Both are manufactured in Silicon-Micromachining. The element for

yaw rate is a combination of Bulk- and Surface Micromachining. The measuring element for linear acceleration is pure Surface Micromachining.
Mechanical balancing of the sensorelements is avoided by implementation of a new dry etching process and precise photolithography.
Both measuring elements are assembled upon a hybrid, together with the ASIC. This hybrid is assembled with the magnet for the actuation of the yaw rate measuring element in a metal-housing.
All this is packaged into a plastic housing for direct mounting in the passenger compartment of the car. The power supply is 12V. The connector has 6 pins.
The sensor is an absolute measuring device with an analog interface and provides a reference voltage for the A/D-converter of the system. An externally triggerable built-in-test (BITE) tests the measuring element and the evaluation curcuit of the yaw rate sensor and adds an offset of 25 deg/s to the yaw rate.

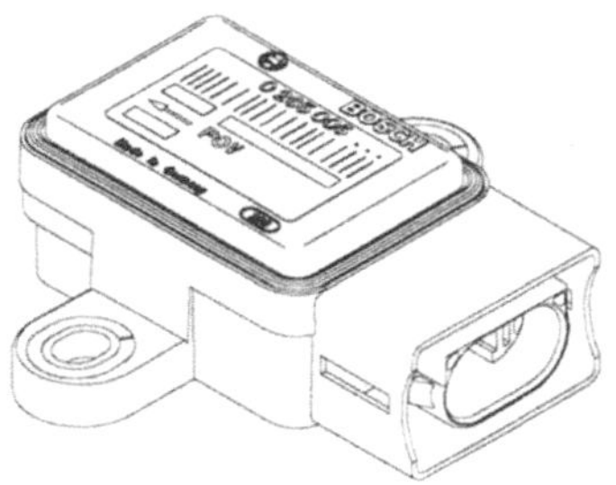

Figure 1: Housing of the sensor

2. Function Principle and Design

2.1 Yaw rate sensor measuring element

The sensor consists of two surface micromachined capacitive accelerometers which are located on their corresponding oscillating seismic masses (Figure 2). The measurement axis of each accelerometer is orthogonal to the direction of oscillation of the seismic masses. A rotation around the third orthogonal axis, a yaw rate, imposes a coriolis force on the accelerometers.

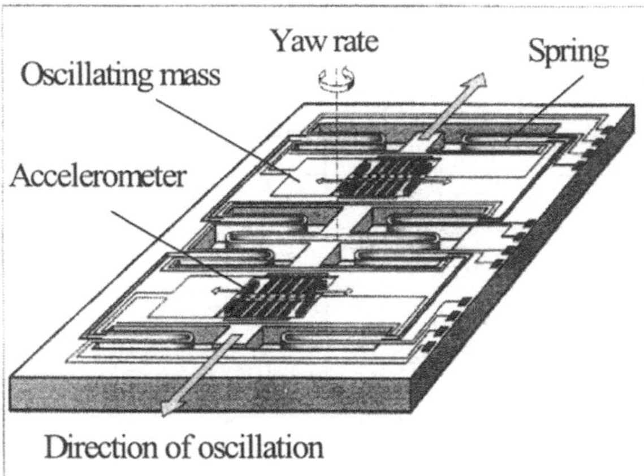

Figure 2: Design of the sensing element

The measurement of the difference between the two accelerometers filters out the linear acceleration and doubles the coriolis signal. Synchronous demodulation using the velocity of the oscillator generates a signal which is proportional to the yaw rate. The two bulk micromachined masses oscillate with a frequency of 2 kHz and a amplitude of 35 μm. Common published sensor designs of micromachined yaw rate sensors [1-6] employ one element as the oscillator and the accelerometer. The disadvantage of such systems is that the oscillator and accelerometer are mechanically and electrically coupled, hindering independent optimization of both elements. Our design, on the other hand, physically separates oscillator and accelerometer which enables independent optimization.

The two oscillating masses resemble and function as a tuning fork. The major advantages of a tuning fork are the stable center of gravity and the compensation of all forces and moments of inertia within the chip. Therefore, there is no need for a special mounting technique of the sensing element.

A one dimensional system with two spring-coupled masses has two natural oscillating modes: in- and out of phase. The system was optimized to separate the two modes by carefully selecting the values for the spring constants and masses.

The system was also optimized with respect to mechanical crosstalk, by decoupling the sensing and oscillating directions. A design in planar technology allows precisely defined structures, in the plane of the chip. On the other hand, dimensions perpendicular to the plane of the chip are process dependent and more difficult to control. Special folded beams increases the stiffness of the oscillator in the direction of the coriolis sensing elements in order to minimize oscillations in this direction.

The Q-value of the system is 1200 at atmospheric pressure, high enough to stimulate the oscillator with small Lorenz forces. Other known designs with vertical detection are dependent on a very narrow gap between the moving plate and a second electrode for detection, which leads to high damping coefficients [7].

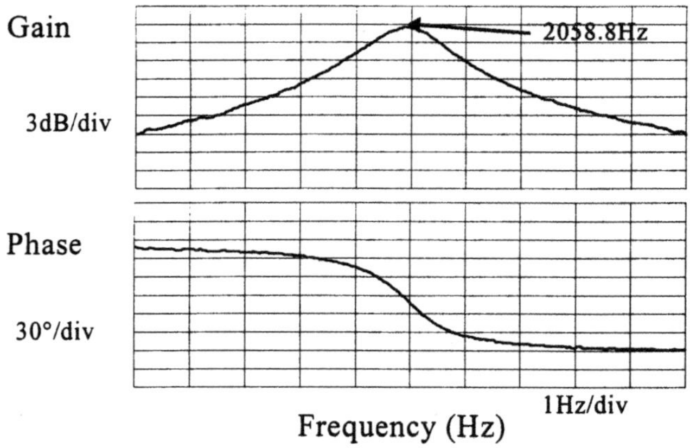

Figure 3: Frequency response of the oscillator

The oscillator is electromagnetically stimulated. The oscillator is mounted under a permanent magnet with a magnetic field through the chip, orthogonal to the surface (Figure 7).

Two conductive paths lead to the masses: one path drives the oscillator via the Lorenz Force [5], while the other provides feedback of the velocity of the masses through the induced voltage. The placement of the conductive paths allows only the stimulation of the out of phase oscillation mode.

The chosen oscillating amplitude of 35 μm leads, at a frequency of 2 kHz, to a coriolis acceleration of 150 mg at 100 deg/s.

The coriolis force is detected with the previously mentioned capacitive accelerometers. Comb structures which are suitable for the measurement of accelerations in the plane of the chip [9, 10], are employed. The accelerometers were optimized with respect to three critical parameters: wide dynamic range, large bandwidth and low cross axis sensitivity to the acceleration of the oscillator masses upon which they are mounted.

The dynamic range was maximized so that the low frequency linear accelerations that occur during normal operation of the car (up to 5 g) and coriolis acceleration (as small as 1 mg at 2kHz) are detected. The linear acceleration is later electrically filtered out. The bandwidth was also maximized to allow detection, in phase, of the 2 kHz coriolis force. The seeming conflict in designing for a high resolution (weak spring constants) as well as high bandwidth (stiff spring constants) was resolved by using weak springs in combination with closed loop position feedback. The natural mechanical resonance frequency was thereby increased from 2 kHz to over 10 kHz. The frequency response of the same sensor under open and closed loop conditions is shown in figure 4.

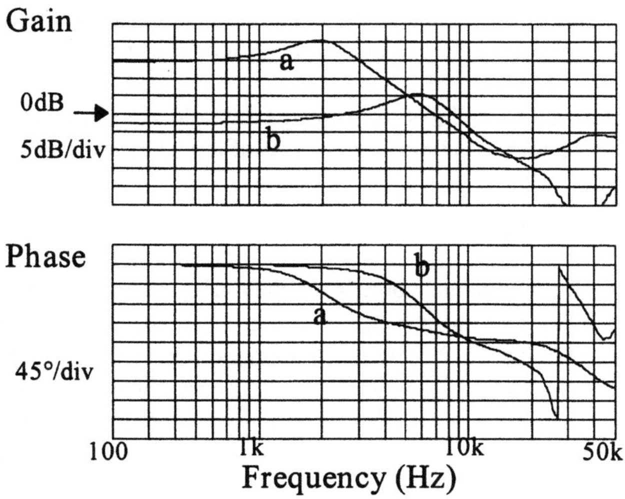

Figure 4: Frequency response of the accelerometer:
a.) open loop b.) closed loop

The cross axis sensitivity, which depends on the error in orthogonality of the measurement axis of the accelerometer to the direction of oscillation of the oscillating mass (see Figure 2), was minimized through stepper lithography.

2.2 Linear acceleration sensor measuring element

The sensor element is designed for using surface micromachining technology. The seismic mass moves as a result of an acceleration. This movement is detected by a differential capacitive comp structure and an open-loop measuring principle.

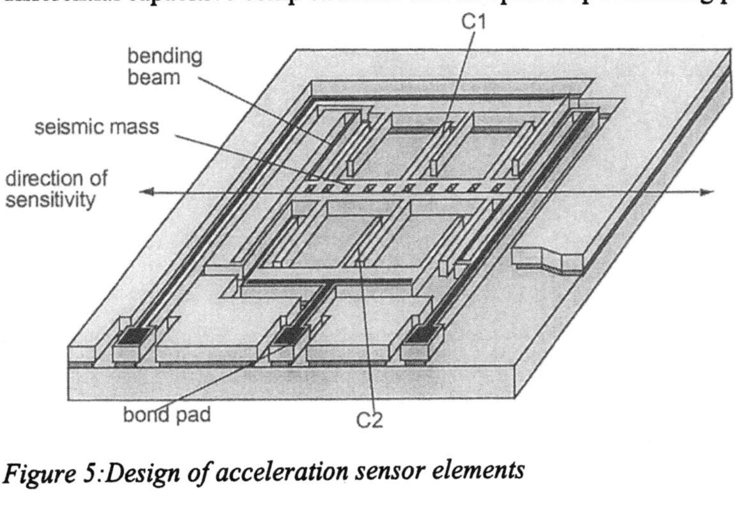

Figure 5: Design of acceleration sensor elements

The design of the acceleration sensor contains a capacitive comp structure with three electrodes: two fixed outer electrodes C1 and C2 and an electrode connected

to the seismic mass CM. If the seismic mass is deflected in response to the linear acceleration, then for examples the gap between C1 an CM increases and between C2 and CM decreases [9, 10, 11].
The offset stability is optimized as follows: the springs are folded to achieve a weak spring constant and to compensate the internal stress of the layer structure.

3. Micromachining Technology

The oscillating masses and accelerometer comb structures are produced on 150 mm silicon wafers using bulk- and surface micromachining, respectively. The process sequence is discussed below. A 2.5μm thick thermal oxide layer is grown and serves as sacrificial layer for the accelerometer and as electrical isolation from the substrate for the contact pads and electrical conduction paths. A 12 μm thick n^{++}-doped polysilicon epitaxial layer is grown and acts as the basis for the mechanical structures of the accelerometer. In the next step aluminum is sputtered and structured. The aluminum serves as a low-resistance electrical path for driving the oscillator masses and makes contact to the electrodes in the accelerometer.
The wafer underneath the oscillator masses is then thinned to 50 μm using KOH-etching in order to simplify the patterning of the oscillator masses. The next step is the dry-etching of the accelerometers using a specially developed dry etching process [8, 10, 11]. The oscillator masses are then etched with the same dry etching process and the structures are released by removing the sacrificial oxide using a gas phase etching process.
The resulting micromechanical structures are protected using two further silicon wafers bonded above and beneath using seal glass. The silicon cap wafer has a recess above the sensor and a hole above the contact pads to allow free movement of the oscillator masses and electrical contact, respectively.

Figure 6 shows a SEM photograph of the oscillator mass, the folded springs and the comb structure of the coriolis accelerometer.

4. Packaging

Intigrated or hybrid versions of inertial sensors have been discussed [1-6, 9]. We chose the hybrid approach in order to decrease the development time and to reduce the interdependence of the fabrication steps for the evaluation circuit and the micromachined sensors.
The metal module contains the hybrid with the sensing elements and the ASIC for the evaluation. The magnetic field is induced by a permanent magnet which is glued into the metal cap.

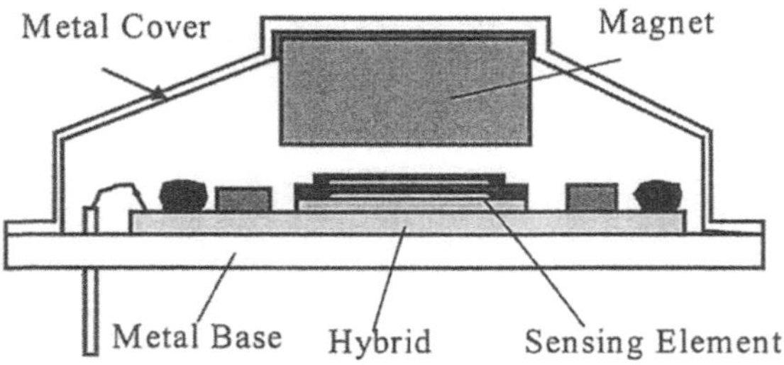

Figure 7: cross-section of the metal module

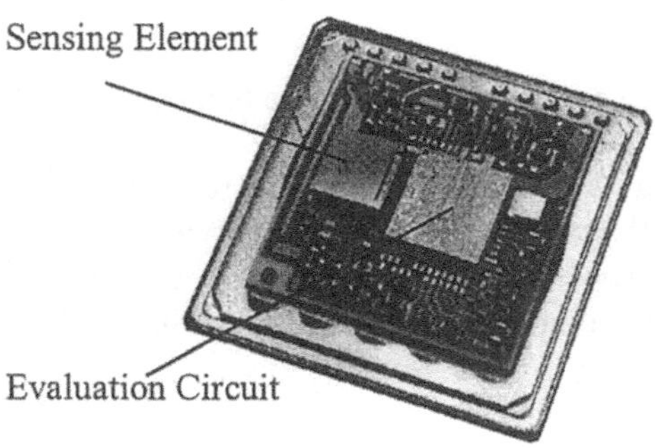

Figure 8: Birds eye view of hybrid on metal base

The plastic housing is designed for mounting the sensor into the passenger compartment. The packaging includes protection against EMC (SAE J 1113), reverse voltages of up to 16V and short circuits to 16V or GND. These features are designed on a printed circuit board PCB which connects the pins of the module with those of the connector (see figure 9).

The connector has 6 pins: 1 power supply +12V, 2 Out reference voltage, 3 Out yaw rate, 4 BITE, 5 Out linear acceleration, 6 GND

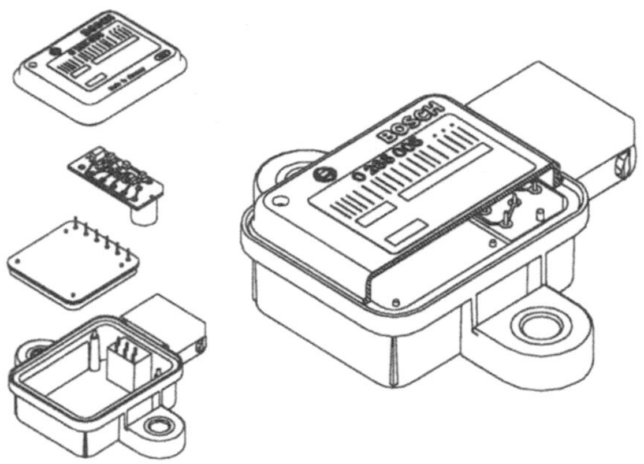

Figure 9: Plastic housing

5. Characterization results

5.1 Yaw rate sensor

The functionality of the yaw rate sensor has been measured in a temperature chamber with a centrifuge.
The sensor is trimmed for a measuring range of ±100 deg/s. The resulting sensitivity and resolution are 18 mV/deg/s and 0.02 deg/s/SQRT(Hz), respectivitly.
An important specification of VDC is the temperature dependence of the offset. The temperature dependence of the offset without any electrical compensation is less than ±10 mV or 0.55 deg/s.

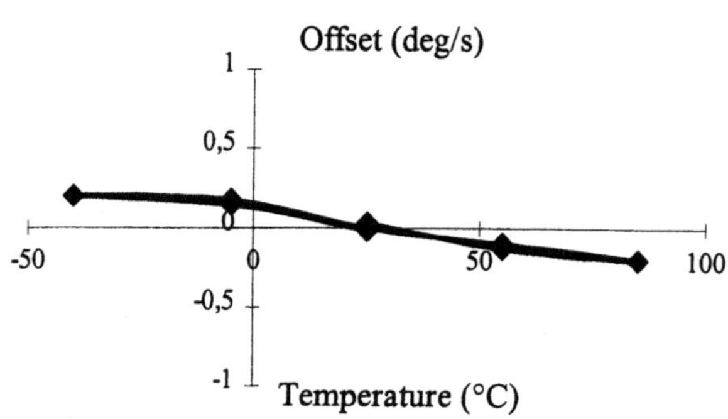

Figure 10: Temperature dependence of the offset

5.2 Acceleration sensor

The acceleration sensor is characterized in a temperature chamber with a turntable to rotate the sensor within the gravitational field. The absolute offset and the temperature dependence of the offset of the acceleration sensor are independently adjusted.
The temperature dependence of the offset is less than ±10 mV. The sensitivity is trimmed to 1 V/g. The sensitivity error of the signal over the temperature range of -40 °C to +85 °C is less than 0.1 %. Therfore there is no need for compensation. The dependence of the nonlinearity is less than 0.2 %.

6. Conclusions

An overview of our new yaw rate sensor is reported. This design leads to a high performance, temperature insensitive and highly reliable device. The sensor is easily adaptable to any system because of the small and robust plastic housing, analog interface and 12 V power supply. This sensor can be implemented in applications which demand a high level of safety because of the built-in-test (BITE). The viability of this sensor for the VDC has been confirmed. The series production will start in the middle of 1998.

ACKNOWLEDGMENTS

The authors would like to thank J. Artzner, L. Tanten, H.-D. Schwarz and J. Mohaupt for the design of the evaluation circuit and for useful suggestions for the complete sensor system.

REFERENCES

[1] M.W.Putty, K.Najfi, "A Micromachined Vibrating Ring Gyroscope", Solid-State Sensor and Actuator Workshop, June 13-16,1994.

[2] J. D. Johnson, S. Z. Zarabadi, D. R. Sparks, "Surface Micromachined Angular Rate Sensor", SAE Technical Paper Series, 950538.

[3] J. Bernstein S.Cho, A. T. King, A. Kourepenis, P. Maciel, M. Weinberg, "A Micromachined Comb-Drive Tuning Fork Rate Gyroscope", 0-7803-0957-2/93, 1993 IEEE.

[4] K. Funk, A.Schilp, M. Offenberg, „Surface-micromachining of Resonant Silicon Structures“, Transducers `95, 519-News, page 50.

[5] M. Hashimoto, C. Cabuz, K. Minami, M. Esashi, "Silicon Resonant Angular Rate Sensor Using Electromagnetic Excitation and Capacitive Detection", Technical Digest of the 12th Sensor Symposium, 1994.

[6] K. Tanaka, Y. Mochida, M. Sugimoto, K. Moriya, T. Hasegawa, K. Atsuchi, K. Ohwada, " A Micromachined Vibrating Gyroscope", Sensors and Actuators A 50 (1995).

[7] Y. Cho, B. M. Kwak, A. P. Pisano, R. Howe, "Slide film damping in laterally driven microstructures", Sensors and Actuators A, 40 (1994).

[8] M. Offenberg, F.Lärmer, B.Elsner, H.Münzel, W. Riethmüller, "Novel Process for a Monolithic Integrated Accelerometer", Transducers 95, 148 - C4.

[9] K. H.-L. Chau, S.R. Lewis, Y. Zhao, R. T. Howe, S. F. Bart, R. G. Marcheselli, "An integrated Force-Balanced Capacitive Accelerometer for Low-G Applications", Transducers 95, 149 - C4.

[10] M. Offenberg, B. Elsner, F. Lärmer, Electrochem. Soc. Fall-Meeting 1994, Ext. Abstr. No 671.

[11] M. Offenberg, H. Münzel, D. Schubert, “ Acceleration Sensor in Surface Micromachining for Airbag Applications with High Signal/Noise Ratio, SAE Technical Paper, 960758.

AWARE A Collision Warning and Avoidance Radar System

Henrik Lind [1], Andrea Saroldi [2], Magnus Kamel [3] and Gerard Delaval [4]

[1] Volvo Technological Development AB , S-412 88 SWEDEN
[2] Centro Ricerche Fiat, Strada Torino 50, 10043 Orbassano (TO), ITALY
[3] CelsiusTech Electronics AB, 175 88 Järfälla, SWEDEN
[4] United Monolithic Semiconductors S.A.S, Route Départementale 128, BP 46 , 91401 Orsay Cedex, FRANCE

1 Introduction and Background

The project Anticollision, Warning and Avoidance Radar Equipment, AWARE, will develop a high performance on-board vehicle system for forward looking Collision Warning and Avoidance, CW/A.

During the project two demonstrator vehicles will be equipped with the AWARE CW/A radar sensor and system. Evaluation and verification of the CW/A function will be made and demonstrated.

Rear-end and stationary object collisions in the front part of the vehicle are the most common types of road traffic accidents. This type of accident often causes severe damages and injuries.

One method to reduce the damages, and even avoid collisions, is to equip the vehicles with radar based CW/A systems.

In such a system the radar detects and measures position and velocity of vehicles and other obstacles on the road ahead. On-board vehicle systems process signals from the radar and vehicle sensors, decide if and how the driver should be warned and, in critical situations, automatically act on the brakes of the vehicle to reduce the velocity.

The objective for the AWARE CW/A system is to warn and brake if no evasive action has been taken by the driver in a critical situation. The system must be able to judge when to warn and when to break in order not to disturb the driver's normal driving behaviour.

2 User and Market Requirements

The consumer and society perspectives of a CW/A system have been investigated in a limited study.

The consumer perspectives are based on individual user aspects, desired functionality and expected use.

The derived requirements are that the system should be functional in all traffic scenarios independent of weather conditions and for all speeds. It should provide an early warning with a low false alarm rate. False brake interventions are not accepted.
The AWARE project will cover part of the possible traffic scenarios.

The system impact on traffic safety and economy has been studied. Also legislation issues have been partly covered.

More than half of the fatal accidents have an impact zone in the front of the vehicle. By using statistics it can be shown that there is a potential for the reduction of the total cost for accidents and damages of more than 30 % when applying a forward looking CW/A system.
The possible cost reduction within the EC alone is more than 16 billion ECU per year.

At the present moment there are no explicit laws or rules regulating the operation and function of CW/A systems. However individual countries might have laws restricting the use of vital parts of the CW/A system. The legislation issues will be addressed within the EC funded project RESPONSE starting in 1998.

3 System Level Requirements

Based on the user requirements, the system level requirements phase intends to translate the user needs into function description and technical requirements. The user requirements are combined with knowledge about what is technically feasible and previous experience in order to define a system that is both useful and possible to produce at the expected market price.

The definition of the system level requirements and the splitting of the function into its modules represent the basis for the specification and development of functional modules.

3.1 Functional Definition

One of the main points in the definition of the Collision Warning and Avoidance function is the definition of the operative limits, that is where the system has full functionality and where functionality is reduced. When these limits have been defined the technical parameters can be derived.

In the development of the Collision Warning and Avoidance function, vehicle path prediction is one of the main issues to be covered. Vehicle path can accurately be predicted if road geometry can be detected from the sensor [4]. However a good estimation can be made using the path of other vehicles together with stationary objects off the road.

The definition of the requirements is primarily made for the highway scenario, followed by other scenarios. In a highway-like scenario, the Collision Warning/Avoidance radar system must be able to avoid a collision with a slower or stationary obstacle in the vehicle lane by warning the driver in time and also automatically braking the vehicle when necessary.

The CW/A function should be fully operative on highways for vehicle speeds from 0 up to 130 Km/h, which includes the highest speed limit for most European countries. From the function specification and the geometry of road constructions the requirements for the system can be derived in a top-down process.

On smaller roads the vehicle speed is generally lower, and therefore less precision is needed for road geometry estimation and path prediction, however the road structure is less visible. This means that the system might be useful and functional also on smaller roads than highways.

3.2 System Decomposition

The CW/A system can be divided into functional modules according to figure 1.

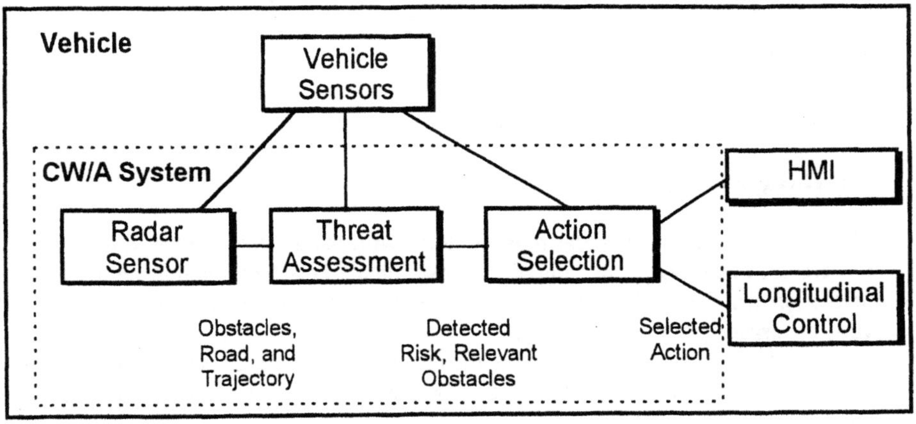

Figure 1 CW/A System Decomposition

The goal of the *Radar Sensor* module is to detect and track obstacles and to estimate road geometry and current vehicle trajectory. Based on these results, the *Threat Assessment* module has to evaluate if the current situation represents a risk. If a risk is detected, then the *Action Selection* module has to decide what is the best evasive action and if this action is best performed by the driver or by direct vehicle control. As a result warnings are generated for the driver and commands are generated for vehicle control.

Only longitudinal vehicle control will be used in the AWARE project. If the best evasive action to be taken is a lateral manoeuvre, then this has to be performed by the driver.

3.3 Compatibility with ACC Functionality

The system level requirements considered within the AWARE project are derived from the CW/A function definition. The requirement for the Adaptive Cruise Control, ACC, sensor is a sub set of the requirements for the CW/A sensor.

The developed radar sensor can therefor be used also for the ACC function or for a combination of ACC and CW/A.

4 Radar Sensor Specification

The radar information has to be accurate enough so that the following systems for threat assessment have the possibility to evaluate the threat severity for each particular traffic scenario. The selected type is a 77 GHz mechanically scanning radar sensor.

The existing ACC radar reports range and angle to the tracked vehicle together with vehicle relative speed. All stationary objects, bridges as well as overhead traffic signs are excluded. Associated with the move from an ACC to a CW/A application, is the consideration of stationary objects. This implies that a number of hardware solutions and algorithms have to be established regarding:

- Identification of overhead objects
- Path prediction
- Classification of stationary objects

4.1 Threat Assessment Needs

The information from the radar must conform with the accuracy required to decide if there are obstacles in the lane of the host vehicle at a long enough range. Path prediction will be used in order to estimate the path of the host vehicle by using yaw rate sensor and road prediction.

If there are obstacles in the path of the host vehicle, the radar sensor information is used in order to decide whether there is space for an evasive action performed by the driver, or not.

In order to warn at an early stage the operative range of the radar must be considerable higher than the ACC radar.

4.2 Path and Road Prediction

The objective with this kind of system is to warn and occasionally also brake when there is an appreciable risk for the host vehicle to run into an obstacle or vehicle approaching the host vehicle path.

In order to accurately predict a potential collision to happen, the system has to make assumptions about where the host vehicle is heading. This is made by the path prediction algorithms. The driver intention is most of the time to continue on the road in the same direction as before. By estimating the road path, the vehicle path in the near future can be obtained.

By using the information provided by the radar sensor concerning road edges, road structures, illumination equipment and moving vehicles ahead, qualified estimates about the road path can be made. To decide to what extent the driver intends to follow the road, the movements both lateral and longitudinal, of the host vehicle are considered, figure 2.

A low false alarm rate has to be obtained in order to make the CW/A system comfortable. This implies that the path prediction has to be very accurate. The acceptable error of the prediction, expressed as lateral distance (predicted path compared to real path), has to be lower than a meter within the operative range.

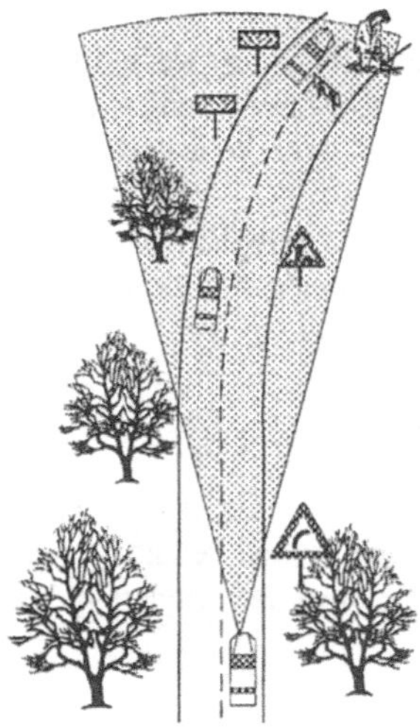

Figure 2: Illustration of the problem of path prediction. The path prediction algorithm shall predict the host vehicle path in the near future in order to decide which ones of the objects that are relevant threats.

4.3 Cost Aspects

The design has to focus on a low cost realisation. Due to expected large-scale production substantial investment is justifiable in order to obtain a low cost per unit. This has already been considered for the ACC sensor and will be for the sensor intended for CW/A applications. A number of approaches will be considered to cope with the market cost expectations:

- Use of comprehensive signal processing in order to use low-cost, medium performance components.
- Maximum integration in ASICs, trading high investments for low cost per unit. To maintain flexibility the design must still provide a certain degree of SW modifications. This is obtained by using a low-cost DSP.
- Antenna design using low scanning mass and limited scan angles to cut motor costs
- Use of MMIC based front end. MMICs are expected to be favourable in cost within the next 3 years.

5 GaAs MMICs for 77 GHz Automotive Radar

In order to reach a good trade-off between the radar range and spatial resolution, and a reasonable antenna size, the radar has to operate at millimetre-wave frequency. The frequency band 76-77 GHz has been selected world-wide for this application.

Prototype versions using hybrid assemblies of millimetre-wave diodes have been demonstrated for a long time. It will be very difficult to reach the required low prices for the millimeterwave front end, due to the necessary tunings, interconnection problems and GUNN diode cost.

GaAs MMICs are today the best candidates to reach good performance at low cost thanks to the availability of millimeterwave processes and the possibility to design highly integrated multi function.

5.1 GaAs MMIC Technology

To develop the necessary low-cost GaAs MMIC technology for such a high frequency is extremely challenging and requires leading edge expertise in several domains, within the millimeterwave frequency range up to 77 GHz, such as:

- Circuit design.
- GaAs processes.
- On wafer measurements.
- Functional validation in text fixture representative of the real environment.

The GaAs processes named PH25/15 have been developed by UMS[4] for this application using pseudomorphic HEMT (High Electron Mobility Transistor) epitaxial structures for ultra low noise performances.

Due to the high current capability in the pseudomorphic GaInAs channel, medium power amplifiers can also be designed and this process can address all microwave functions at millimetre-wave frequencies, up to 100 GHz.

The material structure is grown using molecular beam epitaxy, and has an etch-stop layer for a precise control of the recess and improved manufacturing process. The Al T-gate is precisely defined by e-beam lithography with two options for the gate length:

- 0.25μm for application up to 60 GHz (PH 25 process)
- 0.15μm for applications up to 100 GHz (PH15 process)

5.2 GaAs MMIC Chip Set

The AWARE project includes the design of a millimeter wave front end incorporating an already existing MMIC chip set [2]. The MMIC chip set incorporates the following MMICs:

- 38 GHz Oscillator
- 38 GHz Mixer for source performance optimisation via PLL
- 77 GHz Transmitter multifunctional chip
- 77 GHz Receiver circuit

During the course of the AWARE project, the millimeter wave front end will be incorporated in an AWARE prototype.

5.3 GaAs MMIC Future

To deal with the low cost demands of automotive applications the GaAs MMIC development in the future will address possible solutions to reduce further manufacturing and assembly costs to stay compatible with large volume production of millimeter wave front end:

- Chip size reduction by exploiting new circuit topologies
- Increased integration through design of highly multifunctional chip
- Process simplification via coplanar structure, optical lithography
- Automatic test using high speed on wafer measurements

6 AWARE Partners

The partners in the AWARE project are:

Volvo[1], Project manager, developer and user
Has a ten year experience on specification, design and testing of ACC and Anticollision Warning and Avoidance systems.

CelsiusTech Electronics[3] , Developer and supplier
Developer and manufacturer of a 77 GHz radar sensor for ACC [1].

Centro Ricerche Fiat[2], Developer and user
Has a ten year experience on Collision Warning radar systems acquired in the design, realisation, and testing of several prototypes and vehicle systems [3].

United Monolithic Semiconductors[4], Developer and supplier
Manufacturer of high performance GaAs MMICs, operating at frequencies up to 100 GHz, for the commercial automotive and communication markets, covering also the military and space applications.

The AWARE project is funded by the partners and the European Commission. The project is carried out under the ESPRIT program of the European Commission, DG III.

The AWARE project started in July 1997 and ends in December 1999.

7 References

[1] L.H. Eriksson and B-O Ås, "A High Performance Automotive Radar for
Automatic AICC", Proc. IEEE Int'l Radar Conference - 95, Washington DC 1995, pp 380-385. Reprinted in: IEEE Aerospace and Electronic Systems Magazine, Dec 1995, pp 13-18.

[2] High Performances MMICs for Automotive Radar Applications at 77 GHz" M.Camiade, C.Dourlens, V.Serru, P. Savary, J. C. Blanc
GaAs Application Symposium (GAAS 96), Paris, France, 1996.

[3] Saroldi, "Implementation of Active Safety Systems to Avoid Frontal Collisions", ATA Conference on Active and Passive Automobile Safety, Capri (Italy), October 10-11, 1996.

[4] Saroldi, Rebora, *et al.*, "Relevance of Road Recognition to an Anticollision Radar System", EAEC'95 Congress, Strasbourg, 21-23 June, 1995.

8 Acknowledgements

The support of the European Commission to the AWARE project through the ESPRIT program is greatly appreciated and hereby acknowledged.

Multifunctional Radar Sensor for Vehicle Dynamics Control Systems

M. Wollitzer [1], J. Büchler [1], J.-F. Luy [1], U. Siart [2], J. Detlefsen [2]

[1] Daimler-Benz Research, Wilhelm-Runge-Str.11, 89081 Ulm, Germany
[2] Laboratories for High Frequency Technology, Technische Universität München, Germany

1. Introduction

Important aspects in future automotive concepts are comfort, reduced energy consumption and particularly improved safety. To optimize these points a precise sensing of the vehicle parameters and the environmental conditions is necessary. For example the velocity of a vehicle, the tilt angle, the height above ground and especially for active distance control and for autonomous driving the road condition have to be known. Well suited to carry out these tasks are radar sensors in the millimeter wave range. However, separate sensors for each purpose are uneconomic. Thus the merging of such sensor types to one multifunctional sensor system is needed. Fig. 1.1. illustrates the impact of the measured parameters on vehicular control systems.

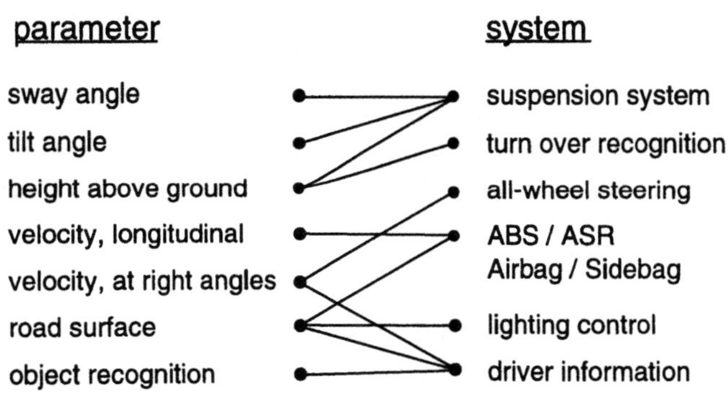

Fig. 1.1. Impact of measured parameters on vehicular control systems

2. Sensor Principle

The sensor is based on a radar system with an operating frequency at millimeter-waves (61 GHz). The small wavelength results in

- compact physical dimensions of the sensor
- high resolution of distance and velocity

- high inertness against pollution, compared to optical systems.

Two sensor modules are necessary to measure the desired parameters. Fig. 2.1. displays two possible arrangements of the modules. The sensor system is realized in the bistatic configuration.

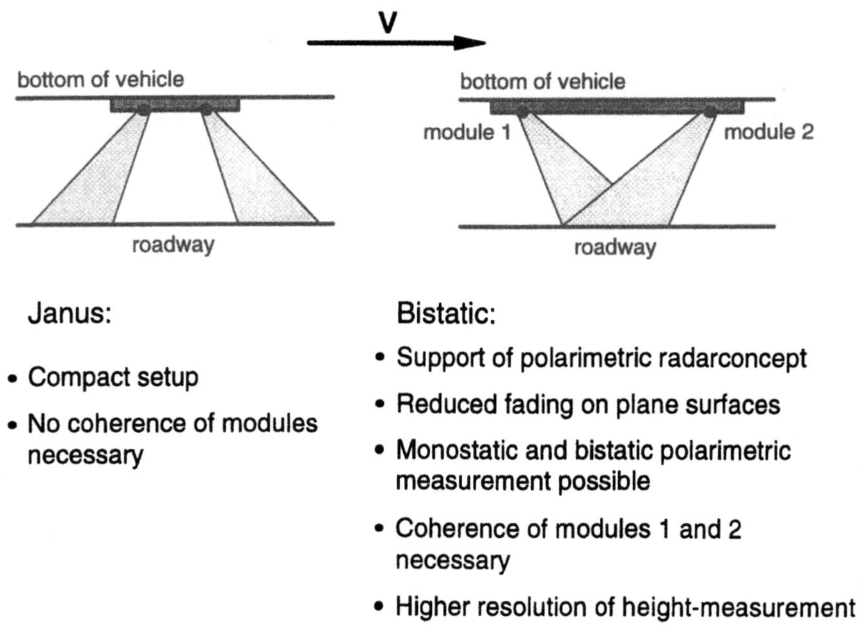

Fig. 2.1. Comparison of Janus and bistatic arrangement of the sensor modules

The sensor concept supports two operating modes: The surface condition of the roadway is determined by a bistatic polarimetric measurement. Height above ground, tilt angle and velocity is measured in a frequency modulated operating mode of the sensor.

2.1. Polarimetric Mode of the Radar Sensor

For the polarimetric measurements, horizontally and vertically polarized waves are transmitted sequentially by module 1. The horizontal and vertical polarized components of the backscattered wave are received in parallel by module 2. Thus, the scattering matrix **S**

$$\mathbf{S} = \begin{pmatrix} S_{HH}\, e^{j\phi_{HH}} & S_{HV}\, e^{j\phi_{HV}} \\ S_{VH}\, e^{j\phi_{VH}} & S_{VV}\, e^{j\phi_{VV}} \end{pmatrix} \qquad (2.1)$$

of the surface is determined. Information on the road condition is contained in the magnitude and phase of the elements of the scattering matrix which describe the polarization properties of the backscattered wave. For illustrative purposes, the polarization state of the backscattered waves is projected on the Poincaré-sphere as is depicted in Fig. 2.2. Linear polarizations are projected on the equator, circular polarizations are found at the poles of the Poincaré-sphere. The transition from smooth to rough surfaces results in a broadening of the cluster centers, changes in the surface material result in a shift of the cluster centers. Even the presence of thin ice-layers on a roadway is dectectable by application of this principle [1].

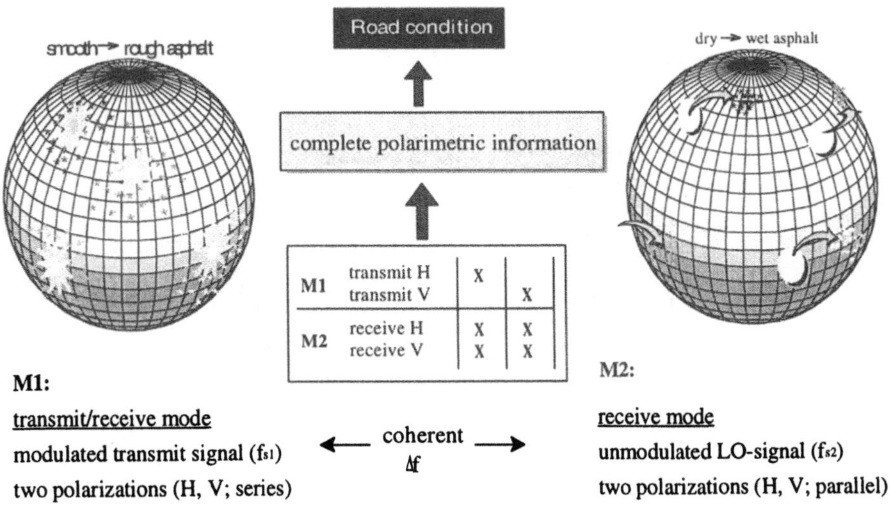

Fig. 2.2. Change of the appearance of cluster centers on the Poincaré-sphere for transitions from smooth to rough and dry to wet road surfaces.

2.2. Frequency Modulated Mode of the Radar Sensor

In a cw mode, the unambiguous range r_u of a radar is $r_u = c_0/2f_0$, which is approximately 3 mm at a carrier frequency of 61 GHz. To assure ambiguity of the measured distances, the carrier is frequency modulated with the modulating frequency f_m. The unambiguous range of the sensor is then given by $r_u = c_0/2f_m$. With a modulating frequency of 80 MHz, the unambiguous range is 1.875 m, which is sufficiently large even for turnover-detection.

The radar modules are mounted at a distance *d*. The angles of the beams with respect to the axis through the modules are ϑ_0 and γ_0 respectively. The beamwidths are ϑ_A and γ_A. The tilt angle of the vehicle is ζ and the height with respect to ground is *h*. Module 1 is radiating a signal at frequency f_{s1} and also

receives the back scattered signal. Module 2 also working at f_{s2} separated by δf from f_{s1} receives the back scattered signal.
The transmitted signal u_{s1} of module 1 is sinusoidally modulated [2]

$$u_{s1} = U_{s1} \cos\left[\omega_{s1} t + \eta \sin(\omega_m t - \varphi_m) - \varphi_{s1}\right], \quad \eta = \frac{\Delta f}{2 f_m} \tag{2.2, 3}$$

the LO signal u_{s2} of module 2 is a CW wave

$$u_{s2} = U_{s2} \cos\left[\omega_{s2} t - \varphi_{s2}\right]. \tag{2.4}$$

The sensor arrangement is depicted in Fig. 2.3.

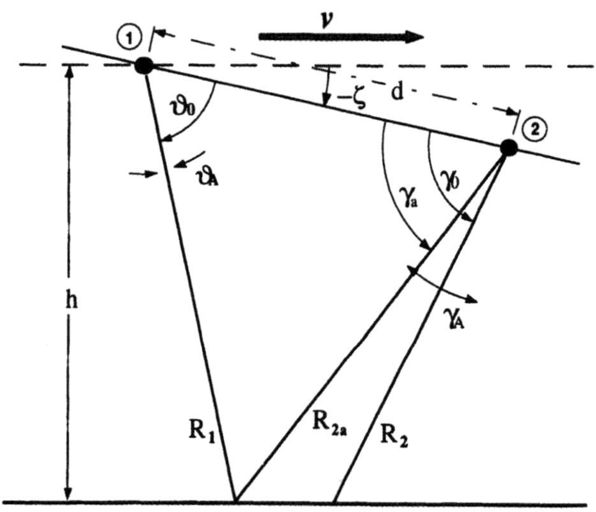

Fig. 2.3. Sensor arrangement

The time delays of the back scattered signal into module 1 and of the scattered signals into module 2 are

$$\tau_1 = \frac{2R_1(t)}{c} = \frac{2}{c}\left[r_{01} \mp v_{r1} t\right]$$

$$\tau_a = \frac{R_a(t)}{c} = \frac{1}{c}\left[r_{01} + r_{02a} \mp (v_{r1} - v_{r2a})t\right] \tag{2.5a, b}$$

$$r_{01} = R_{01} \pm v_{r1} t_0 \qquad v_{r1} = v \cos(\vartheta_0 - \zeta)$$

$$r_{02a} = R_{02a} \pm v_{r2a} t_0 \qquad v_{r2a} = v \cos(\gamma_a + \zeta) \qquad (2.6a\text{-}d)$$

Thus the received signals are

$$u_{e1} = U_{e1} \cos\{\omega_{s1}(t - \tau_1) + \eta \sin[\omega_m (t - \tau_1) - \varphi_m] - \varphi_{s1}\}$$
$$u_{e2} = U_{e2} \cos\{\omega_{s1}(t - \tau_a) + \eta \sin[\omega_m (t - \tau_a) - \varphi_m] - \varphi_{s1}\} . \qquad (2.7a, b)$$

and after down conversion

$$u_{M1}(t) \propto \cos\left[\omega_{s1}\tau_1 + \alpha \cos(\omega_m (t - \frac{\tau_1}{2}) - \varphi_m)\right]$$
$$= J_0(\alpha)\cos\Phi_D + 2\sin\Phi_D \sum_{n=0}^{\infty} (-1)^n J_{2n+1}(\alpha)\cos[(2n+1)x_1]$$
$$+ 2\cos\Phi_D \sum_{n=1}^{\infty} (-1)^n J_{2n}(\alpha)\cos(2nx_1) \qquad (2.8)$$

$$\alpha = \frac{\Delta f}{f_m}\sin(\frac{\omega_m \tau_1}{2}), \quad x_1 = (\omega_m \pm \frac{\omega_{Dm}}{2})t - \frac{\omega_m r_{01}}{c} - \varphi_m \qquad (2.9a, b)$$

$$\omega_{Dm} = \frac{2\omega_m v_{r1}}{c}, \quad \Phi_D = \pm\omega_D t - \frac{2\omega_{s1} r_{01}}{c}, \quad \omega_D = \frac{2\omega_{s1} v_{r1}}{c} \qquad (2.9c\text{-}e)$$

and

$$u_{M2}(t) \propto \cos[\delta\omega t - \delta\varphi - \omega_{s1}\tau_a + \eta \sin(\omega_m (t - \tau_a) - \varphi_m)]$$

$$= J_0(\eta)\cos\Omega_1 - 2\sin\Omega_1 \sum_{n=0}^{\infty} J_{2n+1}(\eta)\sin[(2n+1)x_a]$$
$$+ 2\cos\Omega_1 \sum_{n=1}^{\infty} J_{2n}(\eta)\cos(2nx_a) \qquad (2.10)$$

$$x_a = (\omega_m \pm \omega_{Dma})t - \frac{\omega_m (r_{01} + r_{02a})}{c} - \varphi_m \qquad (2.11a)$$

$$\omega_{Dma} = \frac{\omega_m (v_{r1} - v_{r2a})}{c} \qquad (2.11b)$$

$$\Omega_1 = (\delta\omega \pm \omega_{Da})t - \frac{\omega_{s1}(r_{01} + r_{02a})}{c} - \delta\varphi \qquad (2.11c)$$

$$\omega_{Da} = \frac{\omega_{s1}(v_{r1} - v_{r2a})}{c} \tag{2.11d}$$

The frequencies f_{Dm} and f_{Dma} are in the range of 10^{-2} Hz and therefore negligible. The principle spectra of the mixer signals in channel 1 and 2 are shown in Fig. 2.4. In module 1 the Doppler signal with frequency f_D is available in the base band and symmetrically around the harmonic frequencies of the modulation frequency. Two sidebands are located at the harmonics of the modulation frequency of the mixer signal of channel 2. From there the Doppler difference frequency f_{Da} can be obtained. The distance to the ground r_{01} can be extracted from the Bessel spectra.

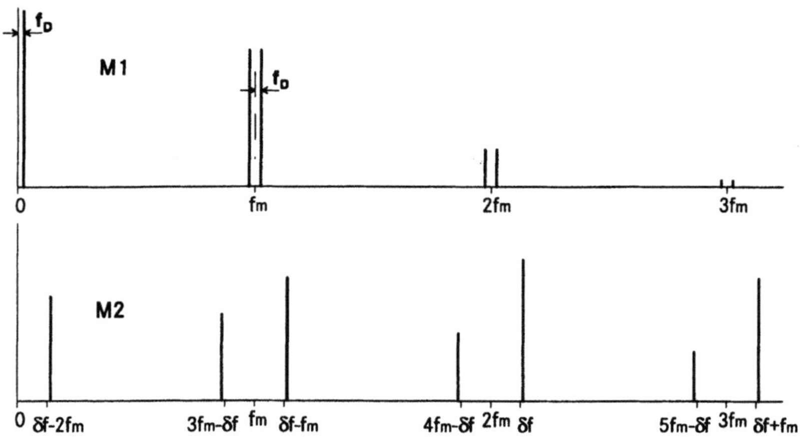

Fig. 2.4. Spectra of the mixer signals, df>2f$_m$

The parameters to be determined ζ, h, v are related to the measured values by

$$r_{01} = \frac{h}{\sin\vartheta} \quad , \quad \vartheta = \vartheta_0 - \zeta \tag{2.12, 13}$$

$$f_D = \frac{2}{c} f_{s1}\, v \cos\vartheta \tag{2.14}$$

$$f_{Da} = \frac{1}{c} f_{s1}\, v(\cos\vartheta - \cos\gamma) \quad , \quad \gamma = \gamma_a + \zeta \quad , \tag{2.15}$$

$$\gamma_a = \operatorname{arcctg}\left(\frac{d\cos\zeta - h\operatorname{ctg}(\vartheta_0 - \zeta)}{h + d\sin\zeta}\right) - \zeta \tag{2.16}$$

By solving equation (2.17)

$$F_\zeta = (1 + AW)\cos(\vartheta_0 - \zeta) - Q\cos\zeta = 0\,, \tag{2.17}$$

$$A = \sqrt{1 + Q^2 - 2Q\cos\vartheta_0}\,,\quad W = 1 - \frac{2\omega_{Da}}{\omega_D}\,,\quad Q = \frac{d}{r_{01}} \tag{2.18a-c}$$

follows the tilt angle ζ. Examples for different W are given in Fig. 2.5.

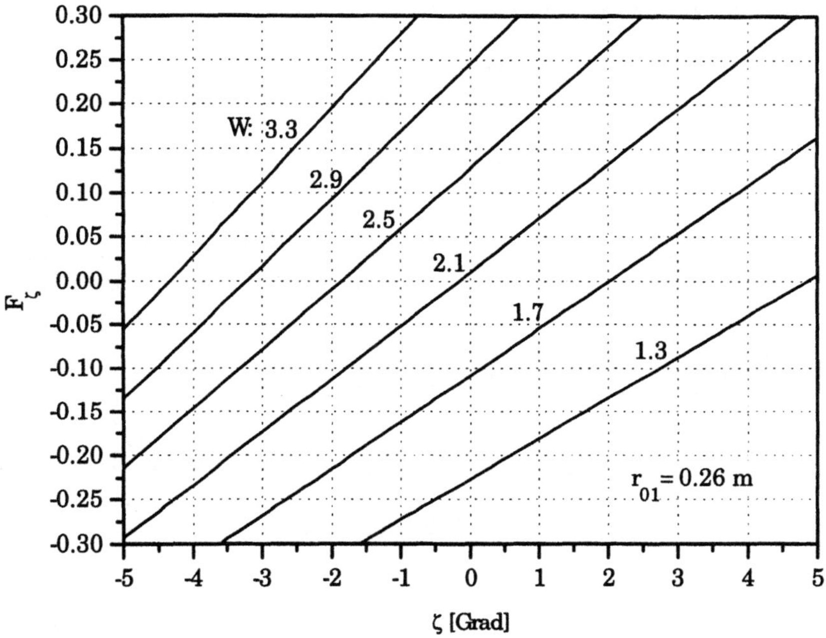

Fig. 2.5. Function F_ζ for different W, h= 25 cm, d/h= 0.9

The height above ground and the velocity can then be determined by

$$h = r_{01}\sin(\vartheta_0 - \zeta) \tag{2.19}$$

$$v = \frac{c\,\omega_D}{2\,\omega_{s1}}\frac{1}{\cos(\vartheta_0 - \zeta)}\,. \tag{2.20}$$

3. Realization of the Sensor

A schematic blockdiagram of the sensor is depicted in Fig. 3.1. Both, transmit- and receive module are fed with reference signals at the n-th subharmonic of the operating frequency. The reference signals of the frontends are provided with a frequency offset $\delta f/n$. The reference signals are phase coherent, thus the received signals are converted to a first IF at δf. If the frequency source of the transmit module is sinusoidally frequency modulated at f_m, IF signals appear at the Doppler frequency and multiples of the modulation frequency f_m in module 1. These signals are amplified and converted to a second IF with an analog bandwidth of 100 kHz. The signals which are received by module 2 are also amplified and converted to the second IF. Polarimetric measurements are possible by alternating the polarization of the transmitted signal of module 1 while receiving both polarizations in module 2 separately.

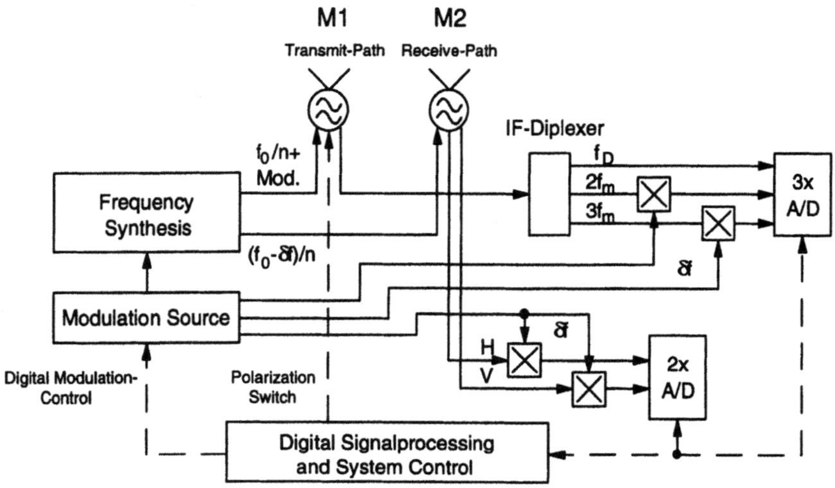

Fig. 3.1. Blockdiagram of the sensor

3.1. Sensor Modules in Waveguide Technique

To be able to make measured signals available for the optimization of the signal processing algorithms in an early stage of developement of the sensor, a waveguide setup of the modules has been realized. These waveguide frontends support the full functionality of the final sensor. The frontends are fed with reference signals at approx. 15 GHz, which enables to guide these signals by coaxial cables. The frequency conversion to the transmit frequency of 61.25 GHz is accomplished by a varactor quadrupler. Downconversion of the received signals is realized by harmonic mixers. The polarization of the transmitted waves is altered by a pin-switch. Corrugated feed horns are used as antennas. They are fed by orthomode transducers. Fig. 3.2. shows a picture of the transmit frontend. The inset depicts the blockdiagram of the module. The receive-frontend is depicted in Fig. 3.3.

Fig. 3.2. Fotograph and block diagram of the transmit frontend

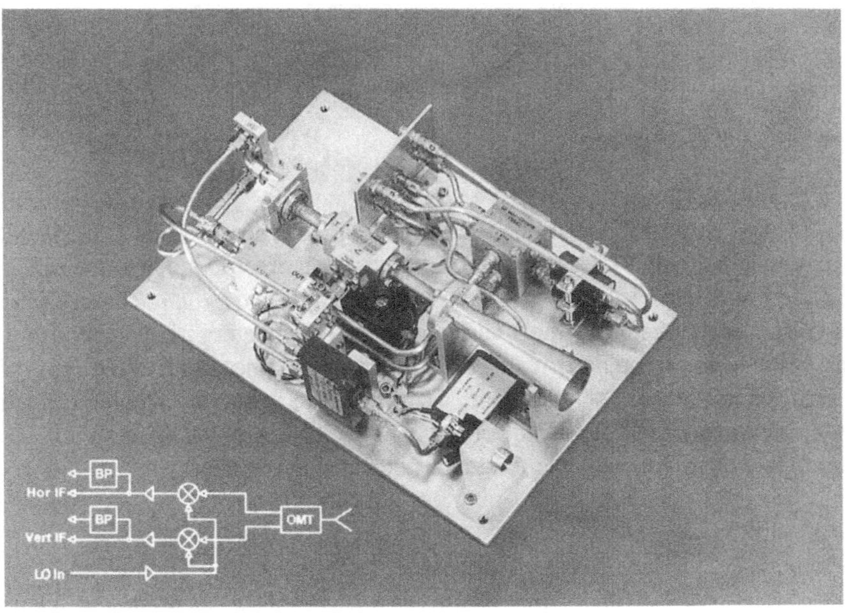

Fig. 3.3. Fotograph and block diagram of the receive frontend

3.2. Integrated Frontend with Silicon MMICs

In order to serve the high volume automotive market with its stringent price targets, it is necessary to apply monolithic integration techniques for the realization of the rf components of an automotive radar system. Silicon MillimeterWave Integrated Circuits (SIMMWICs) [3] which are used in this system offer many advantages over other integration techniques. The thermal conductivity of silicon is three times higher in comparison to GaAs, silicon substrates exhibit low losses in microwave and millimeterwave frequencybands and can withstand high mechanical stress. Therefore and due to the high maturity of silicon processing techniques, SIMMWICs are a solution with an interesting perspective to serve the needs of high frequency mass markets.
The outlines of the sensor, which are determined by the size of the antenna diameter and the base width of the two rf-frontends are shown in Fig. 3.4. The dimensions of the sensor make its mounting below the bottom of a vehicle easily feasible.

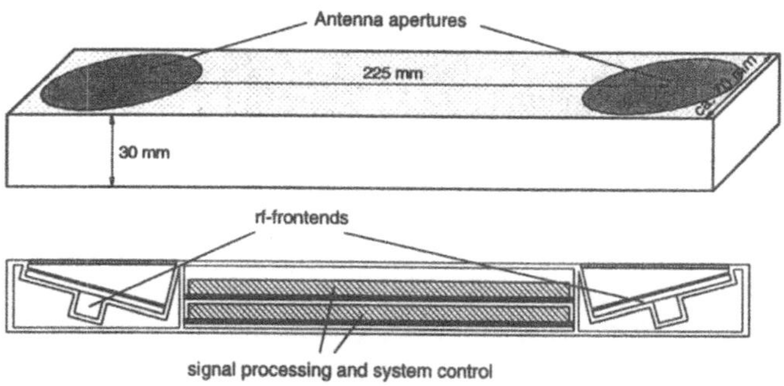

Fig. 3.4. Outlines of the radar sensor

The monolithically integrated SIMMWIC oscillators [4] generate the transmit signal. Each chip transmits one linearly polarized wave. Two SIMMWIC transmitters are therefore necessary to support the polarimetric mode of the radar. The detection of the back scattered signals is accomplished by application of the selfmixing properties of the millimeterwave oscillators [5]. Frequency stabilization of the frontend is realized by subharmonic injection locking [4,6] of the frontend at the third subharmonic. Therefore, the integrated frontend is fed with a reference signal at a frequency of 20.4 GHz. A block diagram of the injection locked millimeterwave frontend is depicted in Fig. 3.5.

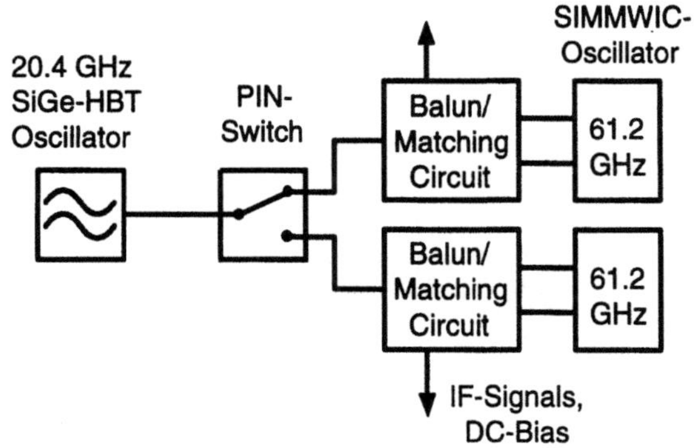

Fig. 3.5. Block diagram of the integrated radar-frontend

In contrast to the measurement system of Chap. 3.1., where signals are transmitted and received via horn antennas, beam forming is realized by a planar antenna [7] in the case of the integrated frontend. A model of the planar antenna has been milled in brass. It is fed in the center of the ground plane by a square waveguide which conducts waves in orthogonal polarizations. Radiation occurs perpendicular to the antenna aperture for both polarizations. Fig. 3.6. shows the milled model of the planar antenna on the left, a realization of the structure in PIM (**P**owder **I**njection **M**oulding)-technique is shown on the right.

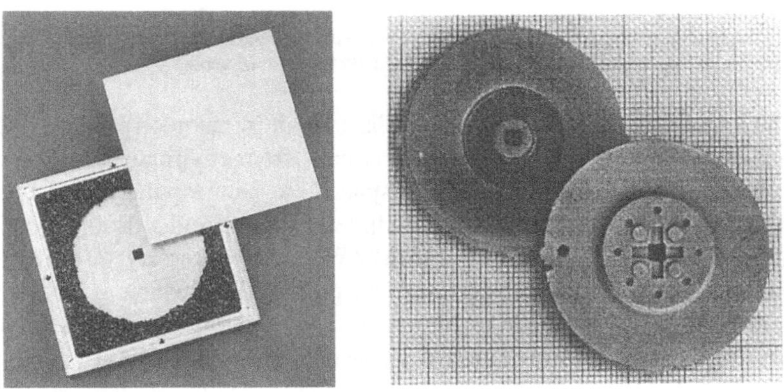

Fig. 3.6. Milled model of the planar antenna and realization in PIM-technique

SIMMWIC transmitters have been integrated with the planar antenna in order to determine the radiation pattern of the whole structure. The frontend consists of parts which have also been milled in brass. Thus, changes in the length of the square waveguide section and the size of the coupling iris can be done easily. Fig. 3.7. gives a schematic cross-sectional view of the realized frontend structure. Calculated and measured antenna diagrams are shown in Fig. 3.8. The measured antenna diagram is well predicted by the theory.

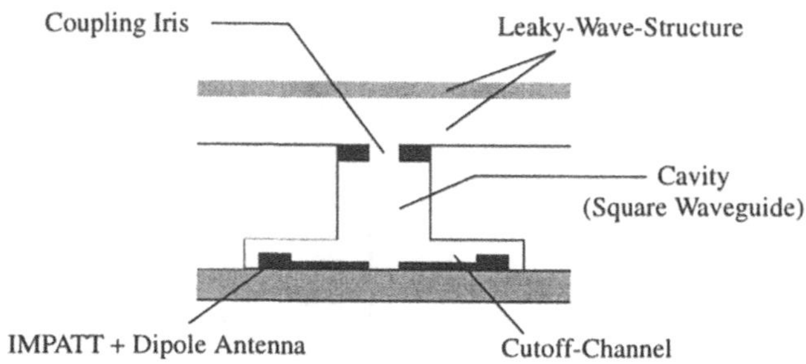

Fig. 3.7. Coupling structure (cross-sectional view)

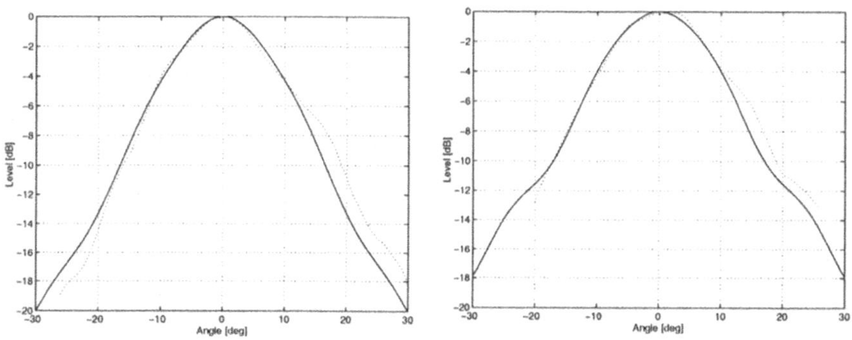

Fig.3.8. Antenna diagrams of the integrated frontend with planar antenna. Radiation pattern in E-plane on the right, H-plane on the left. Straight lines: Calculated, dotted lines: Measured

It is inevitable for the operation of a radar system in automotive applications to hermetically shield the millimeter wave circuits from environmental factors like dust, water spray and corrosive atmosphere. A new topology was therefore investigated for the mounting of the millimeter wave circuits. Both SIMMWICs are located in a cutoff tunnel. Two SIMMWICs are necessary to transmit and receive signals in horizontal and vertical polarization, which is necessary to support the polarimetric mode of the sensor. The radiating structure is coupled to the waveguide section with square cross-section. This cavity itself is coupled to the planar leaky wave antenna by an iris. To hermetically seal the structure, the iris is realized by a slab of ceramic which is metallized on both sides. Both metallizations are shorted by metallized via holes at the edges of the iris to prevent the excitation of parallel plate modes in the ceramic substrate. The waveguide section consists of a metallized PIM part. The bottom of the waveguide is closed by a multilayer· LTCC (Low Temperature Cofired Ceramics)-substrate on which the millimeterwave circuits are mounted. This substrate also provides a feed-through for DC and RF-signals below the walls of the PIM structure in a deeper wiring layer, thus hermetically shielding the rf section. The three parts of the frontend housing are assembled by solder techniques. A schematic view of the integrated frontend is shown in Fig. 3.9.

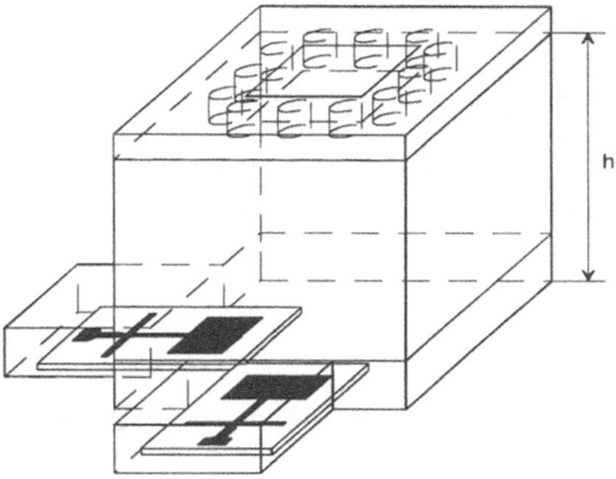

Fig. 3.9. Schematic view of the integrated SIMMWIC-frontend

4. Measurements

Acquisition of polarimetric measurement data is preliminary carried out under application of the waveguide-based measurement system. First evaluation of measurement data indicates a very high sensitivity and phase stability of the setup. The characterization of various roadway surfaces under dry, wet and icy condition shows differing polarimetric signatures which is necessary for the classification of the road condition. If the setup is set in motion, the elements of the polarimetric scattering matrix are Doppler shifted as is depicted in Fig. 4.1.

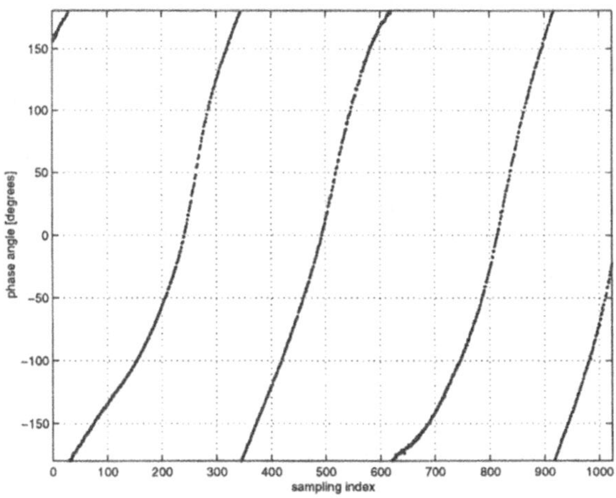

Fig. 4.1. Phase of S_{hh} caused by doppler effect

5. Acknowledgement

This work was funded by the federal ministery of research and education (BMBF) under grant 16 SV 417/0. Support from our SIMMWIC team (Dr. Strohm, Mrs. Lindenmaier, Mr. Eisbrenner) and the radar signal processing team (Mr. Finkele, Mr. Schreck, Dr. Wanielik) is gratefully acknowledged.

References

1 R. Finkele: 'Detection of ice layers on road surfaces using a polarimetric millimetre wave sensor at 76 GHz', Electron. Lett., vol. 33, 1153-1154 (1997)

2 J. Buechler: 'Two Frequency Multifunctional Radar for Mobile Application',IEEE MTT-S `95,Orlando, Florida, May 17-19, 125-128

3 J.-F. Luy, K. M. Strohm, H.-E. Sasse, A. Schueppen, J. Buechler, M. Wollitzer, A. Gruhle, F. Schaeffler, U. Guettich, A. Klaaßen, 'Si/SiGe-MMICs', IEEE Trans. MTT-**43**, 705-714, Apr. 1995

4 M. Singer, K. M. Strohm, J.-F. Luy, E. M. Biebl:'Active SIMMWIC-Antenna for Automotive Applications', IEEE-MTT-S '97, Conf. Dig. 1265-1268

5 M. Claassen, U. Guettich:'Conversion Matrix of Self-Oscillating Mixers', IEEE Trans. MTT-**39**, 25-30,(1991)

6 M. Wollitzer, J. Buechler, J.-F. Luy:'Subharmonic Injection Locking (of) Slot Oscillators', Electron. Lett., vol. 29, no. 22, 1958-1959, (1993)

7 H. Ostner, E. Schmidhammer, J. Detlefsen, D. R. Jackson, , 'Radiation from Dielectric Leaky-Wave Antennas with Circular and Rectangular Apertures', Electromagnetics, vol. 17, No. 5. Sept/Oct 1997, 505-535

Forward Looking 2- dimensional Laser Radar for ACC-Systems

Takeshi Matsui

Denso Corporation, R&D Dept. 1, 1-1 Showa-cho, Kariya-shi Aichi-ken, 448 Japan

Abstract.
In the adaptive cruise control system, the scanning laser radar is widely used to detect the preceding car even on the winding road. The conventional scanning laser radar scans only horizontally , utilizing a fan beam which is horizontally narrow and vertically wide. Since the conventional system scans only horizontally, it has the following drawback: To detect the preceding car even on the highway with up/down hill, it is necessary to utilize a vertically wide fan beam. Therefore, while the car is traveling on a flat road, the scanning laser radar may detect the objects above the road, such as traffic signs, guideboards and bridges etc., in stead of the preceding cars.

To overcome this drawback, we have developed 2-dimensional laser beam scanner with a rotating polygon mirror. Using this laser beam scanner, we have realized the new 2-dimensional scanning laser radar system characterized in that it has six lines raster scan by vertically and horizontally narrow beam.

By this system, it is made possible to differentiate the preceding cars from the objects above the road as well as to detect the preceding car even on the highway with up/down hill.

The complete manuscript was not available by date of print.

Surface Micromachined Sensors for Vehicle Navigation Systems

Christophe Lemaire and Bob Sulouff

Analog Devices, Inc., Micromachined Products Division, Cambridge, MA 02139, USA

Abstract. Today, most vehicle navigation systems use a GPS receiver as the primary source of information to calculate position. While the systems have become increasingly accurate in generating coordinates under ideal conditions of operation, it is a widely acknowledged fact, and studies have shown, that such systems can become highly unreliable in urban environments. Recently, much work has been accomplished and many solutions have been proposed to improve accuracy and minimize problems associated with poor signal reception. Nevertheless, dead reckoning relying on inertial navigation has become the de facto back up method to generate position in the absence of GPS signals. Together with gyroscopes which are an intrinsic part of such inertial navigation systems, accelerometers are now being considered for the function of measuring velocity, one of the requirements of dead reckoning. Recent progress in the development of Integrated MicroElectro Mechanical Systems (*i*MEMS) using surface micromachining technology are now enabling new form factors and price points for this function. In this paper, we introduce the concept of dead reckoning and its requirements, describe the process of surface micromachining and propose a cost effective solution for velocity measurement in an inertial navigation system with the use of the ADXL202, a low-cost, low-g dual axis accelerometer from Analog Devices.

Keywords. Accelerometer, Surface Micromachining, Navigation, Dead Reckoning

1. Introduction

Global Positioning by Satellite (GPS) receivers have rapidly emerged as the standard engines at the core of most navigation systems found on the market today. The system relies on a network of 24 low earth orbiting satellites and associated ground stations to determine longitude and latitude coordinates based on triangulation techniques.

However, like any other radio-communications system, GPS is subject to interference and signal reception quality can be affected by the location of the receiver with respect to the environment. Specifically, there are two types of limitations affecting GPS based navigation systems: selective availability and multipath reflections. Selective availability refers to the fact that GPS signals are being deliberately altered by the military who control the system and for whom it was originally intended, in order to limit its accuracy when used in civil or commercial applications. It is expected that selective availability will be abandoned by year 2005 [1]. Multipath reflection describes what happens to the satellite signals as they bounce off of reflective surfaces such as tall buildings and other structures in the "urban canyon", introducing errors due to inaccurate distances and times used in the computation of position. In some instances, when fewer than three or four satellites can be "seen", the receiver in unable to compute a position altogether. While the coverage can be improved by the use of a dual GPS and GLONASS receiver (GLONASS - GLObal NAvigation by Satellite System - is the competing Russian equivalent to the U.S.-designed GPS), and the effect of selective availability minimized by the introduction of differential GPS (DGPS), most manufacturers of GPS-based systems generally provide GPS back-up navigation systems which rely on the technique known as "dead reckoning" [2, 3].

Dead reckoning is an ancient method used to derive position on the basis of three distinct inputs: a set of starting coordinates, the direction of travel, and the speed of travel. In a GPS-based system, the set of starting coordinates is usually given by the last known position recorded by the GPS before loss of the satellite signals. The direction of travel, often referred to as azimuth, can be provided by an electronic compass, although direction is usually derived from the angle rate measurement or rotation input provided by a gyroscope. Finally, in the case of an automobile, the speed of travel is typically recorded by the odometer and made available to the user via the vehicle's electronic bus. Alternatively, it can be calculated as the integral of an acceleration measurement.

2. Accelerometers in Navigation Applications

Cost and space constraints are driving manufacturers of vehicle navigation systems to investigate new approaches to solve the problems of azimuth and velocity measurement to perform dead reckoning. As mentioned earlier, velocity is the integral over time of acceleration. Incidentally, position is the integral of velocity over time. If one were to accurately measure acceleration of the vehicle from time t_1 to time t_2, one could calculate the speed of displacement and eventually the position of the vehicle at time t_2.

Recent progress in the development of semiconductor processing technology has led to the introduction of new sensors using microscopic electromechanical structures, known as Integrated Micro ElectroMechanical Systems (*i*MEMS). The process, referred to as Surface Micromachining, is being used to develop tiny sensors surrounded by signal conditioning circuitry, all on the same chip. At the same time, this technology is also being considered for the development of miniature gyroscopes. Accelerometers built using surface micromachine technology are becoming increasingly attractive to manufacturers of navigation systems because of their small size, low cost and ruggedness. In addition, they provide a means to measure velocity that is an integral part of the navigation system unit, thereby eliminating the dependency from the vehicle's odometer unit. A self-contained system is easier to install, making it attractive to after-market suppliers.

In addition to being used for velocity measurement, micromachined accelerometers also measure the degree of tilt or inclination of gyroscopes used in navigation systems. Surface micromachined gyroscopes measure yaw or angular rate by picking-up a signal generated by an electromechanical oscillating mass as it deviates from its plane of oscillation under the Coriolis force effect when submitted to a rotation about an axis perpendicular to the plane of oscillation. These sensors are subject to gravity and therefore, can introduce errors if the oscillating plane deviates from the horizontal. As vehicles roll and pitch, due to traveling conditions, the gyroscope is likely to generate erroneous readings unless the error introduced by its inclination can be compensated for.

A compensation algorithm can be generated, provided that tilt can be accurately measured. A pair of surface micromachined accelerometers can be set-up to provide distinct measurement signals for tilt and acceleration [4]. Incidentally, accelerometers using other technologies such as piezo-ceramic film or piezoelectric material cannot be used in this application as they do not provide a DC response and therefore are not capable of measuring a change in gravity resulting from some degree of tilt.

3. Surface Micromachining Fundamentals

As its name implies, surface micromachining differs from bulk micromachining in that the sensor element is built onto the surface of a silicon wafer, rather than etched into the bulk of the wafer, as is the case for a bulk micromachined sensor. In surface micromachining, the sensor is surrounded by signal conditioning circuitry. To build a surface micromachined sensor, a 1.6 μm thick layer of sacrificial oxide is deposited on the passivation layer of a silicon wafer, in which n^+ wells have been previously diffused. Openings are then etched through both insulators to the diffused areas in the substrate. A thick layer of polysilicon is subsequently deposited over the entire sensor area, filling the openings and establishing both a mechanical and electrical bond with the n^+ diffused areas (see Figure 1.)

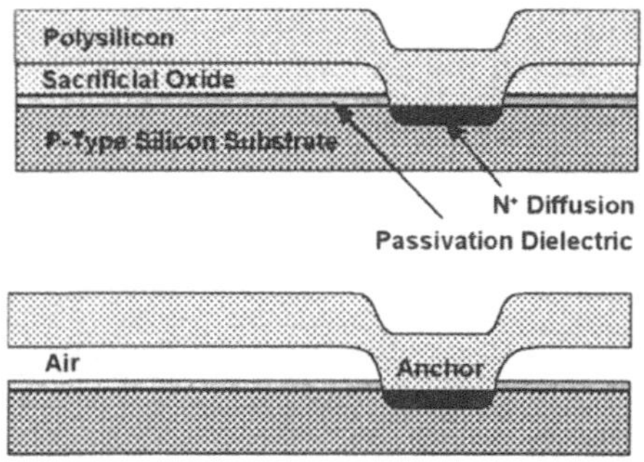

Fig. 1. Surface Micromachined Structure

The sensing elements are then etched into the layer of "floating" polysilicon. The sacrificial oxide is subsequently removed from under the polysilicon through further etching, leaving the polysilicon layer essentially suspended in mid-air, yet still attached to the substrate via the anchor posts, or pedestals, formed at the points of diffusion.

4. Analog Devices' Implementation

4.1. Sensor Design

An accelerometer or a gyroscopic sensor is a combination of springs, masses, motion sensing and actuation cells. It consists of a variable differential air capacitor whose plates are etched into the suspended polysilicon layer. The moving plate of the capacitor is formed by a large number of "fingers" extending from the "beam", a proof mass supported by tethers anchored to the substrate. Tethers provide the mechanical spring constant that forces the proof mass to return to its original position when at rest or at constant velocity (see Figure 2.) The fixed plates of the capacitor are formed by a number of matching pairs of fixed fingers positioned on either side of the moving fingers attached to the beam, and anchored to the substrate.

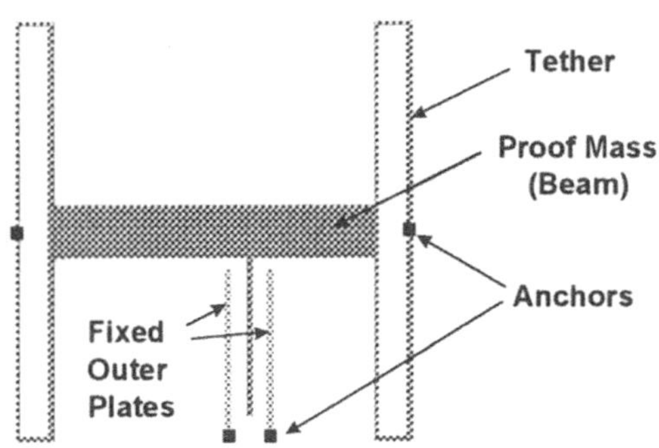

Fig. 2. A Micromachined Sensor Unit

4.2. Sensor Operation

When responding to an applied acceleration or under gravity, the proof mass inertia causes it to move along a predetermined axis, relative to the rest of the chip (see Figure 3.) As the fingers extending from the beam move between the fixed fingers, capacitance change is being sensed and used to measure the amplitude of the force that led to the displacement of the beam.

To sense the change in capacitance between the fixed and moving plates, two 1 MHz square wave signals of equal amplitude, but 180° out of phase from each other, are applied to the fingers forming the fixed plates of the capacitor. At rest, the space between each one of the fixed plates and the moving plate is equidistant, and both signals are coupled to the movable plate where they subtract from each other resulting in a waveform of zero amplitude.

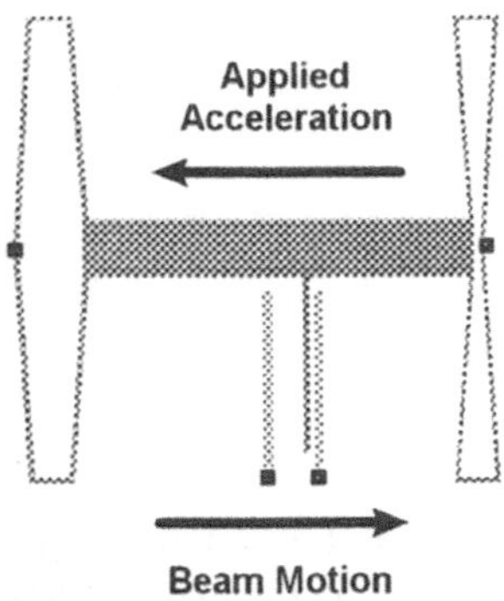

Fig. 3. Sensor Under Applied Acceleration

As soon as the chip experiences acceleration, the distance between one of the fixed plates and the movable plate increases while the distance between the other fixed plate and the movable plate decreases, resulting in capacitance imbalance. More of one of the two square wave signals gets coupled into the moving plate than the other, and the resulting signal at the output of the movable plate is a square wave signal whose amplitude is proportional to the magnitude of the acceleration, and whose phase is indicative of the direction of the acceleration.

The signal is then fed into a buffer amplifier and further into a phase-sensitive demodulator (synchronized on the same oscillator that generates the 1 MHz square wave excitation signals), which acts as a full wave-rectifier and low pass filter (with the use of an external capacitor). The output is a low frequency signal (dc to 1 kHz bandwidth), whose amplitude and polarity are proportional to acceleration and direction respectively. The synchronous demodulator drives a preamplifier whose output is made available to the user.

5. A Low Cost, Low *g* Accelerometer

Figure 4. shows the block diagram of the ADXL202, one of the first implementations of a low *g*, dual axis accelerometer on a single monolithic IC, designed for mass production and commercially available in a space-saving, convenient, 14-pin surface-mount Cerpak package. Typical noise floor is 500 $\mu g\sqrt{Hz}$, allowing signals below 5 mg to be resolved. The ADXL202 is a dc accelerometer with the capacity to measure both ac accelerations, (typical of vibration) or dc accelerations, (such as inertial force or gravity).

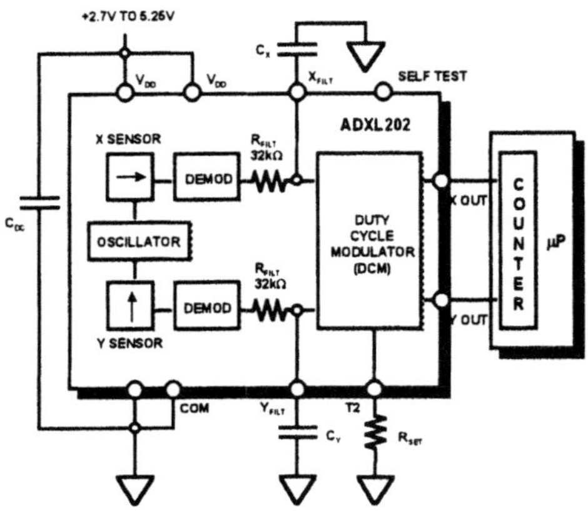

Fig. 4. ADXL202 Block Diagram

One of the distinctive features of the device is the availability of a digital output. The output circuit converts the analog signal to a duty cycle modulated (DCM) signal that can be decoded with a counter/timer port on a microprocessor. This is a particularly convenient feature for car navigation applications as the accelerometer can directly interface to the system's microcontroller without the need for an A/D converter or glue logic.

A single resistor sets the period for a complete cycle (T2), which can be set between 1 ms and 10 ms (see Figure 5.) A 0 *g* acceleration is nominally 50% duty cycle. The acceleration signal is a function of the ratio T1/T2, and can be determined by measuring the length of the T1 and T2 pulses with a counter/timer

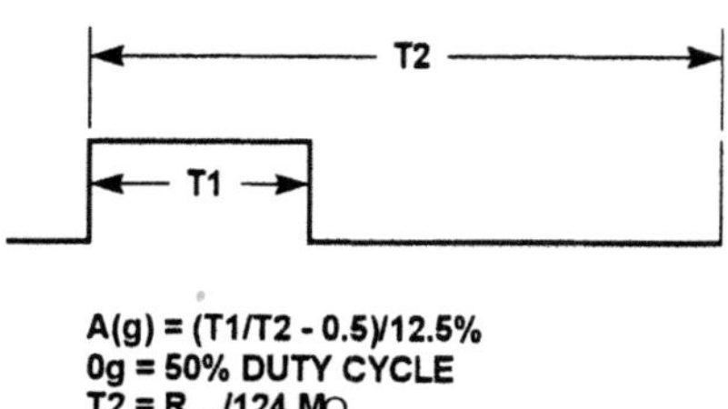

Fig. 5. Duty Cycle Output Signal

To be used in a supporting function to a GPS vehicle navigation system, an accelerometer must demonstrate high accuracy and resolution. These measures are largely determined by the device noise and the measurement bandwidth. The lower the bandwidth, the lower the noise. The lower the noise, the wider the dynamic range and the finer the resolution. With the ADXL202, the user can set the operating bandwidth by easy implementation of a low-pass filter for antialiasing and noise reduction. It is recommended that the user limit bandwidth to the lowest frequency needed by the application to maximize the resolution and dynamic range of the accelerometer.

In car navigation applications, the accelerometer measures inertial forces, acceleration and tilt, which are considered DC signals. So the 3 dB bandwidth can be set around 10 Hz, resulting in an RMS noise figure of 1.9 mg, and less than 5 % probability of a peak-to-peak noise exceeding 7.6 mg. Further reducing the bandwidth will result in lower noise, as will averaging multiple readings.

6. Preliminary Results

As mentioned earlier, accuracy of the sensor will play a critical role in its acceptance as a viable replacement of odometer input in an inertial navigation system. While noise plays a major contribution in reducing the sensor's accuracy, stability of the zero *g* offset over temperature and time are other important characteristics to take into consideration, as drift beyond a certain limit will introduce unacceptable errors. The following charts are indicative of the ADXL202 zero *g* drift over time, as measured every twenty minutes over a twenty four hour period. Both the X and Y channel of the device drifted by a maximum 2.4 mg.

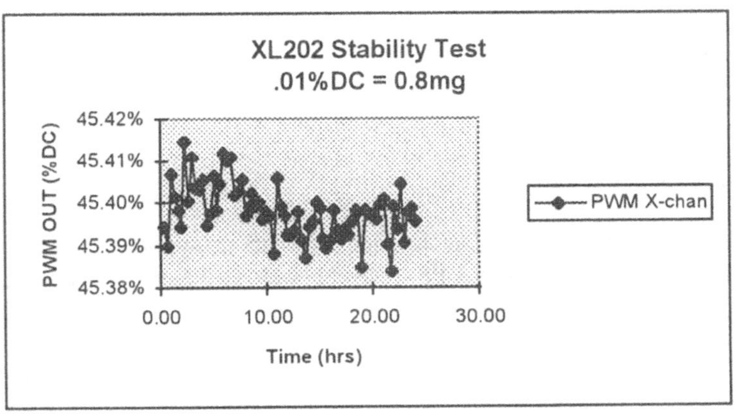

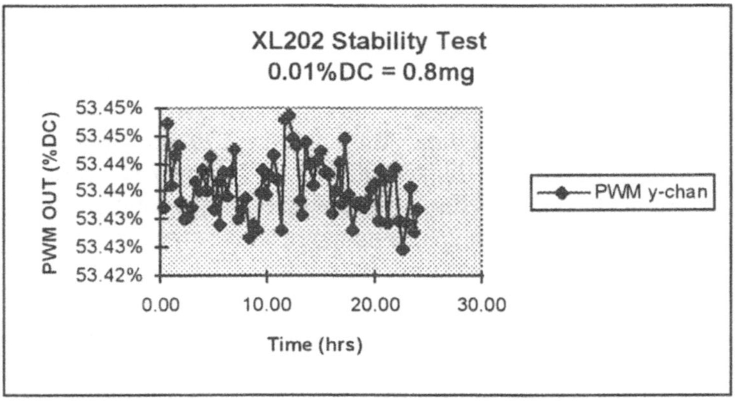

Fig. 6. Zero g Stability Data

During relatively recent field evaluation, the ADXL202 was used in an actual car navigation system, replacing the odometer input in a velocity measurement function (See Figure 7).

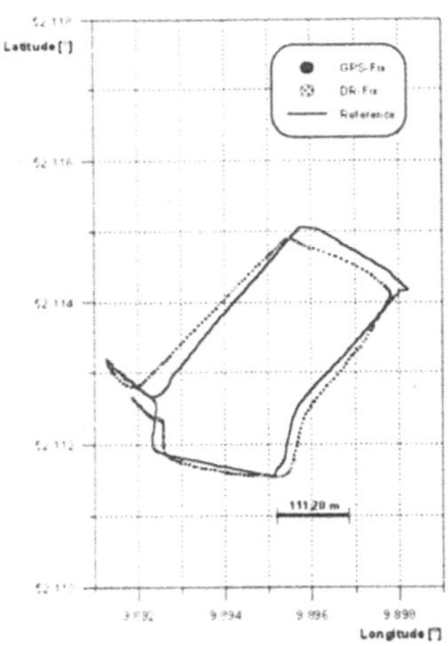

Fig. 7. Actual Navigation Test Results

A conventional gyroscope was used to provide azimuth information. As the vehicle moved around a loop, the GPS receiver was switched on and off, simulating periodic loss of signal as would be the case when driving through tunnels. Dead reckoning was then the only method used to track the vehicle's position. The above plot shows that from start to finish, the system only deviated from the reference by about 30 meters. Each time the GPS signal is recovered, inputs from the receiver are correlated with inputs from the inertial navigation system to ensure a smooth transition, hence the uninterrupted travel path shown on the plot. It must be noted that various degree of accuracy can be achieved depending on the complexity of the correlation algorithms being used to interpret data from the different sources of information, including digital mapping information. When digital maps are used in the system, the function of map matching provides valuable added information to improve position accuracy.

7. Future Developments

While surface micromachined accelerometers are finding their way into vehicle navigation support systems, semiconductor manufacturers are already setting their sight on providing improved performance, enhanced functionality or further integration levels, at costs never thought to be attainable only a couple of years ago.

One can expect to see higher performance (lower noise, higher accuracy) accelerometers being made available for the navigation markets within a few months. Simultaneously, the cost of slightly less accurate accelerometer sensors will drop even further, such that it is not inconceivable that new applications will emerge in the area of personal navigation. Finally, processes are being currently developed that will allow the implementation of gyroscopic and magnetic sensors or electronic compasses in surface micromachine technology in a not too distant future. Ultimately, vendors will likely be offering multiple and complementary functions, by integrating and combining accelerometer sensors with gyroscopes and/or electronic compasses on a single chip.

8. Conclusion

As recent market analysis reports strongly suggest, the GPS-based vehicle navigation system markets are poised for rapid growth, predicted to reach $1.7 billion in semiconductors sales by year 2001. While the Japanese market, forecast to reach $318 million this year, is already well established, Europe and North America still represent enormous growth potential, as they lag behind with forecasts of $41 million and $4 million respectively [1]. Simultaneously, surface micromachining technology and processes are becoming better understood and controlled, allowing increased performance and lower manufacturing costs, making them a viable technical and economical alternative to more traditional devices. In this paper, we implied that micromachined accelerometers can advantageously replace the odometer function in providing velocity information to the inertial navigation system and we showed that current levels of performance achieved by the state-of-the-art may be adequate for today's systems requirements, as suggested by preliminary field evaluation results using Analog Devices' ADXL202. We anticipate that surface micromachined accelerometers and gyroscopes will become more pervasive in navigation systems which rely on dead reckoning as a supplemental function to GPS receivers.

References

[1] Williams, M., "Worlwide Automotive GPS Navigation Markets—A New Direction for Semiconductors in Cars", Dataquest, July 7, 1997.

[2] Bennett, S.M., D.E. Allen, W. Acker and R. Kidwell, "Blended GPS/DR Position Determination System", 9th International Technical Meeting, Institute of Navigation, Satellite Division, Kansas City, MO, September 1996.

[3] Bennett, S.M., D.E. Allen, W. Acker and R. Kidwell, "Improving Position Location by Combining Dead Reckoning and GPS", Andrew Corporation, Orland Park, IL.

[4] Nakamura, Takesi Nagaokakyo-shi, Kyoto-Fu, "Acceleration Sensor" European Patent Application EP 0 744 622 A1, Murata Manufacturing Co., Ltd., May 1996.

MEMS: The New Challenge for the Electronic Design Automation Vendors

J. M. Karam[1], J. Oudinot[2], Dirk Backhaus[2], Ariel Cao[3], Joël Alanis Rodriguez[3]

[1]TIMA-CMP, 46, av. Félix Viallet, 38031 Grenoble Cedex, France
[2]Mentor Graphics , 49, avenue de l'Europe, BP.22, 78142 Vélizy Cedex, France
[3]Mentor Graphics Corporation, 8005 S. W. Boeckman Road, Willsonville, Oregon 97070-7777, USA

e-mail: Jean-Michel.Karam@imag.fr, *Phone:* +33 4 76 57 46 20, *Fax:* +33 4 76 47 38 14

1. Introduction

Mechanical and electronic designs are very tightly linked in the automotive world. Each electronic device is a part of an electrical-mechanical system or sub-system. Micro Electro Mechanical Systems (MEMS) bring this link to the integration scale. The use of MEMS has the potential of a significative cost reduction, which, together with increased performance requirements and possible new function capabilities, is at the source of the current interest on their increasing use in the automotive industry.

A major hurdle to the development for automotive systems resides on the lack of communication link between the mechanical world and the electronic world. This void is felt even more acutely in the development of integrated micro-electro-mechanical systems, where each team handles the tools traditionally used in its disciplines without any common interface. When using field solvers, e.g. finite elements methods (FEM), microsystem engineers identify materials properties and boundary conditions, and build a mesh, so the tool can run a 3D finite elements solution. The tool can predict the amount of stress and strain in the structures, the movement or any other interesting characteristics, but none of this information can be automatically transferred to an IC design tool.

In addition, the straightforward advances within the latest developments of the mainstream semiconductor industry is the use of already available intellectual property (IP) in the development of systems optimally matched to the end product specifications.

These prospects calls for a new generation of design tools that combines aspects of EDA and mechanical / thermal / fluidic CAD. The new product suite developed by Mentor Graphics together with Circuits Multi Projects (CMP), presented in this paper, offers an integration solution allowing a continuous design flow from front-end to back-end. This MEMS Engineering Kit supports both monolithic (1 chip) or hybrid (multi-chips, multi-technologies) solutions, providing a common language and interface for teams working in different fields.

The end objective is to bring to the system level designer of automotive products, a complete design flow, down to the chip level, anchored on design re-use and reliable system-level simulation, thus leveraging standard I.P. products for the realization of sophisticated miniature systems, at low cost.

2. The CAD Structure

The CAD environment considers both monolithic and hybrid solutions, answering the needs of both companies and universities as well as foundries. While the former considers the present state of the art and thus should be available , the monolithic approach aims at the co-fabrication of electronic and non-electronic functions: existing microelectronics production lines are being extended and adapted to allow MEMS production.

The environment allows a continuous design flow which can be seen or considered according to two points of view of a non-specialised *system-level designer* desiring to create a new microsystem with devices from at least two rather different areas, e.g. micro-mechanical and electronics, and the view of a device designer having expert knowledge on his domain of work, knowledge he would like to make available [1]. Efforts to build more general simulation environments have been on going, e.g. to interface process simulators with FE tools, but these tools, besides having their own (non-standard) interface formats, require much expertise from the user, both in simulation and in device modelling. Our approach tends towards more flexibility and compatibility, and is intended for non-expert users as well as for expert users.

The environment hence contains elements for the device designer, enabling him to design modules, to simulate them, and finally to put the knowledge in the form of characterised standard cells in library.

The system level user takes profit of this standard cell library that contains multi-level information (e.g. layout information, behavioural models, FEM-models). He assembles the desired cells, and simulates them at system level. Then, the resulting assembly is handed over to a second set of tools, designed for chip level procedures. Once the final layout is produced, both the system level designer and the device level designer can intervene again to check the features of the resulting microsystem. The figure 1 shows the global CAD principle.

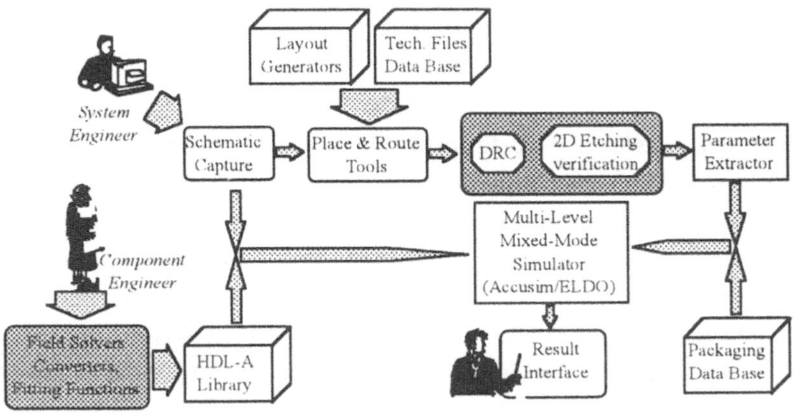

Fig. 1. The CAD structure

3. The MEMS Engineering Kit

The methodologies detailed above were used to develop the Mentor Graphics MEMS Engineering Kit. The resulting tools seeks to bridge the electro-mechanical gap and allows a continuous design flow achieved by the extension of the currently available IC design framework from Mentor Graphics, particularly [2]:

- the layout generation of the whole microsystem (electronic and non-electronic parts) and a design rule verification. This is linked to the specific addressed technologies and can support both monolithic (one technology file) or hybrid (many technology files) solutions,
- layout generators for mechanical structures, such as bridges, cantilevers, membranes and application oriented structures, allowing the schematic driven layout feature,
- a parameter extraction tool from layout level to netlist level enabling post-layout back annotation which will consider the packaging effect on the global system behaviour,
- full verification of the design functionality (ELDO / HDL-A),

- verification of the anisotropic etching procedure (2D and 3D) by the means of the anisotropic etching simulator, ACESIM, developed at CMP, and integrated within the Mentor environment,
- additional features, such as the cross-section viewer,
- a set of parametrized cells described at different levels (symbolic, system / behavioural, layout).

3.1. Extended Design Rules Checker

This section is tightly linked to the technology. The design rules checker can support any kind of monolithic or hybrid technology: bulk micromachining as well as surface micromachining or any other specific process, such as LIGA. The related work consists on the integration of the design rules extracted from specific test structures. The methodology of implementing the design rules into the CAD environment has been applied to the 1.0μ CMOS (from ATMEL-ES2 foundry), the 1.2μ CMOS (from AMS foundry) and the 0.2μ HEMT GaAs (from PML foundry) compatible bulk micromachining technologies available through CMP [2, 3]. These design rules are imposed by the properties of the anisotropic etching. The rest of the design rules strictly concerning the conductive layers such as metal and polysilicon, has been kept unchanged for the simple reason that their electrical functionality is not altered even on suspended parts. Figure 2 shows a DRC example applied to a microsystem integrating microstructures as well as electronics.

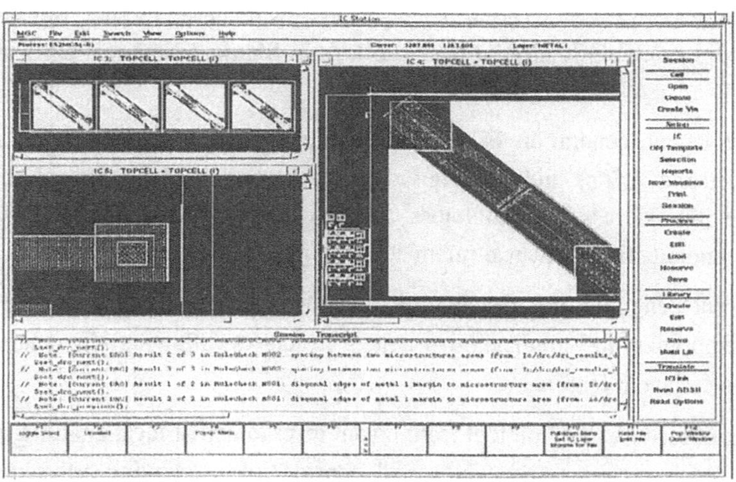

Fig. 2. Example of DRC execution: microsructure rules and electronic rules checked in the same time.

3.2. Parametrized Cell Libraries

A fully characterized set of HDL-A microsystem library elements has been developed. This library includes accelerometers (different types), temperature sensors, pressure sensors, IR detectors, electro-thermal converters, electro-thermo-pneumatic micro-pump, humidity sensor, magnetic sensor (Hall effect) and gas flow sensor. Works towards new elements are in progress. These models are linked to technology independent, application oriented structure generators insuring a schematic driven layout generation. Three family of generators exist: elementary functions (such as zig-zag function), elementary structures (such as bridges, membranes, cantilevers) and application oriented structures (such as IR sensors, bolometers, accelerometer) (figure 3).

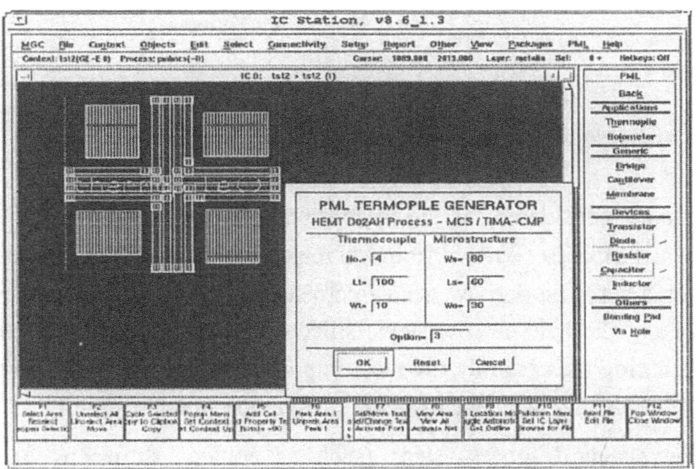

Fig. 3. Seebeck effect IR sensor structure generator

3.3. Cross-Section Viewer

Additional features, useful for the microsystem engineers, have been added. One main function is the cross-section viewer allowing the visualization of the different layers according to a cut-line performed, in any angle, at the layout level. Figure 4 illustrates this function.

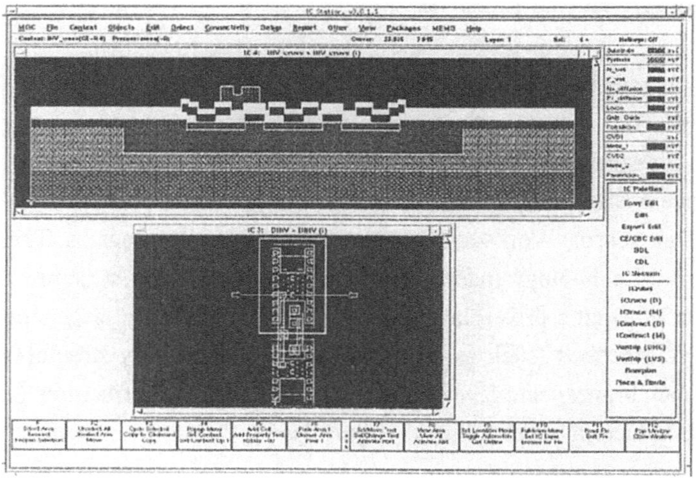

Fig. 4. The Cross-Section Viewer

3.4. The Anisotropic Etching Simulator

Microsystems considered as an extension of microelectronics include all technological processes known from microelectronics, plus new processes which can in some cases be derived from microelectronics by adding post-processing steps. Among the new process steps, silicon bulk micromachining and surface micromachining are certainly the most important. More than fifty percent of all microsystems contain a device made by silicon bulk micromachining.

While the standard process steps (such as epitaxy, diffusion, etc.) can be modelled with existing tools like SUPREM IV the modelling of the new process steps introduced by the micromachining technologies is still not mature.

The ACESIM simulator developed at TIMA is based on a geometrical anisotropic etching model, in contrast with the complex atomistic models that require huge data space. The current version of this tool provides a two dimensional simulation. It has its main potential by the fact that it has been integrated within the Mentor MEMS Engineering Kit allowing an interactive design and dimensioning procedure of the microstructures considered as a post design rules checking operation. In addition, it's obvious that when targeting the monolithic solution where the structures are processed together with the electronics, one of the most important factors is the etching time necessary to liberate the suspended part, because the longer it is, the more the electronics parts can be damaged. This time mainly depends on the characteristics of the etchant, the shapes of the open areas constituting the microstructure and their relative positioning: these physical parameters determine the sequence of planes apparition within the silicon substrate, which is to be known in detail in order to predict the etching time.

It is to be noted that when using the layout generators there is no need to perform anisotropic etching verification since the design rules considered by these generators ensure the design of correctly etchable structures. However, this simulator is very vital for full-custom MEMS designs.

Figure 5 shows an etching simulation performed by ACESIM within the Mentor MEMS Engineering kit. The etch rate diagram of EDP solution is also showed. It represents the etching speed of planes in μm /mn (indicated by graduations) as a function of their angle with the (100) plane.

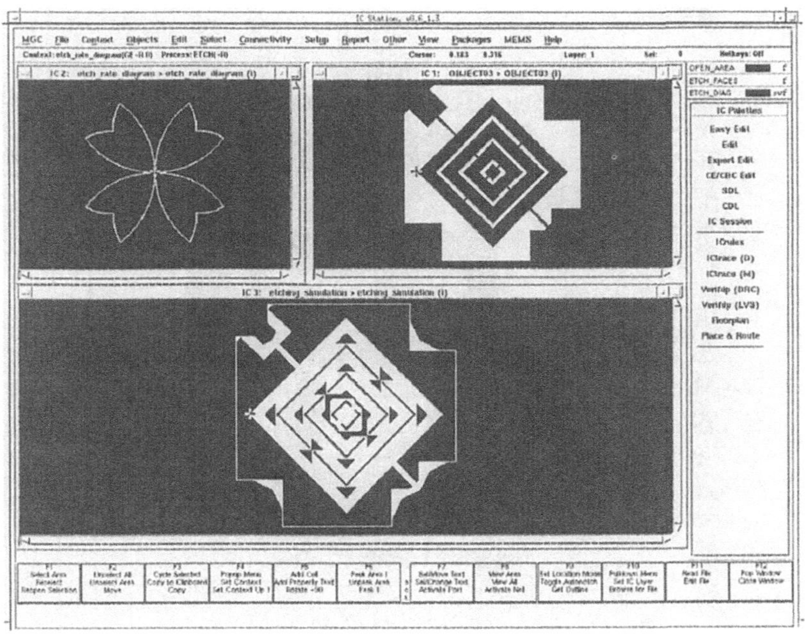

Fig. 5. Etching simulation of an accelerometer structure performed by ACESIM within the Mentor Microsystem Engineering Kit. The etching diagram of EDP solution is also showed.

4. Answer the Need of the Automotive Market

Today's car customers requires more safety, comfort and performance features. This market pull justifies the rapid technological progress and the potential integration of MEMS in the automotive system in the last few years. This increasing leverage of electronics in automotive design is starting to push EDA software in new directions. If available, these design tools will contribute to the cost reduction required by the automotive market, increase the productivity and decrease the time-to-market.

The figure 6 gives a block diagram of an airbag system. It is very easy to see that this system needs a multi-disciplinary team to be achieved. Today, the different designers use a FEM simulator to design the MEMS device and an electrical simulator to design the electrical part.

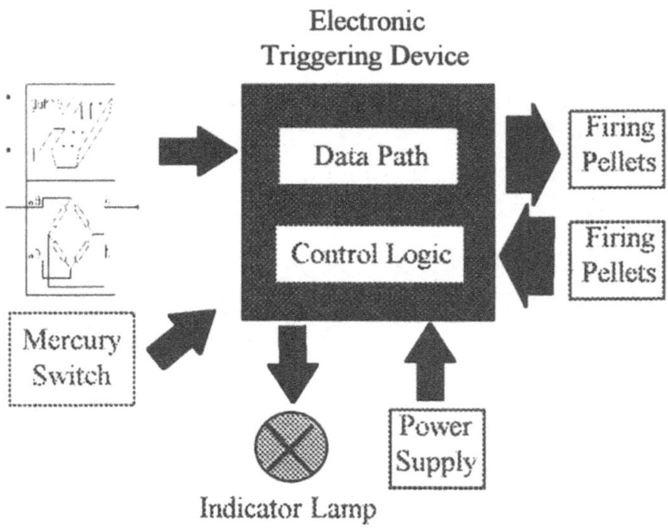

Fig. 5. *The block diagram of an airbag system*

Using the MEMS engineering kit, the simulation of the global system, i.e. MEMS device, analog electronics, digital electronics, will be achieved through the multi-level, mixed-mode, multi-language simulator, ELDO / Accusim, supporting HDL-A, VHDL / Verilog and spice-like languages as well as the emerging standards, such as VHDL-AMS and Verilog-A. Then a complete flow from front-end to back-end can be achieved through the extended Mentor Graphics tools. Figure 6 shows an accelerometer model written in HDL-A as well as a plot of the stimuli and the output signals. The technology as well as functional parameters are considered. HDL-A allows the developers to build models which handle different kinds of signals (electrical, mechanical, etc.) by modelling the physical phenomena using ordinary differential equations.

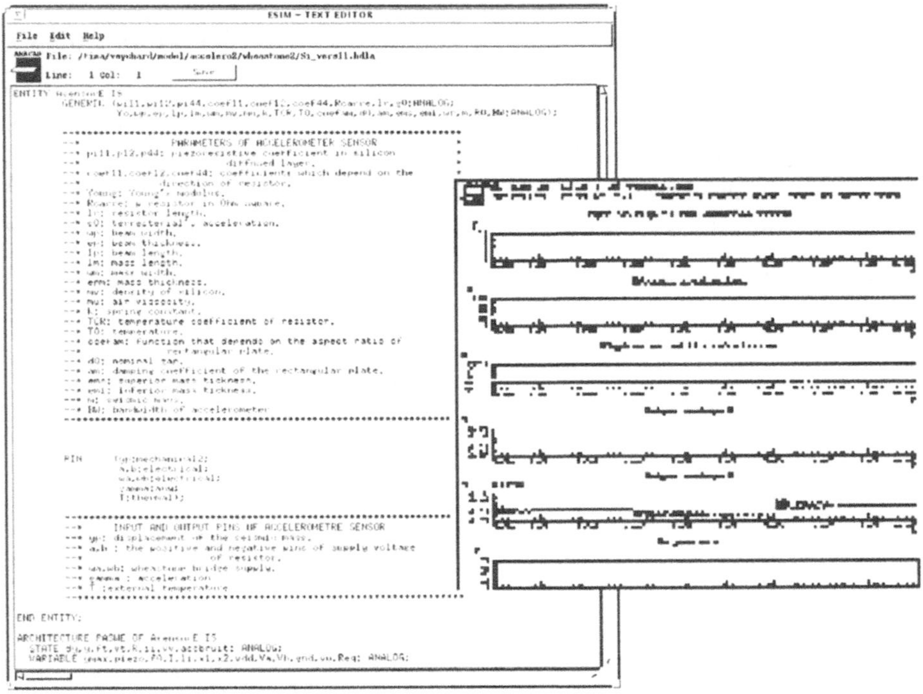

Fig. 6. An HDL-A accelerometer model with the simulation results

5. Conclusion

After a general introduction of the automotive market requirements, the paper shows the approach followed by Mentor Graphics and CMP to provide a solution that bridge the electro-mechanical gap. The MEMS engineering Kit is also detailed as well as its applicability to automotive system design.

References

[1] J. M. Karam, B. Courtois, H. Boutamine, "CAD tools bridge microsystems and foundries", Special Issue of IEEE Design & Test of Computers, June 1997.

[2] J. M. Karam, B. Courtois, H. Boutamine, P. Drake, M. Rencz, A. Poppe, V. Székely, K. Hofmann, M. Glesner, "CAD and Foundries for Microsystems", 34th Design Automation Conference (DAC), Anaheim, California, USA, June 1997.

[3] R. P. Ribas, J. M. Karam, B. Courtois, J. L. Leclercq, P. Viktorovitch, "Bulk Micromachining Characterization of 0.2μm HEMT MMIC Technology for GaAs MEMS Design", Low Dimensional Structures and Devices (LDSD 97), Lisbon, Portugal, 19-21 May 1997.

The Smart Image Sensor Toolbox for Automotive Vision Applications

Peter Seitz

Centre Suisse d'Electronique et de Microtechnique SA, CH-8048 Zurich, Switzerland

Abstract. Advances in semiconductor fabrication technology have made it possible to co-integrate analog and digital circuitry in each "smart" pixel of an image sensor. As a consequence, a wide range of functionality modules can be realized. Their collection represents the smart image sensor toolbox, with which custom image sensors can be tailored to an application. Many types of automotive imaging and vision problems can be solved successfully with the smart image sensor toolbox, among them high-dynamic-range imaging, low-cost color imaging, range or 3D imaging, high-speed imaging, as well as low-cost single-chip vision systems.

Keywords. Solid state image sensors, CCDs, CMOS imagers, cameras, vision.

1 Introduction

The budding engineering discipline *Electronic Imaging* is concerned with all elements and aspects of the imaging chain in a holistic way, starting with the generation of light up to the processing and interpretation of the acquired images, as illustrated in Fig. 1.

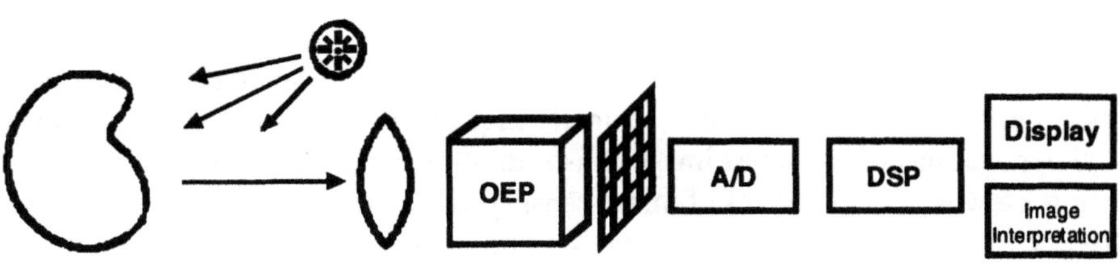

Fig. 1 : Information flow in an Electronic Imaging system, including a source of radiation, the interaction of light with an object, the image formation of the obtained information-carrying radiation, the optoelectronic preprocessing of this radiation (OEP), the photo-transduction in a 1D or 2D image sensor, analog/digital signal conditioning (A/D), digital signal processing (DSP) and the display or the interpretation of the data.

In this view, the traditional role of the image sensor as a simple device for transducing an incident two-dimensional distribution of light into a corresponding voltage pattern has been replaced by a much wider conception : The functionality of an image sensor is not restricted to the radiometric acquisiton of local brightness in a scene; the modern view of an image sensor is a front-end device acquiring and pre-processing a wide range of optical image information [1]. There are two primary reasons for this fundamental change :

- The technology of CMOS image sensors (or APS image sensors) has made it possible to realize photosensors with a performance similar to the one of CCDs, in terms of sensitivity, noise, dynamic range, etc. [2]. Additionally, such an image sensor can be provided with on-chip electronic circuitry such as analog-to-digital convertors, digital processors or complete microcontrollers.
- The advent of the smart pixel concept (see for example [3]), providing each pixel with application-specific functionality — in CMOS as well as in CCD technology – very much widens the scope of the imaging concept, from "image acquisition" to "optoelectronic information extraction".

This development is very desirable for most automotive applications because here one is rarely interested in a faithfully reproduced image but rather in an interpre-ation of the scene contents. Typical examples include questions like "is there a traffic sign in the field of view, and if yes, what does it indicate", to much more elaborate analysis such as "does a dangerous traffic situation evolve now, where certain precautions should be taken by the driver" .

The present contribution puts an emphasis on the second point mentioned above, describing a modular concept (the smart pixel toolbox) for the quick and efficient realization of custom "smart" image sensors based on application-specific pixels with tailored functionality. We will certainly have to go a long way until complex scene interpretations such as the ones described above will be reality; however, many different types of image sensors have already been realized, showing the way to a higher level of image understanding.

In Section 2 of this paper, the technological requirements of smart pixels are described in terms of advances of semiconductor technology. A summary of the smart image sensor toolbox and some of its major modules is offered in Section 3. Practical applications of the smart image sensor toolbox are presented in Section 4 by listing a few generic problems of image sensing for automotive applications and by discussing solution concepts drawing on the smart image sensor toolbox. In Section 5, some practical limitations of the smart pixel concept are reviewed, and possible ways of overcoming these difficulties — effectively a consequence of imperfect semiconductor manufacturing — are suggested. In the concluding Section 6, mid-term predictions concerning the future development of the smart image sensor field are given, including an appraisal of the "seeing chip" concept.

2 Semiconductor Technology for Image Sensors

Semiconductor technology has made significant and surprisingly predictable progress over the past decades. The most obvious measure for this development is the minimum feature size which is available in a certain technology, also called design rule. The evolution of the minimum feature size is illustrated in Fig. 2, from which an exponential trend is easily derived : The minimum feature size is predictably reduced by an average of about 10% per year. In the early 70's, a minimum feature size of 6-8 μm was employed, while today's most advanced DRAM (dynamic memory) chips are fabricated with 0.25 μm design rules. Using the above stated empirical rule, it can be predicted that the minimum feature size will have dropped to well below 0.2 μm by the year 2000 [4].

At the same time that design rules were continuously shrinking, the diameter of the silicon wafers has increased, making 20 cm (8 inches) the standard for many silicon foundries. The technology employed for image sensors is less advanced, making use today of 15 cm (6 inches) diameter silicon wafers, and often employing larger minimum feature sizes. These two developments, firstly, the increase of the wafer diameter, and, secondly, the reduction of the design rules, demand much more expensive fabrication equipment. The investment is worthwhile, however, because more electronic circuitry can be placed on the same chip area, and more wafer area can be processed at a time in the batch processes of semiconductor technology.

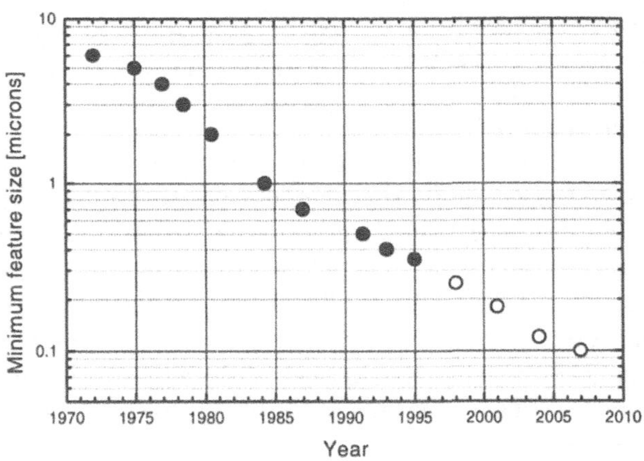

Fig. 2. Evolution of the minimum features size (design rules) in semiconductor manufacturing technology. Values are taken from published data (solid dots) and the predictions made in Ref. [4] (open dots).

One of the most visible consequences of these impressive advances of semiconductor technology is the marked increase in resolution of image sensors (Fig. 3). This graph shows the evolution of the record number of pixels on a CCD image sensor, as a function of the date when the work was published. There is a marked lack of progress in the number of pixels in the years 1975-1983. Our interpretation of this phenomenon is the following:

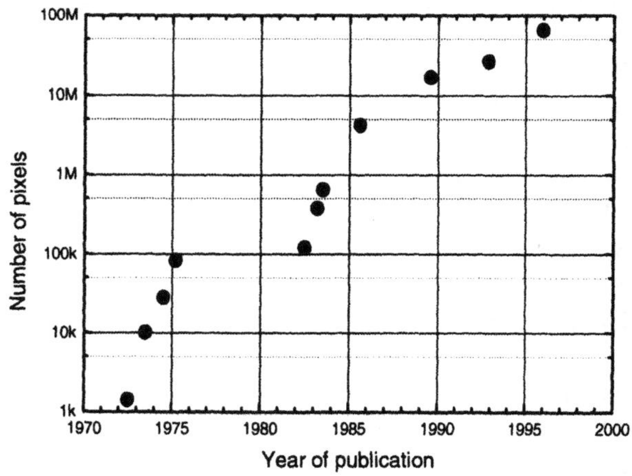

Fig. 3. Evolution of the number of pixels on a CCD image sensor as a function of the year when the record-breaking number was published.

The number of pixels in image sensors was increased rapidly by different research groups, until enough pixels on an image sensor were available for the realization of solid-state video cameras. After that initial period of research activity, it took significant time and effort to develop the semiconductor technology, necessary for the mass-fabrication of these devices with high yield. It was only then, after 1983, that the technology was pushed again, and image sensors with increasingly large numbers of pixels were fabricated. The largest image sensor that has been realized to date is a frame-transfer CCD with 7168×9216 pixels, covering an area of 9×12 cm^2 [5]. This sensor is fabricated using silicon wafers with 150 mm diameter. It is very difficult, however, to produce so large image sensors with a sufficiently high yield.

Another important consequence of the advances in semiconductor manufacture is the shrinking of the geometry of the pixels. A continuous reduction has been observed over the past 25 years. Today's smallest pixels in a CCD image sensor measure just 2.4×2.4 μm^2 [6]. Similar to the minimum feature size shown in Fig. 1, an exponential trend in the decrease of the minimum pixel size is also observable, with an average reduction rate of about 10% per year. For optical reasons, however, the trend of shrinking pixel size will not continue much beyond this value. Fraunhofer diffraction at the usually circular aperture of the optical lens system used for imaging the scene onto the image sensor's surface blurs the images [7]. As an example, consider a diffraction-limited photographic lens with f/2.8, for which a diameter of the Airy disk of d = 3.8 μm is obtained for green light (λ = 555 nm). Actual, low-cost photographic lenses - for example in video applications or for consumer electronic photography - will not be of the highest quality, and they will not reach diffraction limited imaging performance. It quite obvious, therefore, that the pixel size will not drop substantially below about 2-3 μm, a value that has already been reached today.

This situation of continuously shrinking minimum feature size with a lower limit to the minimum dimension of a pixel, has an obvious consequence: It is possible to supply the individual pixels with added functionality in the form of electronic circuitry, without substantially increasing the geometrical dimensions of the pixels.

Image sensors with such pixels offering added functionality are called Active Pixel Sensors (APS) [8]. It is the main purpose of the present paper to investigate the possibilities offered by APS and related technologies, and to discuss the consequences for automotive imaging and vision applications

3 The Smart Image Sensor Toolbox and Its Mayor Modules

The detection of light in a traditional image sensor is restricted to the simplified signal chain illustrated in Fig. 4. A semiconductor device with an associated electric field (usually a photodiode or a precharged MOS capacitor) is employed for the separation of photo-generated charge pairs, resulting in a photocurrent that is highly linear with the incident light intensity over nine orders of magnitude or more [9]. This photocurrent is integrated over a certain time, the so-called exposure time, and the photocharges are retained on a suitable storage device.

Fig. 4. Simplified signal chain in traditional solid-state image sensors. Incident light generates a photocurrent in each pixel. The photocurrent is integrated and stored. During sequential scanning, photocharge is amplified and read out.

The individual pixels are then sequentially scanned with a suitable switching mechanism. The pixels' charge signals are read out, and they are amplified, one by one, to complete the detection process. At the same time, the storage device is cleared (reset), so that it is ready for the next exposure cycle.
Modern semiconductor processes and the reduced feature sizes for electronic circuits are the basis of functionality in the individual pixels that is much increased above what is illustrated in Fig. 4. Some of the possibilities and novel functionalities offered at the different stages of the image acquisition chain are symbolized in Fig. 5. This "smart pixel toolbox" offers a wide variety of modules, with which a specific problem in electronic imaging can be solved, where the system task is to acquire spatio-temporal light distributions and process the contained information.
The content of the smart pixel toolbox illustrated in Fig. 5 is discussed in more detail elsewhere, see for example [1]. For this reason, the most important modules are only summarized very briefly in the following :

- Arbitrary geometry of the pixels can be exploited for signal processing in the analog domain, realizing linear and non-linear spatial transformations [10].

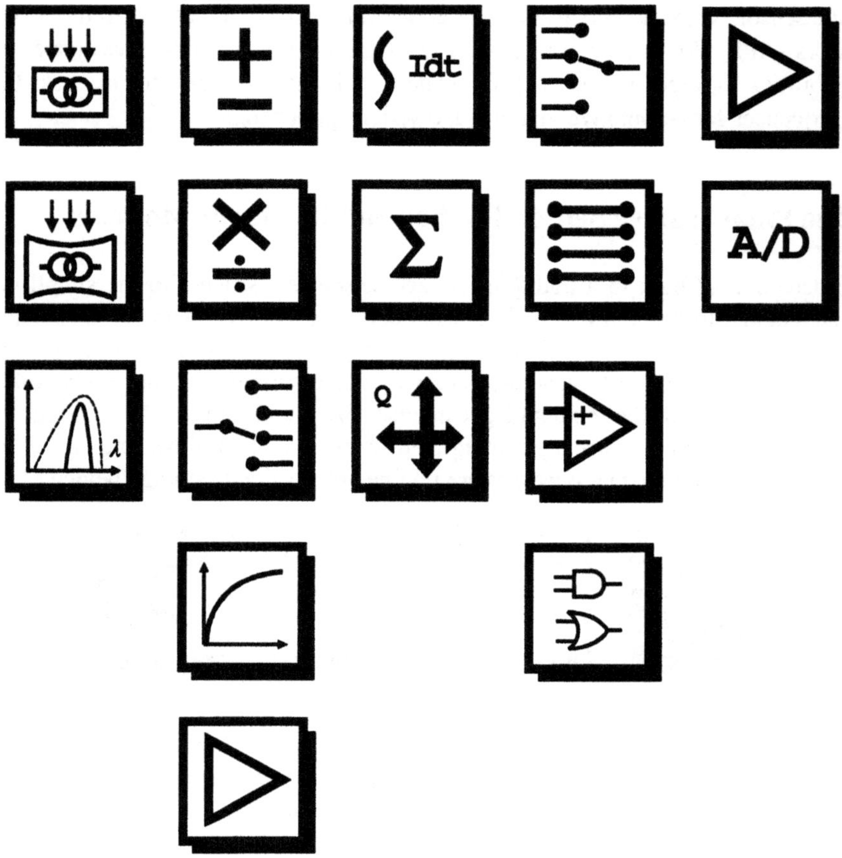

Fig. 5. Illustration of the enhanced image acquisition and processing chain in electronic imaging, the smart image sensor toolbox. It makes use of the capabilities of CMOS and CCD technology for the realization of application specific image sensors with on-chip functionality. This picture is a symbolic and incomplete representation of the possibilities offered by image sensors with smart pixels for novel optical measurement techniques and integrated vision systems.

- The wavelength-dependent absorption properties of silicon can be exploited for the realization of low-cost color-sensitive pixels without additional filters [11].
- Applications in which small (temporal or spatial) modulations must be processed can profit from programmable offset-subtraction pixels. This leads to wide dynamic range image sensors with a D/R exceeding 150 dB [12].
- Fixed or programmable gain over a wide range of at least $10^{-4}...10^{4}$ can be realized with the gain-pixel concept [13].
- Photocurrents can be redirected in sub-microsecond time to different processing or storage channels. This forms the basis of new types of high-speed

imagers and lock-in pixels for the synchronous detection and demodulation of temporally modulated light [14].

- Non-linear compression of the pixel's optoelectronic signal (e.g. logarithmic) for the increase of image sensor dynamic range above 120 dB [15].
- Perfect summation and redistribution of charge signals in CCD structures for high-speed FIR filtering in the analog domain [16].
- One- and two-dimensional photocharge transportation for signal processing applications such as in the convolution CCD [17].
- Monolithic integration of (conventional) analog and digital electronic circuitry in each pixel or on the image sensor for signal pre-processing, including analog-to-digital convertors (ADCs), digital processors, memory blocks, input-output circuitry, on-chip drivers, etc., up to the realization of complete single-chip machine vision systems [18].

4 Example Solutions of Automotive Imaging and Vision Problems with the Smart Image Sensor Toolbox

4.1 High Dynamic Range Imaging

To cover the wide light intensity range of typical traffic scenes, in broad daylight as well as in darkness, image sensors are desired approaching or even surpassing the human eye's dynamic range of almost 150 dB.

The basic APS architecture illustrated in Fig. 6 already indicates a first approach to this problem. Photocharge is accumulated non-destructively, and the pixel can be interrogated repeatedly, therefore, resulting in differing exposure times without any degradation of the signal. This approach to wide-dynamic-range imaging has recently been demonstrated in [19].

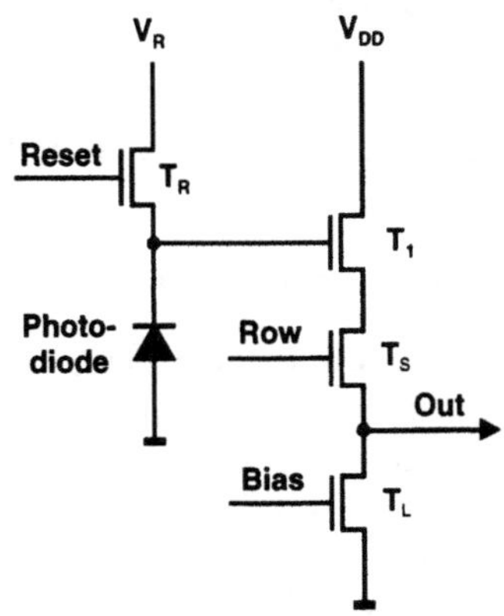

Fig. 6. Illustration of the basic APS (CMOS sensor) architecture

A similar approach is the use of CCD technology for the realization of pixels whose integration times can be programmed individually, as described in [20]. Since this approach requires a relatively large amount of circuitry in each pixel, it is not very practical for the time being.

Both of the above solutions employ a linear light-to-voltage transduction principle. An alternative is to compress the signal range non-linearly. A very simple realization is to use the reset transistor of the standard APS pixel as a logarithmic photocurrent-to-voltage convertor [15], as illustrated in Fig. 7. The physics of MOS-FETs results in a typical light-to-voltage conversion of about 35 mV per decade of incident light and a dynamic range of 120 dB and more [13]. Since the different pixels show marked variations in their conversion slopes and their offsets, suitable additive/multiplicative corrections in hardware or software are necessary for a practical camera solution with high dynamic range.

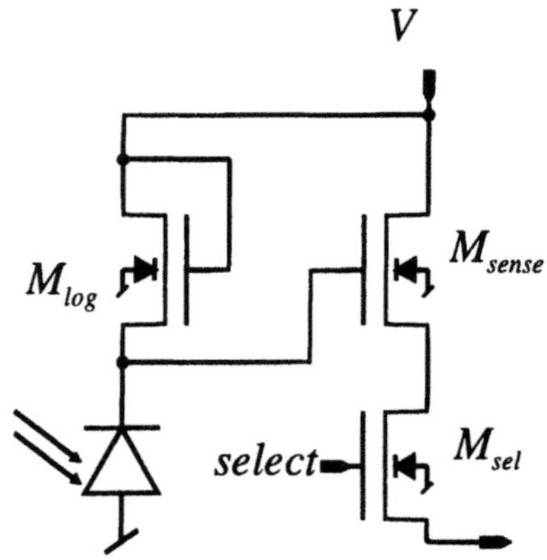

Fig. 7. Schematic of the log pixel exhibiting logarithmic, wide-dynamic range response > 120 dB.

Another approach to wide-dynamic-range imaging is a pixel with programmable gain [13]. This so-called gain-pixel relies on a variation of the current mirror circuit with which a pixel-individual gain between at least 10^{-4} up to 10^{4} can be selected. Depending on the local intensity at a pixel site, the local gain can be adapted to provide output signals in the proper range. Compared to the other pixels described above that offer a wide dynamic range, the gain-pixel is more complex in its architecture as well as in the complete system concept.

4.2 Low-Cost Color Imaging

Conventional color cameras are realized with dye or dielectric filters on each pixel, most often arranged in a mosaic pattern for single-chip color cameras. As described for example in [9] it is possible to use the absorption properties of silicon itself for the realization of a color pixel. Blue light is absorbed in the first few 100 nm of silicon while red light can penetrate more than 10 μm into the silicon. By realizing a stack of several photocharge-collecting structures on top of each other, a pixel is realized that is sensitive to different portions of the spectrum in the same place. Color pixels with reasonable color reproduction properties can be realized with such an approach [9].

By varying the voltage conditions in the photocharge collecting structures, it is even possible to change the spectral characteristics of a suitable color pixel in real-time [21]. Again, this comes at the cost of increased circuit complexity in each pixel, and it might not be necessary for most standard color-camera applications.

4.3 High-Speed Imaging

In automotive applications, many processes occur in time frames well below 10 ms. For this reason, they are not accessible any more to the human eye or to standard video cameras with their 50/60 Hz field rate. Image sensor architectures have been proposed with which image sequences can be acquired at very high frame rates.

An obvious solution is to provide an image sensor with several output ports all working in parallel. Such high-speed image sensors have been available for several years now, reaching repetition rates of several 1000 frames per second [22].

A new type of CCD image sensor offering short-time frame rates of 1 million frames per second has recently been demonstrated with intermediate photocharge storage sites at each pixel [23], so that the complete image sequence is read out only at the end of an exposure cycle. While offering ultra-high frame rates, this approach is limited to applications requiring a maximum of a few tens of exposures for each high-speed sequence. Additionally, the fraction of photosensitive to (opaque) storage area — the so-called optical fill-factor — is very low, making it difficult to provide sufficient lighting to expose the photosites properly.

4.4 Robust Range and Velocity Imaging

In automotive applications, it is often more important to know the three-dimensional configuration of an object or a traffic scene than the traditional two-dimensional brightness or reflectance image. For the acquisition of such depth or range images, many different techniques are known from the field of optical metrology. Several such techniques profit substantially from the recent availability of custom smart image sensors :

The oldest type of range imagery is provided by the passive stereo technique, on which also most range image acquisition in nature is based. It consists of two camera systems (two eyes), each looking from a different angle at the same portion of a scene. From the lateral disparity of features with high contrast in the two images, a depth or range map can be determined. An integrated stereo-vision chip has been demonstrated that relies on this principle [24]. It consists of two image sensors and the necessary processing electronics for providing range images in real-time.

An alternative to this passive triangulation technique is active triangulation where a source of light (a single beam or a sheet of light) is projected on an object from one direction, and the reflected light is observed from another direction, as illustrated in Fig. 8. By scanning the light over the scene, a complete range image can be obtained. This has been demonstrated with a time-to-maximum pixel image sensor [25] and a wedge-pixel image sensor with which range images can be acquired at video speed of 50 Hz [26].

These types of triangulation-based range imaging solutions become unpractical for larger object distances because the optical base (the distance between light source and sensor) becomes too large for the distance resolution typically desired. In such cases, time-of-flight techniques are used successfully, measuring the time taken by light pulses between their emission, their reflection at the object and their detection in a suitable photosensor. The high speed of light of 300'000 km/s results in distance resolution of 15 cm for a temporal resolution of 1 ns in the sensor plane. For this reason special signal acquisition and processing techniques are employed that provide the required high temporal detection resolution. The so-called lock-in technique is the most popular of these techniques, and thanks to the smart image sensor toolbox it is possible to realize smart image sensors in which each pixel detects and demodulates temporally encoded light synchronously. A combination of modules from the smart image sensor toolbox has led to the realization of lock-in pixels offering pixel-individual offset-compensation [14] for the practical usage in time-of-flight range cameras. Since such range cameras can be miniaturized and they do not require any moving parts, they might be of large interest in automotive applications for smart cruise controls or collision avoidance systems.

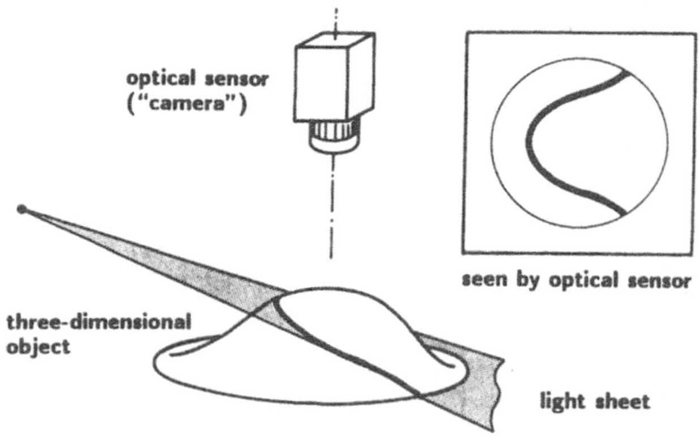

Fig. 8. Illustration of the active triangulation principle for the acquisition of range images. In this implementation, a sheet of light is scanned over the object and the diffusely reflected light is observed by the camera ("light sectioning")

4.5 Low-Cost Micro-Cameras

The smart image sensor toolbox provides all elements for the realization of image sensors combining all necessary circuitry for the acquisition of images and the generation of video signals. It is possible, therefore, to realize a complete video camera on a single chip. In combination with a suitable micro-optical lens system, the complete micro-camera is so small (a volume of a few 10 mm^3 is sufficient) that it can be integrated almost everywhere. This capability is illustrated in Fig. 9 with a video camera contained in the upper third of a conventional pen [27].
Obvious extensions of this technology include single-chip color cameras, single-chip digital cameras with on-chip analog-to-digital convertor or single-chip cameras with on-chip modems.

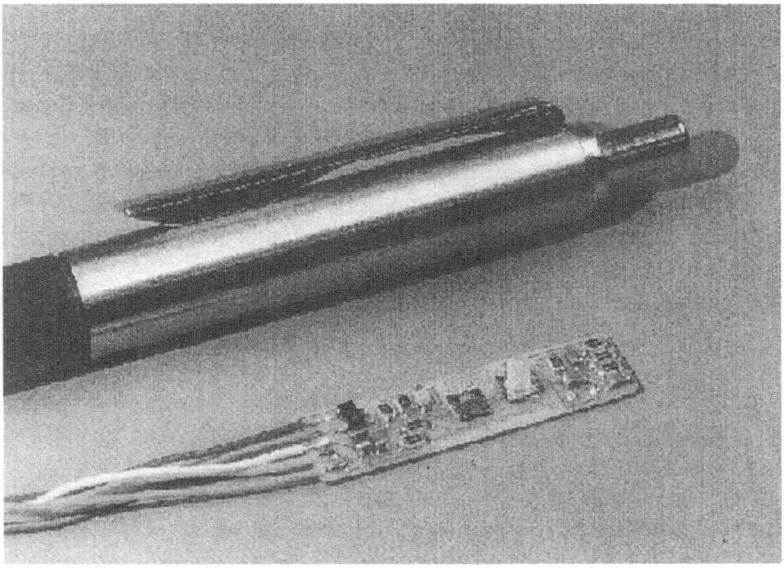

Fig. 9. Example of a miniaturized camera in a pen, realized with a CMOS image sensor, a micro-objective on top and a commercially available microcontroller (PIC) for programming the image sensor and providing data communication.

4.6 Low-Cost Vision Systems

The possibilites described above can be extended even further, giving real meaning to the term "smart image sensor". On-chip circuitry can be combined with the image acquisition part for the immediate processing and interpretation of the acquired scenes. Practical examples are described in [18], one of which is particularly inspiring : An image sensor is combined with ROM, RAM and a custom 2 GOPS digital processor that can compare acquired fingerprint images with stored references prints. Depending on the comparison result, a go or no-go signal is provided, effectively realizing a single-chip access control system as a replacement of the conventional mechanical key.

For automotive vision applications one might envisage improvements of this type of single-chip vision systems for the recognition of traffic signs or to support automatic driving. A few years ago, this development has even led prominent researchers in the field to proclaim the imminence of "seeing chips", single-chip vision systems that can perceive certain visual aspects of their environment [27]. Today such predictions are seen more moderately, and the future of smart image sensors and seeing chips are discussed in a little more detail in Section 6 of this work.

5 Practical Limitations of the Smart Pixel Concept

The smart pixel concept, as powerful as it might be, suffers from several practical problems. The restricted floor-space available for circuitry in smart pixels dictates the use of a minimum number of electronic devices. For these reasons, smart pixels exhibit a range of imperfections, which a potential user should know before expecting too much from a custom photosensor. In the following, a list of typical shortcomings in smart pixels is given, together with possible solutions how to overcome them in the complete optical microsystem.

- Pixel non-uniformity : Differences in the size or radiometric properties of the individual pixels can cause variations in response of 1-5%. These differences can be calibrated out, for example by postprocessing in a microcontroller. In the future, it might become possible to store these calibration values directly in the pixels, by making use of analog EEPROM storage techniques, see for example [29].
- Bad pixels : Imperfections in the fabrication process can lead to individual "bad" pixels that behave significantly different than other pixels. Although the occurrence of such bad pixels is usually rare, of the order of a few bad pixels per million pixels in good CCDs, there values need to be replaced in a postprocessing step, for example in a microprocessor.
- Offset voltage variation and temperature drift. The offset voltages in typical MOS-FET transistors usually differ by a few mV. Even when the transistors are placed very closely to each other, this offset difference is present. In addition, this offset is temperature dependent, of the order of a few mV/K. By using difference methods, such as correlated double sampling as described in Section 3, the influence of offset voltage variations and temperature drifts on photometric measurements can be kept below 1% [54].
- Dark currents : State-of-the-art dark current densities in silicon are between 10-500 pA/cm^2 at room temperature. The dark current can be reduced by lowering the temperature (they are roughly halved for each reduction in temperature of 8-9K), or by reducing the circuit size. Technological improvements might also lead to reduced dark current densities, but they are usually beyond the control of a user. In some cases, charge might be storable in analog EEPROM structures for which charge loss is negligible.
- Ill-matched electronic components: Depending on the electronic device type, the properties of identically designed components can vary by as much as 10%. For this reason, one cannot predict the exact parameters and properties of a circuit. The only way out of this problem is the development of robust circuits, which unfortunately often require an increased area of silicon.
- Non-linearities: Simple electronic components often suffer from non-linear behavior. An example is the charge-voltage characteristic of a photodiode, showing non-constant capacitive behavior. These problems are either resolved within the pixel, by making use of more complex circuitry requiring more space, or by correcting the non-linearities in a post-processing step.

- Switching noise : Reset operations for clearing capacitors or reprogramming electronic devices is always associated with reset or kTC noise. As indicated by its name, the only possibilities one has to reduce this (charge) noise is to reduce the involved capacitances and/or the temperature.
- System complexity : Although the complexity of a highly-integrated custom photosensor might become more manageable by the use of libraries of digital and analog building blocks, it is more practical to reduce the functionality one desires from an image sensor and to work with commercially available components wherever possible. The resulting hybrid solution might be not as appealing as a single-chip system but it is quite possible that it is more economical.

The above discussed imperfections encountered in smart pixels do not mean that smart image sensors cannot perform well. The point to be made is quite different: It is not realistic to expect a custom photosensor to be perfect. Often it makes a lot of sense to correct the shortcomings of an application-specific image sensor with a complementing digital processor, so that the functionality offered by the photosensor becomes precise, reliable and stable. Not an individual image sensor solves an optical measurement problem but rather a complete system does.

6 The Future of Smart Image Sensors and "Seeing Chips"

The past few years have seen tremendous advances in the field of smart image sensors, and electronic circuitry of increasing complexity is being co-integrated with image sensors. Indeed, image sensors become smarter and smarter. Will it take much longer until they can "see" ?

The initial suspicion that "vision is difficult [28]" has been fully verified, and it has become obvious that the early expectations of monolithically integrated single-chip vision systems were too high. The basic philosophy behind the seeing chip is to distribute the processing power over the photosensitve part. This strategy is inspired by the biological concept of highly parallel, low-speed and low power distributed analog computing, which is the basis of nature's marvellous visual perceptive systems, such as our own highly-developed sense of vision. In contrast to the planar, essentially two-dimensional semiconductor fabrication technology, nature realizes fully three-dimensional processing systems, in which each "pixel" is backed by a tremendous number of nerve cells (more than 10^5 in the human visual system [29]) performing the necessary calculation for the sense of vision. In the near future, it is unrealistic to expect that each pixel on a solid-state image sensor will contain more than a few ten transistors, while maintaining a useful pixel size of the order of 30x30 mm^2 and a reasonable fraction of at least 10% of the pixel area being photosensitive.

As a consequence, recent developments in the area of integrated machine vision also consider architectures based on different planes: an image acquisition plane might be followed by several (analog) preprocessing planes, an (essentially digital) classification plane and an output plane, all connected using suitable high-bandwidth bus schemes with an appropriate software protocol. This guarantees a maximum fill factor for the image sensing part and allows to use optimal architectures and technologies for the different parts of the complete system. Such an approach does not necessarily mean that every plane resides on its own chip; different planes can be integrated on the same chip. The technology for stacking and interconnecting silicon chips, so called 3D or z-plane technology, has already been developed [30] but the appealing idea of a low-cost single-chip vision system, a seeing chip, becomes seriously compromised.

It was only about a decade ago that a few researchers started to exploit one of the most exciting capabilities offered by modern silicon-based semiconductor technology, the monolithic integration of photosensitive, analog and digital circuits. Some of the results of these efforts are described in this work, representing just a small fraction of the many applications already demonstrated. They all support the main assertion of this paper, that today's image sensors are not restricted anymore to the acquisition of optical scenes. Image sensors can be supplied with custom integrated functionality, making them key components, application-specific for many types of optical measurement problems. It was argued that it is not always optimal to add the desired custom functionality in the form of highly-complex smart pixels, since an increase in functionality is often coupled with a larger fraction of a pixel's area being used for electronic circuit, at the cost of reduced light sensitivity. For this reason, each new imaging problem has to be inspected carefully, taking into account technical and economical issues. For optimum system solutions, not only smart pixels have to be considered. Functionality could also be provided by separate on-chip or off-chip circuits, perhaps by using commercially available electronic components.
In any case, the freedom offered by the large selection of functionality available in smart pixels and smart image sensors, increases a system designer's options tremendously. Not only is he capable of optimizing existing electronic imaging solutions significantly, the emerging technology of smart image sensors will also be at the heart of manx novel types of optical measurement techniques and vision solutions for automotive applications.

References

[1] P. Seitz, "Smart image sensors : An emerging key technology for advanced optical measurement and microsystems", *Proc. SPIE*, Vol. 2783, 244-255 (1997).
[2] E.R. Fossum, "CMOS Image Sensors: Electronic Camera-On-A-Chip", *IEEE Trans. Electron Dev.*, Vol. 44, 1689-1698 (1997).
[3] P. Seitz, O. Vietze and T. Spirig, "Smart image sensors for optical micro-systems", *Laser und Optoelektronik*, Vol. 28, 56-67 (1996).

[4] "Technology Roadmap for Products and Systems", BPA Technology Management Ltd., BPA House, 250-256 High Street, Dorking, UK - Surrey RH4-1QT. Tel. +44-1306-875-500, FAX +44-1306-888-179.

[5] G. Kreider, J.T. Bosiers, B. Dillen, J.M. van der Heijden, W. Hoekstra, A. Kleinmann, P. Opmeer, J. Oppers, H.L. Peek, R. Pellens and A.J.P. Theuwissen, "An mK×mK Modular Image Sensor Design", *Proc. IEDM 1995*, 155-158 (1995).

[6] H.L. Peek, D.W. Verbugt, M.J. Beenhakkers, W.F. Huinink and A.C. Kleimann, "An FT-CCD Imager with true 2.4×2.4 μm^2 Pixels in Double Membrane Poly-Si Technology", *Proc. IEDM'96*, (1996).

[7] K. Knop and P. Seitz, "Image Sensors", in *Sensors Update — Volume 2 : Sensor Technology - Applications - Markets*, (Eds. H. Baltes, W. Göpel and J. Hesse), VCH-Verlag, Weinheim, 85-103 (1996).

[8] E.R. Fossum: "Active pixel sensors (APS) - Are CCDs dinosaurs ?", *Proc. SPIE,* Vol. 1900, 2-14 (1992).

[9] W. Budde, "Multidecade linearity measurements on Si photodiodes", *Applied Optics,* Vol. 18, 1555-1558 (1979).

[10] P. Seitz, T. Spirig, O. Vietze and K. Engelhardt, "Smart sensing using custom photo-ASICs and CCD technology", *Optical Engineering*, Vol. 34, 2299-2308 (1995).

[11] P. Seitz, D. Leipold, J. Kramer and J.M. Raynor, "Smart optical and image sensors fabricated with industrial CMOS/CCD semiconductor processes", *Proc. SPIE,* Vol. 1900, 21-30 (1993).

[12] O. Vietze and P. Seitz, "Image sensing with programmable offset pixels for increased dynamic range of more than 150dB", *Proc. SPIE*, Vol. 2654A, 93-98 (1995).

[13] O. Vietze, "Active pixel image sensors with application specific performance based on standard silicon CMOS processes", Ph.D. thesis No. 12038, Federal Institute of Technology (ETH), Zurich (1997).

[14] T. Spirig, P. Seitz and M. Marley, "The multi-tap lock-in CCD with offset compensation", *IEEE Trans. Electron Dev.*, Vol. 44, 1643-1647 (1997).

[15] H.G. Graf, B. Höfflinger, Z. Seger and A. Siggelkow, "Elektronisch Sehen", *Elektronik,* Nr. 3/95, 3-7 (1995).

[16] J.D.E. Beynon and D.R. Lamg, "Charge-coupled devices and their applications", McGraw Hill, London (1980).

[17] P. Seitz, T. Spirig, P. Seitz, O. Vietze and P. Metzler, "The convolution CCD and the convolver CCD : Applications of exposure-concurrent photocharge transfer in optical metrology and machine vision", *Proc. SPIE,* Vol. 2415, 276-284 (1995).

[18] P.B. Denyer, D. Renshaw and S.G. Smith, "Intelligent CMOS Imaging", *Proc. SPIE,* Vol. 2415, 285-291 (1995).

[19] O. Yadid-Pecht and E.R. Fossum, "Wide Intrascene Dynamic Range CMOS APS Using Dual Sampling", *IEEE Trans. Electron Dev.*, Vol. 44, 1721-1723 (1997).

[20] S. Chen and R. Ginosar, "Adaptive Sensitivity CCD Image Sensor", *Proc. SPIE*, Vol. 2415, 303-310 (1995).

[21] Q. Zhu, H. Stiebig, P. Rieve, H. Fischer and M. Böhm, "A Novel *a*-Si(C):H Color Sensor Array", *Proc. MRS Spring Meeting*, Symposium A - Amorphous Silicon Technology, San Francisco, April 4-8, 1994.
[22] D.J. Sauer, F.L. Hsueh, F.V. Shallcross, G.M. Meray, P.A. Levine, G.W. Hughes and J. Pellegrino, "High Fill-Factor CCD Imager with High Frame-Rate Readout", *Proc. SPIE,* Vol. 1291, 174-184 (1990).
[23] W.F. Kosonocky, G. Yang, R.K. Kabra, C. Ye, Z. Pektas, J.L. Lowrance, V.J. Mastrocola, F.V. Shallcross and V. Patel, "260×360 Element Three-Pahse Very High Frame Rate Burst Image Sensor: Design, Operation and Performance", *IEEE Trans. Electron Dev.*, Vol. 44, 1617-1623 (1997).
[24] J.M. Hakkarainen, J.J. Little, H. Lee and J.L. Wyatt, "Interaction of algorithm and implementation for analog VLSI stereo vision", *Proc. SPIE,* Vol. 1473, 173-184 (1991).
[25] A. Gruss, L.R. Carley and T. Kanade, "Integrated Sensor and Range-Finding Analog Signal Processor", *IEEE J. Solid State Circ.,* Vol. 26, 184-192 (1991).
[26] J. Kramer, P. Seitz and H. Baltes, "Inexpensive range camera operating at video speed", *Applied Optics,* Vol. 32, 2323-2330 (1993).
[27] N. Blanc, K. Engelhardt, S. Lauxtermann, P. Seitz, O. Vietze, F. Pellandini, M. Ansorge, A. Heubi, S. Tanner, W. Salathé, R. Dinger, M.E. Meyer, E. Doering and H. Mahler, "Microcameras", Paul Scherrer Institute Zurich, Annual Report 1996 / Annex IIIB, 59 (1997).
[28] C. Koch, "Seeing Chips: Analog VLSI Circuits for Computer Vision", *Neural Computation*, Vol. 1, 184-200 (1989).
[29] D.H. Hubel, "Eye, Brain and Vision", Scientific American Library, New York (1988).
[50] J.E. Carson (Ed.), "Materials, Devices, Techniques and Applications for Z-Plane Focal Plane Array Technology", *Proc. SPIE,* Vol. 1097 (1989).

A Low Cost Image Processing Device for Blind Spot Monitoring

G. Burzio[1], G. Vivo[1], E. Lupo[2]

[1] Centro Ricerche FIAT, Sistemi Elettronici
Strada Torino 50, 10043 Orbassano (TO) Italy

[2] Magneti Marelli Spa, Divisione Retrovisori
Viale C. Emanuele II 118/150, 10078 Venaria (TO) Italy

Abstract. Preventive safety is a topic where car manufacturers will spend a great amount of research effort in the next years. In the Vision Laboratory of FIAT Research Centre a specific sensor has been developed to cope with this argument in an innovative manner. The device, named Blind Spot Monitoring Sensor, is based on a CCD micro-camera and on a very compact image processing unit; the provided support function consists on the activation of a suitable warning indication when overtaking vehicles, moving in the "blind zone", are detected by the system. The described sensor, that improves significantly safety and comfort during overtaking manoeuvres, shows two fundamental characteristics for an automotive device: high performance and low cost. The processing unit of the system has been specifically designed and developed; it is based on a single electronic board, with a PCB size of 80 x 40 mm. and uses, as "core processor", a popular (and inexpensive) MCU. Blind Spot Monitoring Sensor has been installed on different prototype vehicles; the paper describes the developed sensor and some functional evaluations, performed in real traffic and weather conditions, using the equipped vehicles.

Keywords. Vehicle safety and comfort, real time image processing, CCD sensors, optoelectronic devices

1 Introduction

Every year, word-wide about 500.000 people are killed in road accidents; for the same reason in Europe every year about 50.000 deaths and many more serious injuries happens. It is estimated that a significant part of these accidents is related to overtaking manoeuvres; human factors involved in these accidents depend mainly on the incorrect or missed perception of the overtaking vehicle in the blind spots and on a wrong distance estimate. The aim of the Blind Spot Monitoring Sensor, showed in figure 1.1, is to contribute to the reduction of the given accident numbers. This hope is supported by many tests, carried out using vehicles equipped with the Blind Spot Monitoring Sensor, indicating that drivers perceive a real improvement of safety and comfort during overtaking manoeuvres when they are supported by the described warning support function.

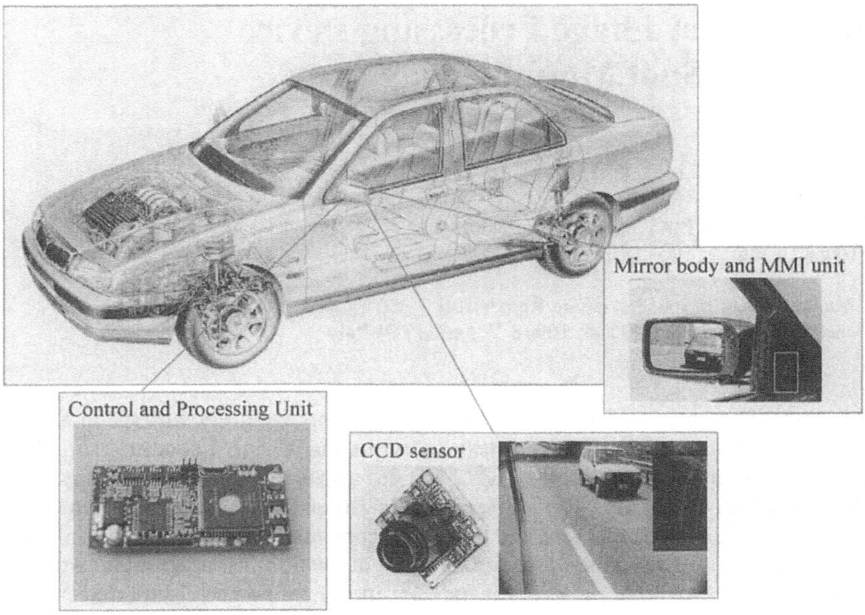

Fig 1.1. General layout and placement of Blind Spot Monitoring Sensor components.

2 System Function Description

Blind Spot Monitoring Sensor is composed by three parts: a CCD micro camera, a real time processing unit, a driver interface unit. The system is continuously monitoring, at the video field rate of 50 Hz, the lateral area of the equipped vehicle. When overtaking manoeuvres are performed by other vehicles, a warning information is passed to the driver.

Using a CCD micro camera with a suitable optics as imaging sensor integrated in the lateral mirror, it is possible to monitor a lateral area much larger than the one perceived by the driver with a normal rear view mirrors. While standard mirrors produce view angles around 20°, the horizontal view angle of a CCD camera can be tuned.

For all prototypes (six vehicles) we have equipped we used lenses producing a viewing angle of 45°. With this angle a trade off is obtained between a large angle (with the possibility of covering a wide area, reducing the blind zone) and a narrow angle (augmenting the pixel resolution, with better performance in term of maximum distance covered by the sensor).

In figure 2.1 a comparison between the area observed through the lateral mirror and the micro camera field of view is shown.

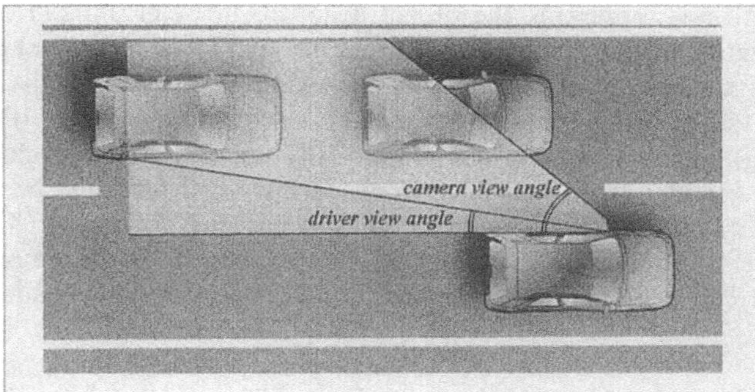

Fig 2.1. Comparison between the area covered by the driver through the lateral mirror and the micro camera field of view.

The working principle of the Blind Spot Monitoring Sensor is based on the real time analysis of objects present in the images during the motion of the vehicle. When whatever objects (horizontal road markings, other vehicles, trees, shadows, bridges over road way ...) are traced in the CCD images, their apparent position is moving towards the vanishing-point in the images, which is placed on the horizon line. Only objects with a positive speed towards the sensor camera have different behaviour: their apparent position is moving away from the vanishing-point in the images. So it is quite easy to filter out all information in the image sequences, keeping a high degree of robustness in the detection of overtaking vehicles. In the following figure the Blind Spot Monitoring Sensor processing task is showed.

Fig 2.2. Blind Spot Monitoring Sensor processing task.

In this image it is possible to see the shadow of a bridge over road way and an overtaking vehicle. In the diagram placed inside the picture, the horizontal axis represents the last three seconds of the performed processing.

The vertical axis represents the object distance, in pixel units, from the ego-vehicle baseline (about 2 meters away from the micro camera). As showed in this figure, the shadow of the bridge is producing two traces with a particular ascending shape (the object distance is augmenting when the time is progressing); a trace with a completely different shape is produced by the overtaking vehicle: in this case the produced low level features consist in many small and fragmented traces with a typical "descending shape". This kind of shape is the discriminating characteristic that permits, in the image sequences, to distinguish between potentially dangerous objects (overtaking vehicles) and all other ones.

3 The Electronic Processing Unit

Figure 3.1 represents the electronic processing unit and the micro camera module of the Blind Spot Monitoring Sensor. All the system processing activities are carried out by this dedicated processing unit, designed and built by the electronic design group of Centro Ricerche FIAT.

Fig 3.1. Picture showing the Blind Spot Monitoring Sensor electronic processing unit (PCB size is 80 x 40 mm.) and a micro camera; the round "object" is the Italian coin of 500 lire.

Blind Spot Monitoring Sensor electronic processing unit is physically composed by a single printed circuit board of 80 x 40 mm.; all the electronic components needed for image acquisition and processing are included in this device, that provides also the processing power for all other computational tasks of the sensor. The processing unit has been designed and built with some application dependant constraints in order to obtain the maximum efficiency from a standard and inexpensive MCU: in all effects the developed device is a miniaturised but powerful real time vision system. The core of the processing unit is a MCU (Micro Controller Unit) Motorola 68HC11 running with a crystal frequency of only 16 MHz. Using some analogic pre-processing before transmitting the video stream information to the core MCU, it was possible to process in real time the images produced by the micro camera, that generates a standard CCIR video signal. Another major difference with other vision systems is the absence of a frame store memory: the electronics of the processing unit extracts the clock signals of field_start and line_start from the video source and passes these signals to the MCU; all the processing activities, including A/D conversion, image filtering etc, must be executed via Interrupt Service Routines and the vision algorithms must be designed and written to work "on the fly" and in synchronous with the field and line clock signals.

The MCU RAM memory (only 768 bytes) is completely used to store A/D converted values and some other program data. The executed program is writted in a segment of about 20 Kbytes of EPROM.

As showed in fig. 3.2, both the electronic processing unit and the CCD micro camera are completely integrated into the lateral mirror.

Fig 3.2. The electronic processing unit and the micro camera unit are totally integrated in the lateral mirror of the showed demonstrator vehicle.

A team of specialists of Magneti Marelli did the design, the development and the building of the integrated mirror. The integration of a TV camera in the lateral mirror of a vehicle is not a trivial task: nowadays mirrors are often complex systems that can be composed of many mechanical, electrical, and electronic parts, including two motors for adjusting the alignment of the mirror, another motor for the capsizing unit, a heating subsystem and sometimes (as happened in the showed demonstrator vehicle mirror), also a temperature sensor for the vehicle air-conditioning system and a demultiplex unit for simplifying the electrical connections of the mirror. The camera module is really compact; this permitted to develop a mirror whose original features were all kept and the style characteristics were very similar to the production mirrors. Also for some environmental problems related to the camera module, specific solutions were found; for example considering the mirror exposition to all weathers and dirtiness, the camera module was made water resistant and a specific protection for the lens was designed and built.

It could be interesting to point out someone of the most important benefits obtained as a consequence of such a deep level of integration. From the assembling point of view, for example, if the system is as much as possible independent from the rest of the car body structure, it is easier, for the car component supplier, to furnish the given function at a reasonable cost (because the links, the connections and the interfaces with the rest of the vehicle are reduced to a minimum level). In addition, from the marketing point of view, it is also feasible, to cope with a low volume market (at least in the initial stages of the mass introduction of the lateral support function) to provide the function as an "after-market" optional. From the functional point of view the integration of the sensor into the lateral mirror guarantees an "automatic" solution to the problem of providing a satisfying enclosure for the micro camera (and this improves also the style and the mechanical robustness of the device). The integration of the micro camera module provides also benefits to the problem of protecting the component against the rain and the dirty: also in low speed driving the air flow produced by the mirror body acts continuously to keep the micro camera lens clean.

4 Decision Procedure and Driver Interface

No visual data or other complex information is presented to the driver. Driver-sensor interface, which is simple and easily understood, is composed by a light source (a red LED) and an acoustic source (a "beeper") integrated in a single, very small device placed near the driver's side lateral mirror.

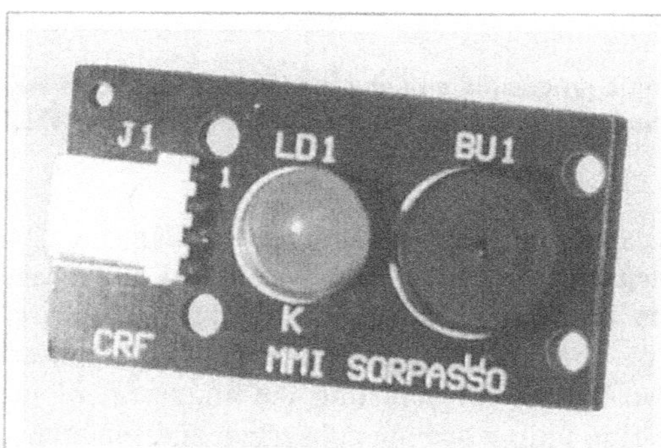

Fig 4.1. Driver interface for the Blind Spot Monitoring Sensor.

The function carried out by the Blind Spot Monitoring Sensor is aimed at integrating and improving the "natural task" of the mirrors, not at replacing them. The system interaction with the driver is very soft: during an overtaking manoeuvre, if the driver can perceive the overtaking vehicle in its lateral mirror, the system is silent; when this vehicle begins to disappear because it is entering the blind spot area, then the red LED is automatically switched on; the red LED keeps this alarm state during the entire crossing of the overtaking vehicle in the blind spot area; only when the driver can see again the vehicle, because it has concluded its overtaking manoeuvre and it is directly visible straight forward, then the red LED is turned off.

Normally, when the driver behaviour is regular, the alarm light indication is not enforced by any acoustic signal; in this way the driver is not disturbed by unnecessary warnings and the only interaction with the system consists of the red LED that flashes in a lateral position when overtaking is in progress. This kind of warning has no effects on the driver concentration because, in average, its focus of attention is on the road, which is pointed by the fovea, in the retina centre. The acoustic signal is turned on only when there is evidence of a potentially dangerous manoeuvre during overtaking. For example this evidence is clearly stated when the driver is declaring explicitly an intention to change the lane, acting the direction indicator. In a future release of the sensor, the actual lane change information can be estimated, using the same CCD micro camera, from the observed position of the road lane markings.

5 Related Projects and Technology Comparison

Until now no many prototypes have been developed all around the world with characteristics, performance and costs similar to one presented in here. There are clear indications into the market that automatic systems providing to the driver the ability to detect objects in the lateral part of the equipped vehicle will become popular in the next years. Nevertheless, only demonstrator prototypes, using different sensing technologies, have been proposed in the last years. Some ones, for example, are:

Demonstrator Name	Producer	Sensing technology	Year
Side Minder	Autosense (U.S.A)	Multi beam infrared	1994
Demonstrator vehicle	Siemens	Infrared	1994
Demonstrator vehicle	PSA	Infrared	1994
Demonstrator vehicle	Mercedes Benz	Radar, 3 beam, 77 GHz	1994
Demonstrator vehicle	Jaguar GEC Ferranti	Scanning laser radar	1993
Electro optic obstacle detection	GEC Marconi Avionics	Radar, 5 beam	1994

Blind Spot Monitoring Sensor represents a very innovative and economic enhancement of the mentioned devices. Due to the adopted sensing technology (vision) and to the wide availability of inexpensive CCD camera modules, the described device represents an optimal solution both for the obtained sensing performances (missed alarms rate near to zero) and for the very low cost figure. Good quality CCD cameras are now available for less than 30 ECU; it is now entering the market a new generation of CMOS cameras, with a good quality and the potential cost of few ECU; consumer technology for this kind of sensors is growing in an impressive way, mainly because of the large expansion of the multimedia application market.

Adopting vision as sensing technology, imply some further advantage respect to the mentioned other ones. CCD cameras are passive sensors; this means that no emission of any kind is released in the environment. There are no important legal aspects nor special cares to cope with for installing and checking the device, for using it in the equipped vehicles, and for destroying it at the end its life cycle. Other typical potential problems of active sensors, are thus eliminated by definition: there are no complex standardisation needs (regarding the emitted power and the wave lengths, for example) and there are no interference problems between vehicles equipped with the same device.

Definitely, we think that for the specific characteristics of the sensor described in here, the competitiveness of the vision based technology shows a performance much better at a cost that is a fraction respect to all other currently available technologies.

6 Social Expected Benefits

Social aspects involved with the developed device are very important. Blind Spot Monitoring Sensor is aimed at reducing the chances of missed detection of overtaking vehicles. The ambitious goal of the system is to provide a substantial contribution to the reduction of road accidents related to the driver behaviour (e.g., the incorrect or missed perception of the overtaking vehicles in the blind spots or a wrong distance estimate).

The given support will be especially useful for people affected by reduced capabilities of lateral perception: although Blind Spot Monitoring Sensor has been developed to provide a general and complete support to the entire population of drivers, the very initial concern in the system definition and design was directed to studying and considering social benefits and driving quality improvement for elderly people and drivers with limited visual acuity.

7 Experimental Tests

Some initial experimental work regarding a camera based overtaking detection system started after the conclusion of the European PROMETHEUS Programme in the Vision Laboratory of Centro Ricerche FIAT; first tests were carried out using a commercial vehicle (a FIAT Ducato Van) specifically equipped to develop and test functions and systems based on optoelectronic sensors. With this facility it was possible to collect videos and other data with scenes related to different road types and lighting conditions. At the end of this feasibility study a first autonomous prototype car (a Lancia Thema) was developed. After these preliminary steps, the sensor algorithms, running with a general purpose processing unit, were widely used and tested. An extensive test (50.000 Km. travelled) was then carried out installing the sensor on another car (a Lancia Kappa). The actual sensor release was designed and built with two general targets: firstly to reduce as much as possible the processing unit costs and volumes; secondly to meet some heavy constraints of industrial production and compatibility with the automotive requirements.

Many demonstration vehicles are actually running with the Blind Spot Monitoring Sensor: a Lancia Thema 3000 V6 an Alfa Romeo 164 a Lancia Dedra 2000 equipped with two Blind Spot Sensors two Lancia Kappa, a commercial vehicle IVECO TurboDaily. Globally the Blind Spot Monitoring Sensor has been tested for many thousand kilometres on different road types, weather and light conditions, by day and by night. Extensive system tests have been done at all speeds up to 130 Km/h; there is no evidence that failures could happen at higher speeds. The system works well on all road types and it is very precise: the detection geometry permits to exclude from the analysis all the vehicles and objects present in external lanes, considering only objects moving in the near lane.

8 Installation, and Marketing Issues

It is not easy to give precise forecast about the introduction into the market of devices, like the one described in this paper, that are at the present time, produced only as demonstration prototypes. On the other hand the peculiar characteristics (extremely low cost and very easy installation) of the described device permit to draw, in general, an optimistic scenario.

Inside to the FIAT group, a limited and preliminary production, for client tests and evaluation, could start in 1998. Depending on marketing strategies, the Blind Spot Monitoring Sensor could be introduced, as an optional accessory, just with the next generation of cars; large scale production could start before January 2001.

The described sensor is easy to install on all type of vehicles, produced by all car manufacturers. One of the most important points in the component supplier marketing strategy is to sell a useful and economic device with the hope to introduce it initially in Europe and then all over the world. Depending on the specific politics of car manufacturers, the sensor could be first introduced into the markets that show more sensitivity regarding aspects like safety, driving comfort and product quality. Many of the European countries, especially the German one that is also the widest, are receptive to these aspects. U.S. and Japan markets have been considered too. In these last countries, where the presence of the most important car manufacturers in the world has the power of closing the market, there is always the possibility of selling the production licence to a third component supplier.

9 Future Planned Activities and Conclusions

Blind Spot Monitoring Sensor has been developed by Centro Ricerche FIAT and Magneti Marelli, with support from FIAT Auto; the sensor showed good performance regarding the levels of false and missed alarms. Also the "driving impression" is very nice. This good performance has been maintained quite constant in all conditions tested during the experimentation of the system; such conditions included also poor visibility situations (as example by night, in foggy days, in sunny days with shadows on the road, with the camera pointing against the light). With the adopted camera lens (focal length of 6 mm. on a 1/3" sensor) the alarms are produced when the overtaking vehicles are about 20 meters away from the camera; the "driving impression" with this set up is really good.

The future prosecution of the described activities is addressed towards the engineering and the production of the Blind Spot Monitoring Sensor. One of the possible technology improvements will concern the micro camera module: suitable CMOS cameras will be available in the next months and this will probably permit a further reduction of the overall sensor cost. To obtain a significant reduction also for the processing unit cost, one possibility is the design and the production of a dedicated ASIC; but only significant production volumes (>25.000 units/year) will make this choice feasible.

There is no doubt that the presence of the type of devices described in this paper on the next generation vehicles will produce a significant improvement for the driver's comfort; we hope that the safety contribution to the reduction of road accidents related to the driver behaviour will be even more important.

Continuous Parallel Analogue Image Processing Using Time Discrete Sampling

U. Apel, H.-G. Graf, B. Höfflinger, U. Regensburger*, U. Seger

Institute for Microelectronics Stuttgart Allmandring 30a 70569 Stuttgart
*Daimler-Benz AG F1M/IA 70546 Stuttgart

Abstract

Image processing systems for special purposes gain high performance by executing standardised operations such as filtering or edge extraction on specific chips. The architecture of the presented chip set is the starting point for a massively parallel image processing directly coupled to the image acquisition in space as well as in time domain. The presented architecture reduces the necessary bandwidth for signal transmission and allows a small and power saving module to be integrated in the camera system without the heavy power and chip size load associated with fully focal plane integrated approaches [1,2].

1 System Concept: Image Acquisition and Analogue Preprocessing with Separate Chips

Video systems are dedicated for future driver assistant systems to increase the safety in road traffic. The prediction of lane curvature, obstacle detection, and interpretation of traffic signs are major tasks that have to be performed by camera systems. A high performance and reliability as well as a compact and cost saving architecture are prerequisites for customer acceptance.

The presented co-designed image sensor and retina chips are part of a multiple stage analogue pre-processor system. They have been developed in the framework of the research program „Electronic Eye" to study the feasibility of a spatial filtering operation close to the image capturing but not within the focal plane array. Shifting these procedures from the digital processors to the front-end devices has numerous advantages: The edge enhancement as one example of low- or high-pass spatial filtering [3], which is a time consuming task on digital processors, can be prepared massively parallel on a resistive network acting analogous to the human retina [1]. The last chip in the analogue processing pipeline is a network of fuses [4] that extracts objects with closed contour lines from the edge-enhanced image at the retina output. Condensing the image information in this way the bandwidth is reduced, as a result cheaper and more flexible bus systems can be used. Furthermore our approach opens the door to even more effective parallel processes because intermediate informations from each processing level is accessible for post-processing modules. E.g. it is possible to use logarithmic grey level information (directly from the acquisition module) for monitoring while dedicated recognition algorithms run on edge enhanced and smoothed retina images using the retina output.

Although the information flow is pipelined, only a short delay between original and processed data has to be taken into account. Due to the quasi-continuous processing and multi accessible processor-array structure, the system behaves quit similar to its time constant counterpart, the focal plane array. At the same time it consumes only a small fraction of the silicon area and is more flexible in system design.

Both sensor and retina chip are fabricated in a 0.8 µm standard CMOS technology. Architecture and electrical characteristics of the imager will be given in section 2, the implemented retina will be presented in section 3, in section 4 a detailed description of the „rolling-retina" concept which is applicable to nearly all analog post-processing hardware, will be shown.

2 HDRC Sensor for Image Acquisition

Acquisition of images with clear and robust detectable objects in it is the most important requirement for the sensor. Neither reflections of the head lamps of approaching vehicles or from the sun close above the horizon nor sudden changes of the brightness as appearing at tunnel entrances are allowed to disturb the image information. The High Dynamic Range CMOS (HDRC®-) sensor architecture with its logarithmic conversion of photo current to output voltage within each sensor pixel is the superior solution for such scenes and is a commonly accepted input device for at least 5 years [5]. The random, non-destructive readout feature without any integration time control necessary is the base for a good co-operation with the analogue post-processors. A constant contrast resolution of about 3% over several magnitudes of intensities guaranties the stable operation of edge enhancement effectively providing a first image intensity compression.

The pixels of the sensor as well as the related cells of the 2 post-processing chips Retina and Fuse are placed in a hexagonal structure like a honeycomb, yielding an identical relationship to all direct neighbour cells. While using this arrangement for the centre of the photodiodes, the device structures are kept strictly orthogonal, yielding a pixel pitch of 37 µm in lateral and a pitch of the rows of 32 µm. Every other row is shifted by half a pixel pitch. The fill factor of the image array is about 48%.

The resolution of the sensor is 384 x 256, resulting in an image diagonal of 16.4 mm. The sensor array is multitapped and allows an access to 6 independent blocks each of 64 pixel width and equipped with a video output of its own. Propagating the image data in parallel to the subsequent analogue image processing chips with a clock rate of 1 MHz (this limit is set by the recent version of the Fuse-Chip) allows a full frame rate of about 60 Hz. Accelerated readout up to 6 MHz per channel (36 Mpixel) of the HDRC imager is possible, enabling the system to serve as a imaging module for high speed post-processing.

3 Retina Chip Architecture

The function of the retina, a high-pass spatial filtering of the image information, is implemented on a regular resistive grid (see Fig. 1). Each node of the grid is mapped to a pixel of the sensor. The output voltage values of the corresponding pixels, representing the local irradiation, are stored at each new frame in the sample-and-hold buffer. The first amplifier of the cell tends to pull the potential of the central node near the irradiation level stored at S&H, acting against its six direct neighbours via the resistors.

While at slight spatial variations of the irradiation level the net potential follows identically, coarse deviations appearing at edges in the scene will be smoothed. The output signal gained from the retina chip is the difference between the stored grey level and the potential on the net, delivering a high-pass filtering of the scene. Only grey-level gradients that occur locally will be enhanced, while slopes constant across a number of nodes are responded with a zero output. To avoid a clipping of the signal even at high amplifications, the second amplifier has a slowly saturating gain characteristic.

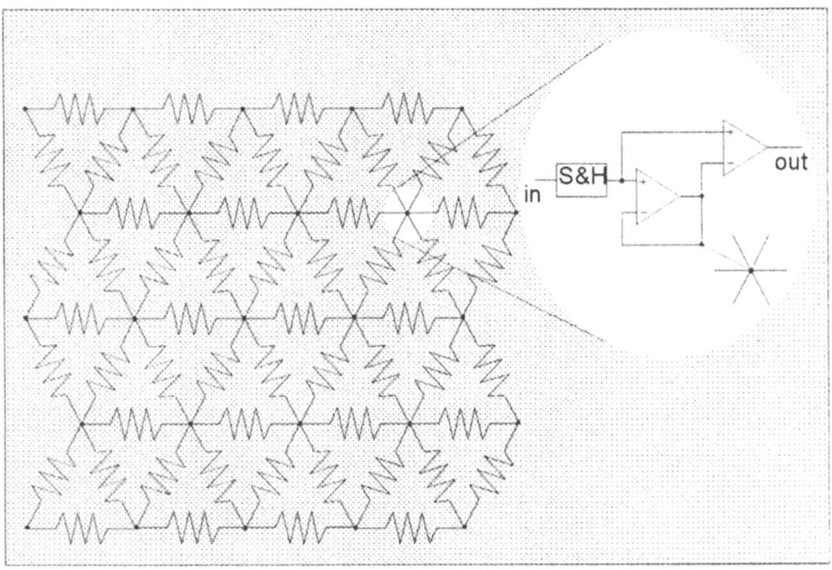

Fig. 1: Configuration of the resistive grid in the retina chip

Depending on the transconductance of the first amplifier and on the resistance, the net responds to edges with a specific coherence length κ. This number describes the distance from an edge the net needs to come close to the image grey level. κ is defined by the transconductances g_m of the node amplifiers and by the branch resistance R, a being the distance between two nodes of the net.

$$\kappa = \sqrt{\frac{3}{g_m \cdot R}} \cdot a$$

On-Chip resistors with required high values at about one Mega-Ohm have to be built with MOS transistors working in subthreshold operation. A control circuit has been implemented within each cell which regulates the gate voltage of the subthreshold MOS-resistor branches keeping a constant offset to the net node, ensuring a homogeneous resistance value of the net. A resistor branch is a symmetric serial connection of 2 transistors, each of them being controlled from the corresponding cell. The global adjustment of the gate offset voltage yields the advantageous property that net parameters can be trimmed for better operation stability of the image processing modules following the retina chip.

The high flexibility of the first prototype retina chip results in a circuit complexity of about 60 active transistors per cell, occupying an area of 66x100 μm². The packaging density of the actual layout can be seen in the microphotograph in Fig. 2. The core size of the fabricated prototype array with a net of 80 rows and 140 columns is 74 mm², adding peripheral control, interface circuits and pads, the chip size rises to about 120 mm². This numbers elucidate that the spatial resolution of resistive grids for analogue image processing can't keep pace with the growing resolution of image sensors. An alternative solution will be discussed in Section 4.

Fig. 2: Chip Microphotograph showing details of the retina chip

Because the imperfect termination of the array border discontinuities in the net behaviour will appear there. To reduce the impact, only the data from the central part of the retina were transferred to the subsequent Fuse-Chip. A frame of 8 rows respectively 8 columns will be loaded with the analogue image data, ensuring correct results in the centre of 128 by 64 cells. The signal lines for reading and writing images are independently in direction of the columns. Also the address generators for load- and read- operations are built separately, using individual shift registers. This architecture allows to adjust the net settling time before readout but at the same times guarantees a strictly time-correlated readout.

The measured power consumption of the retina chip is 400 mW at 6 MHz operation (6 channels a 1 MHz). A non-negligible part is caused by the loss of the video line buffers that supply the output pads and the internal signal distribution lines to the sample-and-hold circuits of the cells. The power consumption of an individual cell is assumed to be in the range of 3 - 5 microamps. A further reduction of this value will hamper the read-out speed capability of the retina due to the limited driving force of the second amplifiers of the nodes.

4 Rolling Retina Operation

Circuit simulations show that the settling time of internal nodes of the retina is in the range of 10 µs, due to the inherent small node capacitance. This time is less than the period to load a complete row of the retina, even at a pixel clock of about 6 MHz. It can be seen clearly that the region of the retina grid that is currently adapting will be limited to a few lines around the row address loaded at the moment. At a spatial distance of 5 times the coherent length (of the processing kernel), the net can be assumed as time continuously working and settled. Addressing the retina grid in an appropriate manner, an actual net size of 32 rows and any numbers of columns can be used to operate on any image resolution.

For the given case (the retina)the output from the grid must be read 16 line cycles after the load process occurs. A further period of 16 lines later the top row of the retina can be used to load the next partition of the image while the settled first partition is being read out at the same time. The retinal grid is supposed to be arranged on the surface of a cylinder that is rolling over the image (see Figure 3) this is where the name of the „Rolling Retina“ comes from.

The address controller of the current retina can be switched to a mode that uses the core centre of 64 rows periodically. In this mode the margin of 8 rows above and below the centre will be loaded in parallel to the corresponding lines to avoid distortions caused from the array borders. A redesign with the proposed row number of 32 can be supplied with interconnects that will close the retinal grid to an actual cylindrical configuration.

The cell size of the presented retina prototype, which has been designed in a 0.8 µm-CMOS-process may be reduced by omitting some of the circuitry for flexible adjustment of the net parameters. Also the transfer of the design to a shrinked process with 3 metal layers gains a relevant size reduction. A retina chip with a full image size capability is possible on economical chip sizes, utilising the rolling retina concept.

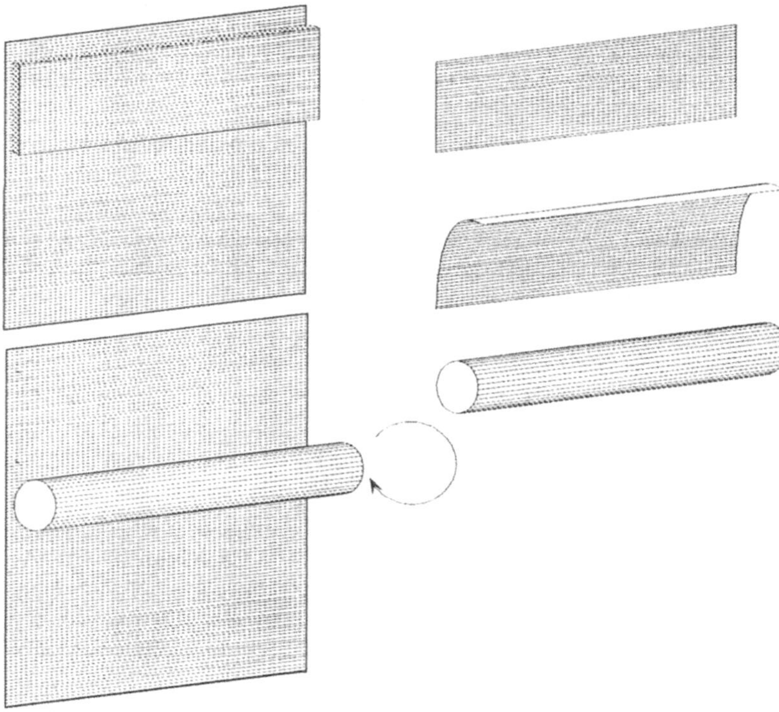

Fig. 3 Schematic of Rolling Retina Concept

5 Summary

Approaches for focal plane image processing functions as low or high-pass filtering, Laplace filtering resistive grids, binary weighting etc. up to now had one dramatic disadvantage. Even usage of deep-subμ-CMOS technologies lead to high power dissipation, high area consumption, and low fill factors for the active optical system. A system using sampled and pipelined analogue processing combined with a multiple read and write structure shows how to overcome these problems while maintaining the advantages of a time continuous analogue processing area. Advantages of our approach over the conventional focal plane approach or pipelined CCD-processing networks are:

- small processing arrays even for high resolution images
  -> reduced Si-Area and power-consumption
  -> independence of spatial resolution of imager and complexity of processing circuitry
  quasi continuous operation mode
- variable cell parameters
  allows variable coherence length constants for filter kernels

- modular, pipelined structures
  -> for minimum pipeline delay
  -> for even recursively usage of filter kernels
  -> for direct access to results of intermediate stages

E.g. spatial filtering of pictures is performed by resistive grids with user defined net coefficients to enhance robustness of subsequent edge detection. The settling time of internal nodes is in the range of 10 μs, so different regions of an analogue processing array separated by at least 5 coherence-length constants are read out and reloaded at the same time. This allows analogue processing hardware to be used recursively and eliminates the necessity of using analog processing arrays of the same arrays size as the high resolution imager chips supplying the image information.

A system of a hexagonal sensor and an analogue processing array for edge enhancement have been demonstrated. It can be shown, that a rolling retina processing unit with only 32 * 384 elements is capable to process an image of 384 pixel width and any pixel in height.

Our approach of „rolling" analogue processing can be used for any similar processing algorithms as long as the settling time of the necessary circuits is short in comparison to the image acquisition

[1] M.A. Sivilotti, M. A. Mahowald, and C.A. Mead: „Real-Time Visual Computations Using Analog CMOS Processing Arrays", *Stanford Conference VLSI, Cambridge, MIT Pres.* pp.295-312, 1987

[2] H. Kobayashi, and J.L. White, and A.A. Abidi: „An Analog CMOS Network for Gaussian Convolution with Embedded Image Sensing", *IEEE International Solid-State Circuits Conference,* pp.216-217, 1990

[3] Rafael C.Gonzales, Paul Wintz, „Digital Image Processing", *Addison-Wesley Publishing*, 1977 London

[4] H.Zinner, P.Nothaft, „Analogue Image Processing for Driver's Assistant System", *Proceedings „Advanced Microsystems for Automotive Applications, VDI/VDE* , Berlin 2./3. Dez 1996.

[5] Seger, H.-G. Graf, and M.E. Landgraf: "Vision Assistance in Scenes with Extreme Contrast", *IEEE Micro*, Vol. 13, No. 1, pp.50-56, February 1993.

High Dynamic Range Image Sensors in Thin Film on ASIC Technology for Automotive Applications

M. Böhm[1], F. Blecher[1], A. Eckhardt[1], B. Schneider[1], S. Benthien[2], H. Keller[2], T. Lulé[2], P. Rieve[2], M. Sommer[2], R. C. Lind[3], L. Humm[3], M. Daniels[3], N. Wu[3], H. Yen[3], U. Efron[4]

1 Institut für Halbleiterelektronik (IHE), Universität-GH Siegen, D-57068 Siegen, Germany
2 Silicon Vision GmbH, D-57078 Siegen, Germany
3 Delco Electronics Corp., Malibu CA 90265, USA
4 Hughes Research Laboratories, Malibu CA 90265, USA

Abstract. TFA (Thin Film on ASIC) image sensors consist of an amorphous silicon based optical detector on top of an ASIC which performs signal readout and signal processing. In this paper recent results on TFA prototypes are presented. The detector multilayer was deposited by PECVD (Plasma Enhanced Chemical Vapor Deposition) in a cluster deposition system, whereas for the ASIC a standard 0.7 μm CMOS process was employed. Focus is put on two locally adaptive sensor arrays that allow the control of sensitivity for each pixel individually. These types of sensors are ideally suited for automotive vision systems, since they do not exhibit certain disadvantages inherent to devices using logarithmic data compression or global sensitivity control and image fusion. Moreover, the reduced chip size of vertically integrated TFA devices leads to a significant reduction of manufacturing costs in comparison to CMOS imagers. In addition to the locally adaptive arrays simpler devices with global sensitivity control and a concept for color recognition in TFA devices are discussed.

Keywords. automotive vision system, image sensor, TFA technology, thin film detector, amorphous silicon, ASIC, autoadaptive pixel, dynamic range

1 Introduction

Recent developments in almost all fields of electronic imaging create a need for optical sensors with improved performance and additional functions. In particular, a high dynamic range and optional on-chip signal processing are essential for automotive systems such as lane marker detection. Conventional CCDs are not qualified for this purpose, since they have a limited dynamic range and permit serial readout mode only so that the pixels can perform just limited "intelligent" operations [1]. To overcome these restrictions, CMOS sensors have been developed supplying an ASIC with a photodiode as the optical detector [2]. Since the detector and the pixel circuit share the same chip area, the area fill factor of such devices is limited.

Unlike a CMOS imager a sensor in TFA technology is vertically integrated, thus providing a fill factor close to 100 % for both the detector and the circuitry. Another key advantage of TFA technology is flexibility, since the concept allows the design and optimization of each component independently. Nevertheless fabrication costs are lower compared to hybrid sensors where detector and CMOS electronics are manufactured separately and joined by e.g. bump bonds [3]. Fig. 1.1 illustrates the layer sequence of a TFA sensor. The crystalline ASIC typically includes identical pixel circuitry underneath each pixel detector and peripheral circuitry outside the light sensitive area. An insulation layer separates the optical detector from the chip. The detector consists of an amorphous silicon (a-Si:H) multilayer thin film system which is fabricated in a low-temperature PECVD process [4]. The thin film system is sandwiched between a metal rear electrode, which is usually the third metal layer of the ASIC, and a transparent front electrode. Due to its higher absorption coefficient and its maximum spectral response for green light, an amorphous silicon detector is better qualified for an image sensor than a crystalline silicon detector. Moreover, the a-Si:H deposition sequence is adaptable to the specific requirements of an application. Generally speaking, pin structures are employed for black-and-white detection, whereas nipinin or ni^{x}p structures serve for color recognition [5].

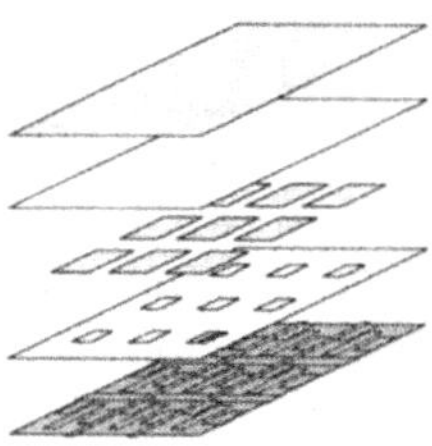

Fig. 1.1: Typical layer sequence of a TFA image sensor

In section 2 various TFA sensor prototypes are discussed which have been optimized with regard to different potential applications. Based on the special requirements of automotive vision systems, two high dynamic range TFA sensors are presented in section 3.

2 TFA Sensor Concepts

2.1 TFA Sensor with One-Transistor Pixel

The **A**na**L**og **I**mage **D**etector ALID provides the smallest possible pixel size for TFA sensors. One ALID pixel consists of an a-Si:H photodiode and a transfer transistor. The photogenerated charge is stored in the blocking capacitance of the photodiode and transferred to the readout column when the transfer transistor is activated [6]. The sensor is addressed by line and column decoders, an amplifier/multiplexer combination provides a serial output signal. The ALID is well suited for illumination levels below 1 lx with 20 ms integration time (video frame rate). It may be operated at higher rates (e.g. 200 Hz), if a reduction of sensitivity is taken into account. The dynamic range of ALID is limited to about 60 dB. Due to the omission of active pixel electronics, the pixel matrix exhibits very low fixed pattern noise and high linearity. On the other hand, the lack of pixel integrated column drivers limits either the maximum line number or the output amplitude. In addition, parasitic cross coupling may reduce local contrast between adjacent pixels. A major advantage is that the extent of integrated peripheral electronics required to operate the pixel matrix is very small. As a result, the ALID is a preferable device if a low resolution is sufficient, as in e.g. a motion control device. Since chip area is a crucial issue with regard to device costs, the described concept allows low cost mass production of sensing devices.

The **V**aractor **A**na**L**og **I**mage **D**etector VALID is similar to the ALID, but contains an additional MOS capacitor inside each pixel. In this way the saturation illumination as well as the column output voltage and the local contrast are increased, since a larger amount of photogenerated charge can be stored. A VALID with comparable pixel geometry may therefore have more pixels than an ALID or it may provide a higher output signal.

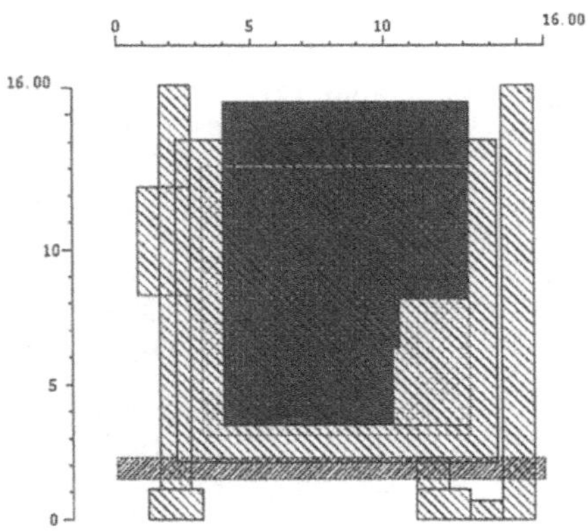

Fig. 2.1 Layout of (16 μm)^2 VALID pixel

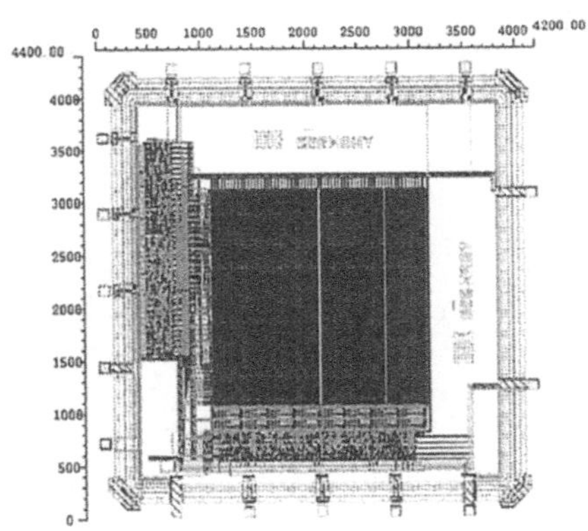

Fig. 2.2 Layout of 128 x 128 VALID array

The current VALID prototype consists of 128 x 128 pixels with an area of 16 μm x 16 μm each. The pixel size may be reduced to about 6 μm x 6 μm in a 0.5 μm process. The layout of a single pixel is depicted in Fig. 2.1. The detector rear electrode is given by a rectangular hatching, the varactor area is shaded. Fig. 2.2 shows the layout of the 128 x 128 AIDA array. Fig. 2.3 and Fig. 2.4 depict images taken with VALID, both without fixed pattern noise correction. The second image shows details in the sky as well as in the shaded building façade.

Fig. 2.3 Image taken with VALID

Fig. 2.4 Another image taken with VALID

2.2 TFA Sensor with Constant Voltage Circuit

In order to achieve the highest possible yield and to lower fabrication costs, the thin film system of a TFA sensor is fabricated in a PECVD cluster tool without temporarily being taken out of the vacuum for lithography. Therefore the pixel is simply defined by the size of its rear electrode. The continuous thin film layer however permits lateral balance currents between adjacent pixel detectors, resulting in a reduced local contrast. One way to suppress this effect is to pattern the thin film system employing a self-structuring technology [7].

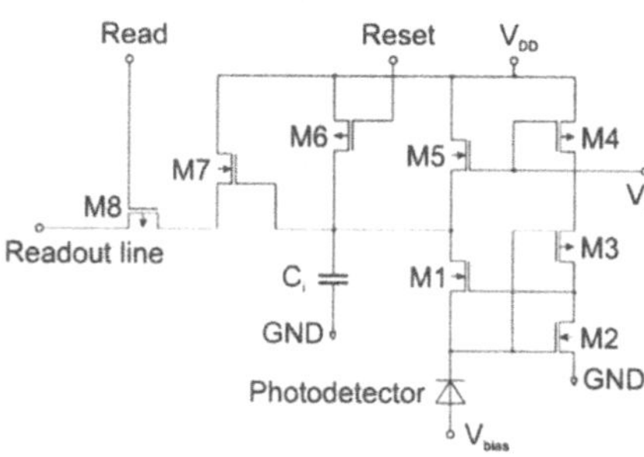

Fig. 2.5 Circuit diagram of AIDA pixel

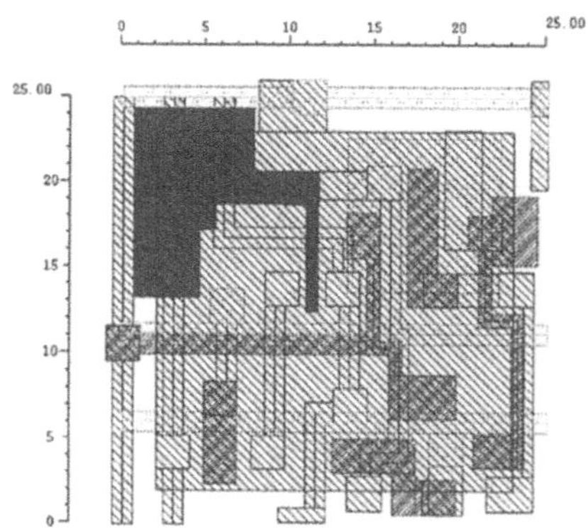

Fig. 2.6 Layout of (25 μm)^2 AIDA pixel

The sensor AIDA is an **A**nalog **I**mage **D**etector **A**rray which overcomes the coupling effect by electronic means. Here, a circuit inside each pixel provides a constant rear electrode potential, whereby the local contrast is significantly enhanced compared to ALID/VALID. The pixel schematic is given in Fig. 2.5. Each pixel consists of eight MOSFETs and one integration capacitance C_i that is discharged by the photocurrent. The inverter M2, M3 and source follower feedback M1 keep the cathode voltage of the detector constant. M4 limits the power consumption of the inverter, whereas M5 restricts the minimum voltage across the integration capacitance in order to always keep the constant voltage circuit working. The integrated voltage on C_i is read out via M7 and switch M8 in the simple source follower configuration. The reset operation is performed as M6 recharges C_i after the end of the integration period. The effective integration time of the pixel is the period between two reset pulses which may be varied according to the brightness of the scene. By this means the sensitivity is controlled for all pixels globally and a dynamic range of far more than 60 dB can be covered.

The prototype consists of a 128 x 128 pixel array with a pixel size of 25 μm x 25 μm. The pixel layout is depicted in Fig. 2.6. In a 0.5 μm ASIC process the pixel will shrink to about 18 μm x 18 μm. Fig. 2.7 shows a die photograph of AIDA. An image taken with the sensor array is given in Fig. 2.8. The device has been tested for illumination levels as high as 80,000 lx and proved to be virtually free of blooming effects or image lag. The maximum illumination level for video rate is about 5000 lx, higher frame rates are also possible. A reduction of the maximum illumination level at video rate is possible using a smaller integration capacitance. This leads to a higher output signal at low illumination levels so that the signal level exceeds the ASIC noise.

Fig. 2.7 Photograph of 128 x 128 AIDA array

Fig. 2.8 Image taken with AIDA

Fast transient response to changing illumination is a crucial requirement of image sensors in automotive systems. First of all, the transient time should be shorter than the frame time. However, in order to allow the evaluation of a portion as large as possible of each frame, it is desirable to have a much faster response. For an estimation of the transient response of AIDA, the sensor was illuminated by a 512 μs LED pulse synchronized to the 40 ms frame. The effective integration time was 64 μs and equal to the duration of one line. Thus eight lines were illuminated by the LED. Since the ninth and the following lines show no visible signal, the response time of the pixels is below 64 μs.

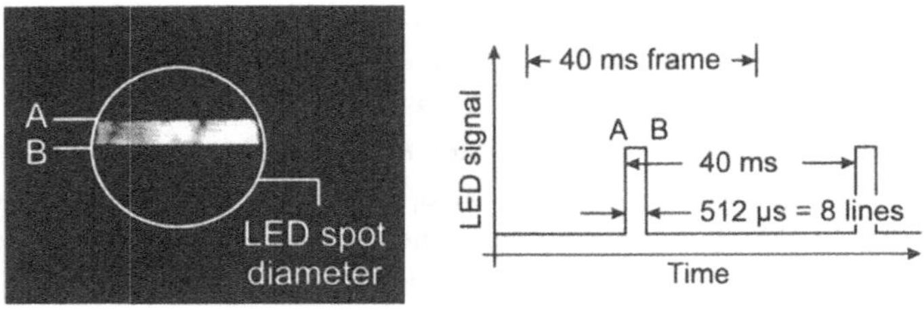

Fig. 2.9 Measurement of AIDA transient response with frame synchronized pulsed LED

2.3 TFA Color Sensor

One of the major benefits of the TFA technology is the possibility to deposit thin film detectors with adjustable spectral sensitivity on top of the ASIC. So for a color sensor array this advantage will preferably lead to a 3-colors-in-one-pixel sensor design. This technology inherently allows smaller pixels even if every pixel is equipped with three information storage and readout units. Such a pixel architecture is well suited for the identification of areas of the same or similar color in automotive systems (color tracking) as well as for single shot flash exposure in still cameras.

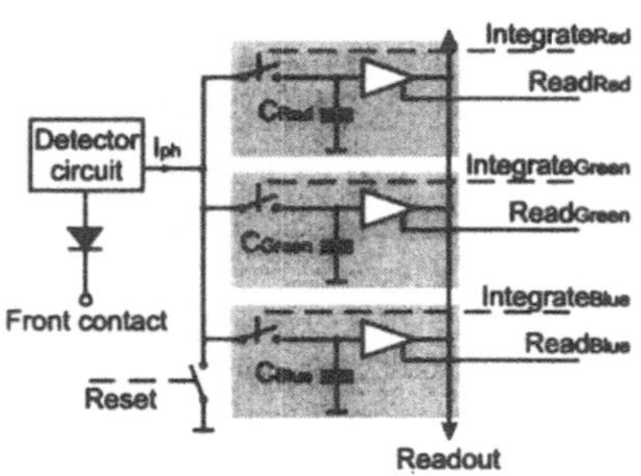

Fig. 2.10 Block diagram of CAESAR pixel

Event	Expected	Future perspective
Switching delay blue	10 ms, during preceding readout phase	
Delay after switching on illumination	300 µs	200 µs
Integrate blue	300 µs	150 µs
Switching delay green	300 µs	200 µs
Integrate green	200 µs	100 µs
Switching delay red	300 µs	200 µs
Integrate red	150 µs	75 µs
Total	1550 µs	925 µs

Tab. 2.1 Timing of CAESAR pixel

Fig. 2.10 shows the pixel block diagram of the color sensor array CAESAR. The detector circuitry keeps the rear electrode at constant potential in order to effectively suppress lateral balance currents. The photocurrent is fed into one of the color integration circuits, one at a time, during the integration phase. Thus the circuitry is able to generate and store the complete RGB information inside each pixel without intermediate readout operation. For readout the integrated color voltages are sequentially applied to the column output line. The CAESAR array has not been fabricated yet. Simulations of the circuitry however show excellent results with a high linearity over more than three decades. Even though random access modes were possible, a fixed line and column address generation scheme will be included on-chip which requires only little external circuitry. Control of spectral sensitivity is carried out globally by varying the front contact voltage.

One important issue of the system design is the minimum time required to integrate all three colors. Tab. 2.1 gives a short overview of the expected values based on current measurements and simulations. The column named future perspectives shows realistic estimates for the near future which will probably result from further research on the thin film system and deeper optimization of the pixel circuitry. The most time consuming step of switching from red to blue sensitivity is done during the readout phase such that the frame time is not unnecessarily extended. The resulting total time, of down to 925 μs, will permit single shot flash operation in the near future.

3 TFA Sensors for Automotive Vision Systems

3.1 Requirements of Image Sensors for Lane Tracking Systems

As new automotive systems such as Adaptive Cruise Control, Collision Warning and Roadway Departure Warning become desirable from a convenience, traffic flow and safety perspective, the need for robust determination of a vehicle's path, i.e. the road becomes essential. Without accurate knowledge of the road ahead of the vehicle it is impossible to determine whether an obstacle directly ahead of the vehicle is actually in its path or whether the road bends before reaching the obstacle. Fortunately there has been extensive effort, on the part of numerous individuals and groups, over the last several years on the process of vision based path determination using video cameras. These efforts have resulted in robust lane tracking algorithms for use in path determination. These lane tracking systems work quite well when provided with a good image from the video camera.

Unfortunately, under real world conditions, video cameras currently are not capable of providing satisfactory images under all conditions. This is directly due to the limited dynamic range of the available video cameras. Most available cameras provide on the order of 48 dB of instantaneous dynamic range. Under constant illumination, this dynamic range is sufficient for real world objects since reflectivities of non-specular objects range from about 0.5 % to 50 % covering a range of only 40 dB. Changes in illumination level account for the extremely large dynamic range encountered in lane tracking situations. The illumination level on a clear sunny day can vary by a factor of over 100 from outside to inside a tunnel for example. This 40 dB illumination level difference combined with the 48 dB of desired contrast resolution defines a requirement of roughly 90 dB of dynamic range. Another limitation of common video cameras is the tendency to bloom when portions of the image are extremely bright, such as for specular reflections and headlights. Under such circumstances, it is acceptable for overilluminated pixels to saturate, since there is no useful information lost, but blooming is very detrimental to the image and therefore must not occur. A camera suitable for lane tracking applications should therefore have a dynamic range of 100 dB or more and a very high resistance to blooming.

Semiconductor image sensors have technology inherent dynamic ranges of slightly more than 70 dB. By means of global sensitivity control the range can be extended to the required 100 dB. However, if this entire range is covered throughout a single frame, global sensitivity control may be ineffective, since saturation as well as signals below the noise level may occur simultaneously. A further demand apart from blooming prevention is therefore that any portion of the image - within the specified dynamic range - can be recognized any time.

A common concept for very high dynamic range image sensors exploits the compressing nature of the logarithmic voltage-current response of diodes. In the standard configuration of such a logarithmic sensor the exposure proportional photocurrent of a photodiode is fed into a second diode which may be a bipolar diode or a MOS transistor in diode configuration [8]. The voltage V response across the diode is logarithmically compressed with regard to the photocurrent I:

$$V = n\frac{kT}{q}\ln\left(\frac{I}{I_0}+1\right)$$

(k: Boltzmann's constant, T: Temperature, n: Diode factor, q: Electron charge, I_0: Reverse saturation current)

This voltage is usually read out in a standard source follower configuration. Logarithmic sensors may seem favorable for cruise control systems since they require only three transistors plus one photodiode per pixel and the compression easily fits 120 dB of intensities into a voltage range of a few hundred millivolts. However, there are some major drawbacks which must be considered. First, these devices are especially susceptible to CMOS inherent fixed pattern noise, since a minor pixel-to-pixel variation in output voltage leads to an exponentially amplified difference in the reconstruction of the photocurrent. Self-offset reduction schemes such as correlated double sampling however cannot be applied since there is no reset level per pixel which may be subtracted for reference. The only successful fixed pattern noise reduction scheme to date is the off-chip compensation by means of storing the per-pixel conversion characteristics into a memory and stripping the pixel-to-pixel variations in a signal processor.

However, matters become even more difficult through the second disadvantage: the strong temperature dependence of the output signal which becomes apparent from the formula above. Temperature also has an exponential influence on the intensity information. Moreover, the off-chip fixed pattern compensation schemes must include this temperature effect. This requires a close control over the actual die temperature, constant temperature across the entire die irrespective of the exposure situation and an elaborate image correction algorithm which demands excessive computation power. Finally, the transient response of logarithmic sensors proves to be very poor at low light intensities due to an unfavorable relation of photocurrents to load capacitance.

3.2 Locally Adaptive TFA Sensor

The TFA sensor LAS (Locally Adaptive Sensor) for automotive applications has been developed in order to overcome the above discussed problems arising with conventional global integration control as well as logarithmic output characteristics. The underlying concept of locally adaptive integration control allows the sensitivity of each single pixel to be adapted to the illumination condition at its respective location in the image. In this way a dynamic range of over 100 dB can be covered throughout the chip at any time [9,10].

The principle of the locally adaptive pixel is explained with the help of its block diagram in Fig. 3.1 and the corresponding timing diagram in Fig. 3.2. Basically the sensitivity of a pixel is controlled by determining the time during which the photocurrent from the a-Si:H multilayer is integrated into an on-chip capacitance C_i. Therefore a second capacitor C_p is included in the pixel which contains the programmed timing information represented by a voltage. Every image frame comprises three phases. First a write signal allows the programming capacitance C_p to be precharged. The higher the illumination intensity on the pixel is, the higher the programming voltage is chosen. Second a voltage ramp is applied to the pixel which is compared to the voltage across C_p. As soon as the ramp exceeds the programmed value, integration of the photocurrent starts. With the falling slope of the ramp the integration is stopped. At the end of integration the voltage across C_i is read out and afterwards reset to its starting value.

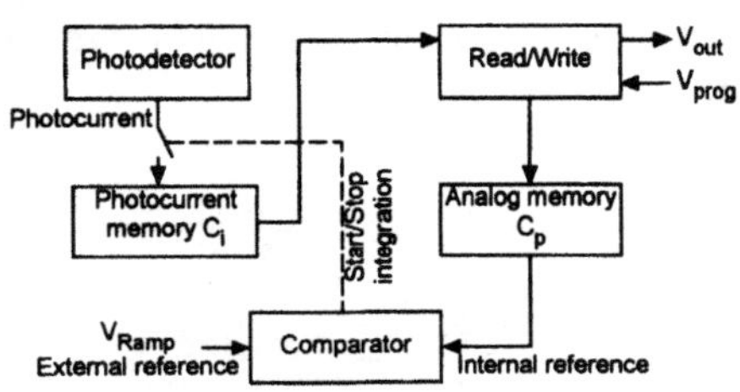

Fig. 3.1 Block diagram of LAS pixel

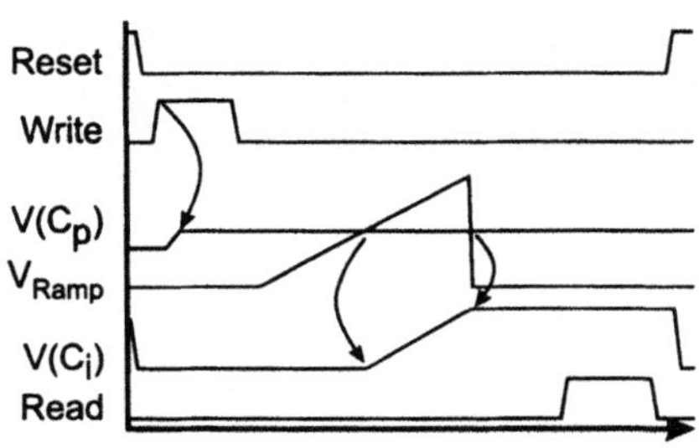

Fig. 3.2 Timing diagram of LAS pixel

The schematic of a programmable LAS pixel is shown in Fig. 3.3. The constant voltage circuit (M20 .. M23) on the right has been described in section 2.2. M19 turns on and off the photo current into C_i. Readout of the integrated voltage is realized by M14 and M16. M15 resets the integration capacitor to its initial value V_{ref}. The integration time control is performed by M1 through M13. The programming voltage is passed through M13 and stored on C_p. M9 and M10 form a source follower which effectively decouples the small programming capacitance of 100 fF only from the following comparator stage (M1 .. M8). The latter compares the programmed value with the ramp voltage V_{ramp} which is applied externally and starts integration when the comparison is positive via M11/M12 and M17/M18. As a special feature the comparator is of the latched type - the recent comparison result is periodically frozen when clock is high. This allows precise control of the final integration period even at higher levels of fixed pattern noise.

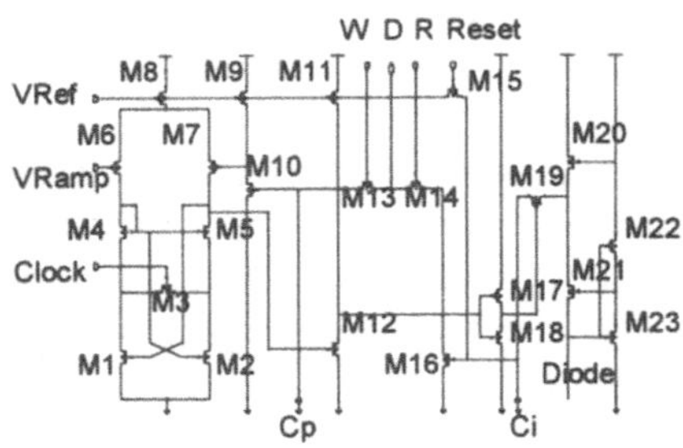

Fig. 3.3 Circuit diagram of LAS pixel

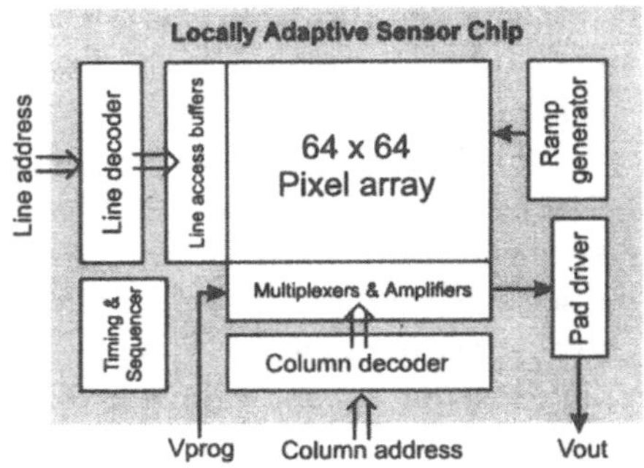

Fig. 3.4 Block diagram of 64 x 64 LAS array

The LAS array block diagram is depicted in Fig. 3.4. In addition to the 64 x 64 pixel array the imager contains line and column decoders for pixel programming and for reading out the integrated pixel voltage. A pad driver makes sure that the output signal can be used in the following external readout stages. The required voltage ramp is generated on-chip and is applied to every line of the sensor array. Finally, a sequencer and timing unit for providing the peripheral circuitry and the pixel array with clock signals is implemented.

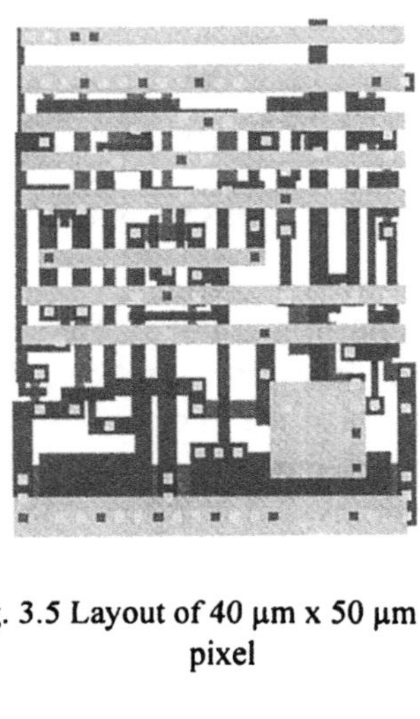

Fig. 3.5 Layout of 40 μm x 50 μm LAS pixel

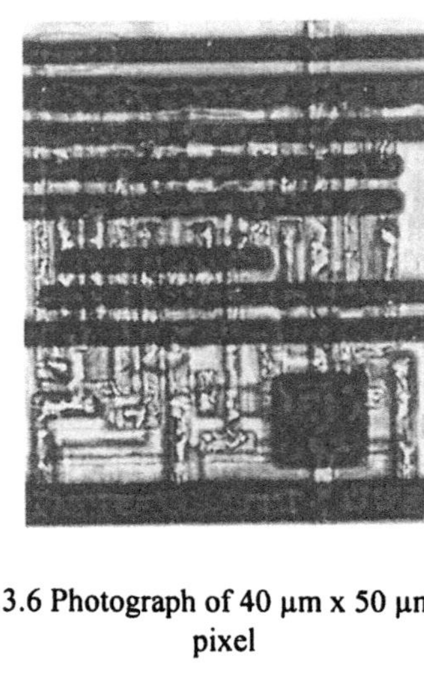

Fig. 3.6 Photograph of 40 μm x 50 μm LAS pixel

The layout in conjunction with the corresponding chip photograph of a 50 μm x 40 μm LAS pixel with programmable sensitivity is depicted in Fig. 3.5 and Fig. 3.6, respectively. The chip was designed and manufactured in a 0.7 μm low power CMOS process. The horizontally running control lines and the contact region between detector rear electrode and pixel electronics can be seen clearly.

Fig. 3.7 and Fig. 3.8 demonstrate the behavior of the integration circuit for two contrary illumination conditions with three different integration times each. The large switching offset of the output voltage at the beginning of each integration period does not limit the dynamic range of the pixel. It just contributes a constant offset which has to be accounted for when designing the readout stages. The measurements prove that the pixel works at 6 lx with a few milliseconds as well as at 80,000 lx with about 10 μs maximum integration time.

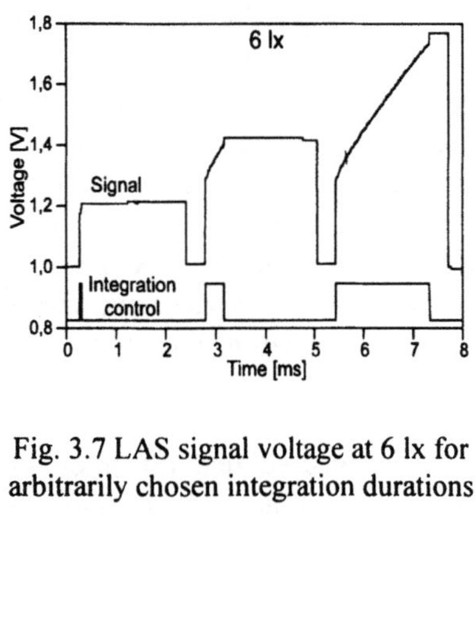

Fig. 3.7 LAS signal voltage at 6 lx for arbitrarily chosen integration durations

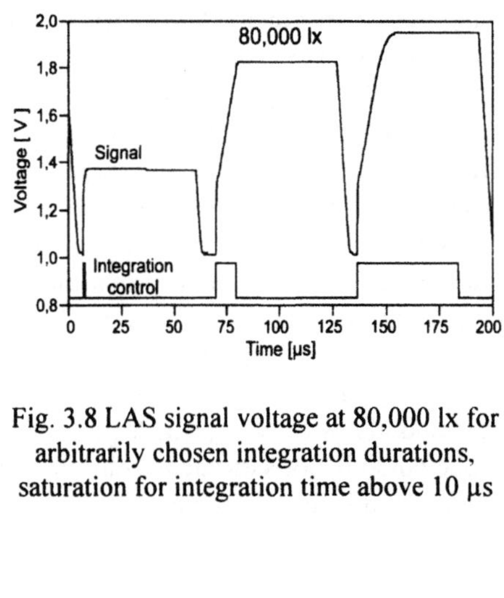

Fig. 3.8 LAS signal voltage at 80,000 lx for arbitrarily chosen integration durations, saturation for integration time above 10 μs

3.3 Locally Autoadaptive TFA Sensor

The locally autoadaptive sensor, LARS, provides a very high global dynamic range by adapting the integration time for each individual pixel according to the local illumination intensity [11]. Unlike the locally adaptive sensor LAS the integration time control takes place in the pixel itself in real time. Therefore off-chip circuitry and additional time for pixel programming are not required. Furthermore, a sudden change to a high illumination intensity is detected immediately so the integration of the photocurrent is stopped before the integration capacitor is saturated. The complex pixel electronics of the LARS concept does create large pixels, however, because of the TFA structure, the photo detector fill factor is still 100 %.

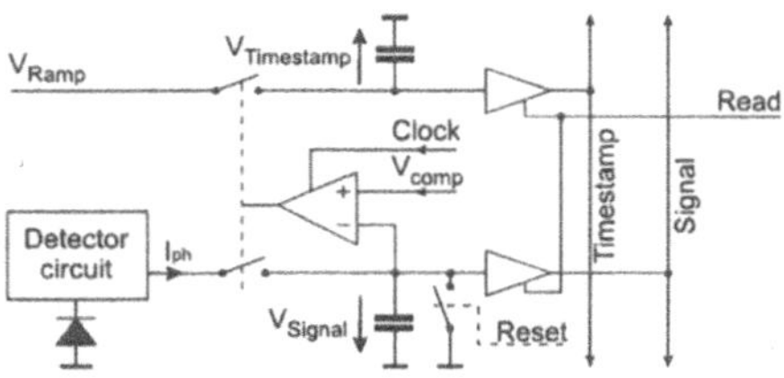

Fig. 3.9 Block diagram of LARS pixel

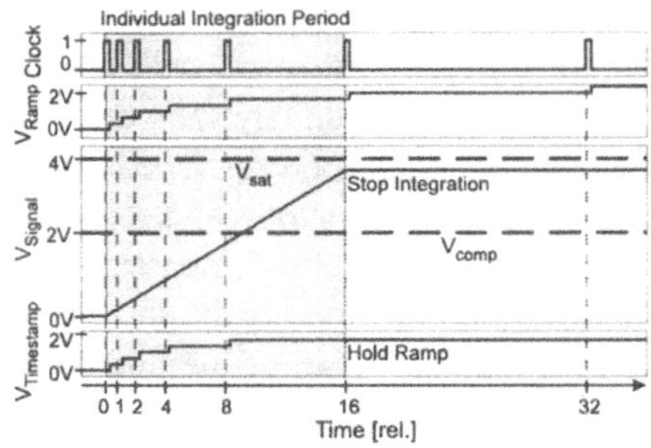

Fig. 3.10 Timing diagram of LARS pixel

Fig. 3.9 shows the schematic of a locally autoadaptive pixel, Fig. 3.10 the corresponding timing diagram. The current of the photodiode is integrated on the integration capacitance to a signal voltage V_{signal}. On every rising edge of the clock input this voltage is compared to a reference voltage V_{comp} which is slightly below half the saturation value of V_{signal}. If the integrated signal is still below V_{comp} the integration time is doubled, whereas the comparator terminates the integration via the switches if the signal exceeds the reference level. With every clock the timestamp input V_{ramp} climbs up one step and is sampled and held in the timestamp capacitance at the moment the integration is terminated. At the end of the integration phase the information stored in every pixel consists of two voltages which are read out: the integrated signal and the timestamp, the latter clearly defines the integration duration over which the pixel signal has been integrated. The binary exponential increase of the integration time steps in the above example corresponds with $V_{comp} \leq ½\, V_{sat}$. In this way it is ensured that the integration capacitance is not saturated within the following step and the range for the signal voltage at the end of the integration time is $½\, V_{sat} \leq V_{signal} \leq V_{sat}$.

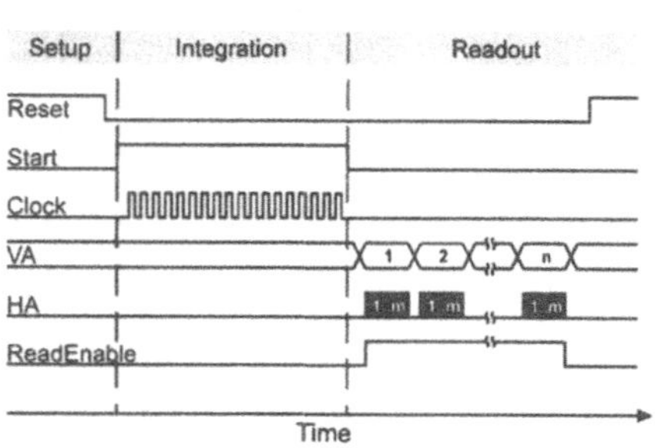

Fig. 3.11 Timing diagram of LARS array

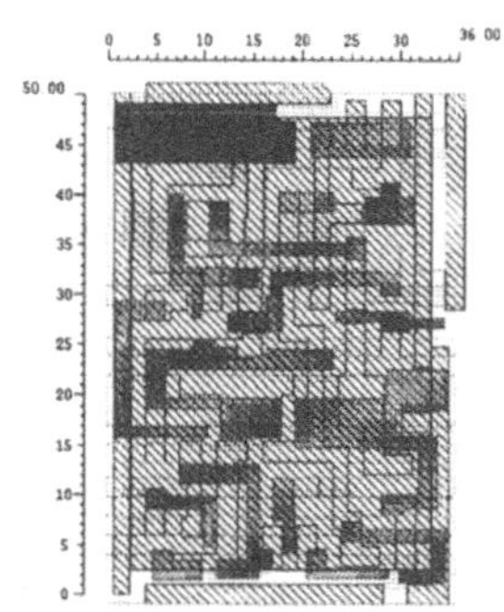

Fig. 3.12 Layout of 35 µm x 49 µm LARS pixel

The timing diagram of the LARS array (see Fig. 3.11) shows a typical setup/integration/readout cycle. During the setup period, all pixels are set to an initial state (reset = high). Then the integration time is globally started with the start signal. After the last pixel has stopped the integration, the pixel data is read out linewise by activating the vertical counter (VA). Within every line the horizontal counter (HA) switches the columns from 1 to m transferring the pixel data to the column multiplexer, where the line data are parallel-to-serial converted. Serial pixel data together with the according timestamps are then transferred to the interface. After the readout period has ended (defined by the read enable signal switching to low) a new setup period begins.

Fig. 3.12 shows a complete pixel circuit in a 0.7 µm low power CMOS technology. The autoadaptive functionality is realized with 24 transistors and two capacitors covering an area of 35 µm x 49 µm. The first 64 x 48 pixel prototype shown in Fig. 3.13 includes the readout periphery and line and column adress generators. For flexibility, timing sequencers and access control were realized off-chip. The block diagram of the LARS array with 64 x 48 pixels is similar to that of the 64 x 64 LAS array depicted in Fig. 3.4. The main difference is that the LARS array includes another analog output instead of the programming input. For reasons of further integration and improved performance, on-chip A/D conversion is aspired. In this case, a bidirectional parallel interface will be employed to export pixel data and to set up global sensor parameters.

Fig. 3.13 Photograph of 64 x 48 LARS array

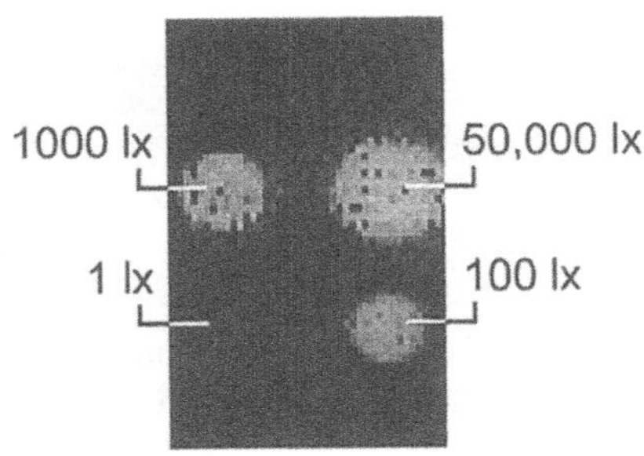

Fig. 3.14 Evaluation of dynamic range of LARS array

Signal (integrated voltage) range	400 µV .. 1.5 V	71 dB
Timestamp (integration time) range	5 µs .. 2,56 ms	54 dB
Illumination range	**55 mlx .. 100,000 lx**	**125 dB**

Tab. 3.1 Exemplary dynamic ranges for LARS

Tab. 3.1 gives an example of the dynamic ranges for the current LARS array. The illumination range (global dynamic) is basically limited only by the detector if any integration time is allowed. With an output swing of 1.5 V and a noise level of some 400 μV_{rms} the signal-to-noise ratio and therefore the range of a single analog voltage is about 71 dB. The additional integration time range depends on the timing and is 54 dB in this example. So the total dynamic range included in the signal and timestamp amounts to 125 dB.

The most obvious demonstration of the capabilites of the LARS array is depicted in Fig. 3.14. The image sensor was illuminated by four spots of 1 lx, 100 lx, 1000 lx and 50,000 lx respectively, or 94 dB altogether. The image shown here reproduces the integrated signal only so the spots show approximately equal brightness. For this measurement timing was designed to allow adaptation to one of nine integration times, ranging from 5 μs to 2.56 ms separated by factors of two, corresponding to Tab. 3.1. The pixels under the 1 lx and 100 lx spots selected the longest integration time of 2.56 ms, whereas the pixels under the 1000 lx and 50,000 lx spot adapted to 320 μs and 10 μs, respectively. The image shows some fixed pattern noise and defect pixels which can be accounted to an improper second metallization of the ASIC and problems with the vias through the insulation layer of the thin film system.

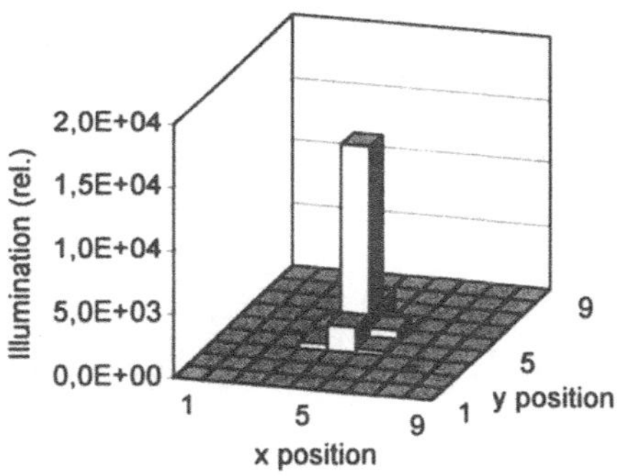

Fig. 3.15 Evaluation of blooming / local contrast of LARS array

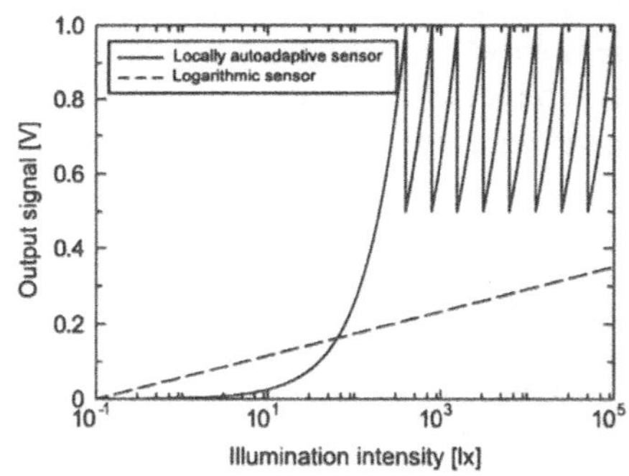

Fig. 3.16 Comparison of logarithmic and linear autoadaptive characteristics

A simple method to evaluate blooming effects in an image sensor array is to illuminate a single pixel with high intensity through an optical fiber and to chart the photoresponse of the pixels. Fig. 3.15 depicts the result for the LARS array, for which both integrated signal and timestamp have been taken into account. The chart demonstrates that the array is virtually free of blooming, since the photoresponse drops significantly outside the illuminated central pixel, which is saturated at an intensity of over 300,000 lx. The slightly raised signals of the adjacent pixels are mainly attributed to light scattering from the fiber cladding so the actual local contrast is still higher.

In contrast to the logarithmic sensors, the integration timing sets the actual working ranges of LARS. Fixed integration times for all pixels are possible, as well as a linear distribution of available integration times just by reprogramming the sequencer e.g. in a PLD. However, the exponential distribution is most favorable, since it spans the widest adaptation range with the fewest ramp steps and clock phases neccessary. For applications with low light conditions the timing can easily be extended to long integration times such as e.g. 5 µs .. 20,48 ms (143 dB total) or even 5 µs .. 81,92 ms which provides an unprecedented dynamic range of a linear image sensor of 155 dB.

This exponential timing distribution leads to some kind of logarithmic compression. The proposed sensor type however has some major advantages over conventional logarithmic sensors. First, there is no diode characteristic involved which alleviates the above mentioned problems of fixed pattern noise. This becomes apparent upon inspection of Fig. 3.16 where the voltage output of a bipolar diode in a logarithmic sensor is compared to the analog signal part of the LARS over intensity. The slope of the curves, a measure for the insensitivity to fixed pattern noise, is higher for the LARS for most of the dynamic range. Only at low light intensities (< 10 lx) does the logarithmic concept give a somewhat lower noise sensitivity, while its transient response becomes insatisfactory as described above. The logarithmic characteristic of the LARS is generated by timing which is much easier to control over excessive dynamic ranges (e.g. 1 ns .. 0.1 s = 160 dB). Furthermore timing driven range compression is inherently immune to temperature drift in strong contrast to the diode compression. Second, the external timing control easily allows switching between fixed and adaptable integration times whenever necessary. Third, transient response is not impaired at low intensities. Finally, the achievable dynamic range is split into two signals, both with moderate demands on noise level (400 µV and 3 mV), whereas 150 dB will not be manageable in the foreseeable future in a linear signal due to the allowable noise level, at the full 10 MHz bandwidth, of only 50 nV_{rms}.

Since maximum integration times are usually limited by external constraints, further measures have to be taken to raise sensitivity at low intensities. A thicker amorphous diode e.g. provides higher sensitivity due to its lower capacitance. Likewise special circuitry such as capacitive transimpedance amplifiers may have the same benefits. In another approach the amorphous multilayer is deposited on the back of a MOS varactor to form a unified detector. Such a device provides the high absorption of amorphous silicon with the possibility of storage and transport like in CCDs. Very high sensitivities can thus be achieved by the use of very low capacitance charge readout stages.

4 Conclusion

TFA technology combines the high-level electrical performance of crystalline ASICs with the excellent photoelectric properties of a-Si:H thin films. Common problems of image sensors as e.g. blooming or smear do not occur in TFA sensors, the transient response is sufficient for moderate readout rates. Apart from the performance, flexibility and low cost fabrication are further benefits of TFA.

Several TFA prototypes with increasing complexity and different optimization criteria have been discussed. While the simpler devices may be fabricated in a large number for multi-purpose use, highly complex pixel circuitries can be designed in order to implement application specific advanced functions. It has been demonstrated that by means of locally autoadaptive sensitivity control, the dynamic range can be expanded beyond the limitations of any other circuit concept or technology. Problems due to exponential amplification of fixed pattern noise and temperature differences as in logarithmic sensors do not arise, since the autoadaptivity is determined by the timing. As a result, TFA sensors are ideally suited for automotive vision systems where images with very high dynamic ranges have to be evaluated.

References

[1] S. E. Kemeny, E.-S. Eid, S. Mendis, E. R. Fossum, "Update on Focal-Plane Image Processing Research," Charge-Coupled Devices and Solid-State Optical Sensors II, Proc. SPIE, vol. 1447, pp. 243-250, 1991.

[2] S. K. Mendis, S. E. Kemeny, R. C. Gee, B. Pain, C. O. Staller, Q. Kim, E. R. Fossum, "CMOS Active Pixel Image Sensors for Highly Integrated Imaging Systems," IEEE J. Solid-State Circuits, vol. 32, pp. 187-197, 1997.

[3] R. Elston (ed.), "Astrophysics with IR-Arrays," Astronom. Soc. of the Pacific Conference Series, San Francisco, vol. 15, 1991.

[4] J. Giehl, H. Stiebig, P. Rieve, M. Böhm, "Thin Film on ASIC (TFA)-Color Sensors - New Applications of Optical Thin Film Detectors," G.Hecht, J. Hahn, DGM Informationsgesellschaft Oberursel mbH, Oberursel, pp. 560-563, 1994.

[5] P. Rieve, J. Giehl, Q. Zhu, M. Böhm, "a-Si:H Photo Diode with Variable Spectral Sensitivity," Mat. Res. Soc. Symp. Proc. 420, pp. 135-140, 1996.

[6] J. Schulte, H. Fischer, T. Lulé, Q. Zhu, M. Böhm, "Properties of TFA (Thin Film on ASIC) Sensors," H. Reichl, A. Heuberger, Micro System Technologies '94, VDE-Verlag, Berlin, pp. 783-, 1994.

[7] H. Fischer, J. Schulte, P. Rieve, M. Böhm, "Technology and Performance of TFA (Thin Film on ASIC)- Sensors," Mat. Res. Soc. Symp. Proc. 336, pp. 867-872, 1994.

[8] B. Dierickx, D. Scheffer, G. Meynants, W. Ogiers, J. Vlummens, "Random Addressable Active Pixel Image Sensors," AFPAEC Europto Berlin, Proc. SPIE, vol. 1950, p. 1, 1996.

[9] T. Lulé, H. Fischer, S. Benthien, H. Keller, M. Sommer, J. Schulte, P. Rieve, M. Böhm, "Image Sensor with Per-Pixel Programmable Sensitivity in TFA Technology," H. Reichl, A. Heuberger, Micro System Technologies '96, VDE-Verlag, Berlin, pp. 675-, 1996.

[10] B. Schneider, H. Fischer, S. Benthien, H. Keller, T. Lulé, P. Rieve, M. Sommer, J. Schulte, M. Böhm, "TFA Image Sensors: From the One-Transistor Cell to a Locally Adaptive High Dynamic Range Sensor," Tech. Digest IEDM 97, pp. 209-212, 1997.

[11] "Bildsensor mit hoher Dynamik," patent pending, PCT/EP 97/05 306, 29.09.1997.

A Wide Dynamic Range CMOS Stereo Camera

Guy Meynants, Bart Dierickx, Danny Scheffer and Jan Vlummens

Imec, Kapeldreef 75, B-3001 Leuven, Belgium

1. Introduction

Novel automotive applications, like automatic car park systems, time-to-crash sensors or autonomous driving systems require an accurate measurement of the environment around the vehicle. Distances to other vehicles and objects have to be known precisely. Stereo vision systems can be used to solve these problems. But high requirements are posted to the stereo camera as well as to the image processing unit. The image sensor has to deal with large variations in operation conditions, like illumination and temperature. The image processing unit needs a lot of computing power : a lot of correlation calculations are necessary to solve the two-dimensional matching problem between both images.

In this paper, we describe our approach to these problems. We have built a stereo camera, made with two dedicated CMOS image sensors. The sensor has a large dynamic range, which is necessary to cope with the large light variations which occur in automotive applications. The camera has a simple digital interface to the image processing unit, with a true random access readout mechanism. Another essential feature is the geometry of the pixels, which makes the alignment of the stereo pair of imagers easy. It limits the matching problem between the stereo images to a one-dimensional problem, thereby strongly reducing the calculation overhead. In stereo vision systems based on traditional camera's, the problem is two-dimensional which complicates the image processing a lot.

2. The Image Sensor

2.1 Image Sensors in Automotive Applications and Stereo Vision

In an automotive application, some of the major operation conditions of the sensor are really pushed to the limit. The variation of the average illumination level in a scene is huge : on sunny days, very bright light may not cause to saturate the image sensor, while at night, an image must be taken with only the illumination of the car's headlights. Not only the average illumination level is varying, but also the illumination in a single image varies a lot.

The maximal intensity variation of an image which can be detected is characterised by the dynamic range of the sensor. It is the ratio between the highest and the lowest intensity possible in an imaging scene. In traditional integrating image sensors, this ratio is limited to about 4000:1, which is too low for automotive applications. Besides that, CCD based imagers have two more problems with large illumination variations in the scene. Vertical white columns appear under and above overexposed pixels. This phenomenon is called smear. It is caused by the illumination of these columns during the transfer of the photocharges. A second phenomenon is blooming. It results in an enlargement of the area of the saturated pixels. If a pixel is saturated, the charges generated under this pixel will be detected by the neighbours until also these pixel saturate. The result is an enlargement of the saturated region of pixels, and in extreme situations, the entire image is saturated. Modern CCDs have provisions for anti-blooming, although these can only reduce the effect, and not eliminate it.

The sensor has to work in a large temperature range. The effect of the temperature is an increase in dark current, which doubles every 8 °C. This dark current is added to the photocurrent and it varies randomly from pixel to pixel. Therefore, the dark current cannot be distinguished from the photo signal and in the image, it is seen as a random temperature dependent pattern which is added to the image. Especially in low light levels, these effects are visible because the dark current is then almost equal to the photocurrent. Typically, the dark current is about 1 nA/cm^2 in a standard CMOS process. For CCD devices, it is one to two magnitudes lower.

Other requirements to the sensor are posted by the application. There is the resolution needed and the pixel geometry. For the envisaged application, namely stereo vision for automotive, the highest resolution is needed in the horizontal direction. In the vertical direction, perpendicular to the ground plane or horizon, the resolution can be much lower, because there is normally no motion in this direction. Later on, we will also show that the accuracy of the stereo computation will be better if the pixels can perform a kind of gaussian filtering in the vertical direction. Because of this, the pixels are rectangular and special-shaped. A second advantage is the better alignment of the two image sensors.

For a vision task like stereo vision, some kind of image processing module (e.g. a microcontroller, DSP or ASIC) is necessary to perform the computations. The interface between the camera and the image processor needs to be as fast as possible. A simple digital interface is therefore the best solution. Windowing functions of the camera can be used to read out only the window of interest, at a higher frame rate. The highest readout speed is however only offered by a true random access image sensor, where only the pixels of interest are read out. The algorithm in the signal processor decides which pixels it wants to read out and then the imager is addressed to read these pixels, in a way comparable to addressing a memory. For this operation mode, the imager should work completely asynchronous, without a fixed integration time or timing scheme.

The latter operation mode is possible, however only in CMOS technology. Before we go more in detail on the actual implementation of the image sensor, we will first discuss how CMOS technology can be used for image sensors.

2.2 CMOS Technology for Image Sensors

The mainstream technology for image sensors is based on the principle of charge-coupled devices or CCDs. The technology is best fitted for high-quality image sensors for video applications. There are however some limitations for the use of CCD technology in vision tasks. The major drawbacks are the lack of the possibility to include logic on the imager chip and the relatively high cost of these imagers. Recently, new attention is paid to the development of image sensors based on CMOS technology, as an alternative for the CCDs [1]. Image quality of these sensors is increasing rapidly, and at this moment, a performance similar to those of consumer CCDs can be reached. There are, however, a lot of other advantages to the use of CMOS technology :

- The possibility to integrate on-chip control and interface logic. An analog-to-digital converter can be added to the sensor output. Control logic can be integrated to obtain a camera-on-a-chip. Intelligent vision algorithms can be included on the imager die to perform certain vision tasks.
- The use of a standard low-cost technology, of which the quality and yield keeps improving.
- Certain performance characteristics for CCDs and CMOS sensors are different. Some are worse for CMOS imagers than for CCDs, like the temporal and fixed pattern noise. For this reasons, CCDs are still superior for high-quality imaging, like in TV broadcast cameras and scientific applications. But for vision tasks and consumer video applications, the image quality of CMOS sensors is satisfactory. And some performance tasks are even better, like the behaviour for large illuminations (smear and blooming). Power consumption of CMOS cameras is also a decade lower.
- CMOS sensors offer the flexibility to use a dedicated pixel type for a dedicated application. A lot of flexibility exists in the pixel design, and the pixel can be optimised for high-dynamic range, image quality or sensitivity. Especially a true random addressable pixel is only possible in a CMOS technology.

These arguments, and especially the last one, convinced us to use CMOS technology for the image sensor.

2.3 Pixel and Photodiode Structure

The pixel is the most important element in any image sensor. For an automotive application, the dynamic range is critical. Therefore, we decided to use a pixel with a logarithmic light-to-voltage conversion [2]. This pixel can convert the light into a voltage over a range of more than 6 decades in light intensity. It is also free of blooming and smear. Its architecture is given in figure 1. The principle of the pixel is as follows: the photocurrent generated by the photodiode, goes through the transistor M1. This transistor acts as a non-linear resistor. Because the photocurrent is small, the transistor is in weak inversion and its gate-source voltage is proportional to the logarithm of the drain-source current (or photocurrent). At the gate of transistor M2, we obtain a voltage which is proportional to the logarithm of the light intensity. M3 acts as a switch, which is closed when the row is selected by the row addressing decoders. When M3 is closed, M2 acts as a source follower and puts the voltage of the photodiode on the column readout bus.

The photodiode is a p^+/n junction, which is identical to the junction used as the source or drain of a PMOSFET. When used in an n-well CMOS process, this junction offers the highest MTF or sharpness.

The envisaged application, namely the use of the sensor in a stereo camera, determines the shape of the photodiodes. Only in one direction, a high resolution is necessary. In the other direction, the resolution may be lower. This makes the alignment between both imagers easier. Therefore, the photodiodes are rectangular, with a dimension of 400 x 10 μm^2. The imager array has 16 rows of each 640 pixels. A second advantage is that only 1D correlation calculations are necessary for the calculation of the distance. In the other dimension, the matching problem is already solved, as it is assumed that both sensors are row-aligned.

Studies have shown that 1D correlations (for optical flow or stereo vision) are calculated with a higher accuracy if a gaussian filter is performed in the direction perpendicular to the direction of the correlation (e.g. the pixel row) [3]. The rectangular shape of the photodiodes does already perform some kind of gaussian filtering, and this filtering is even improved by placing a metal light shield on the corners of the photodiodes. In this way, a light sensitive area as indicated in figure 2 is obtained in the pixels. On this figure, the approximation for the gaussian filter can be distinguished.

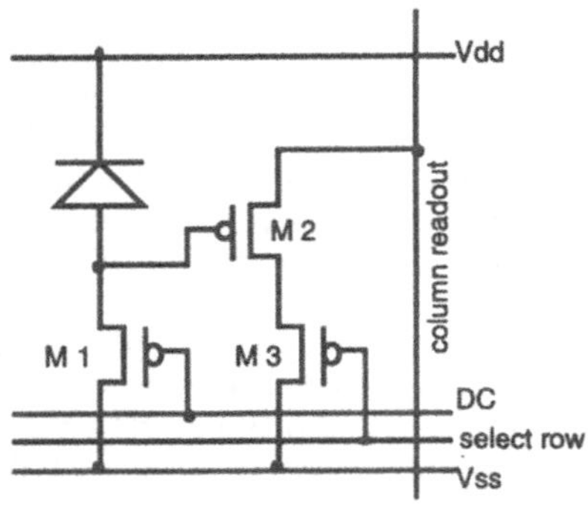

Figure 1 : Pixel schematic

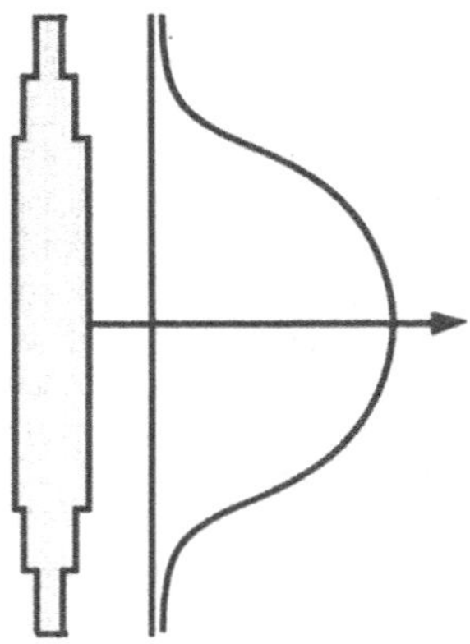

Figure 2 : Shape of the photodiodes and spatial response

As mentioned before, the output of the pixel is logarithmic. The output characteristic on the analog output signal of the imager before amplification is shown in figure 3. The theoretical slope of the curve is 60 mV per decade in light intensity, but this slope is attenuated by the pixel source follower and the column multiplexer circuit. Only 25 mV per decade remains at the output. The minimal light intensity is about 10^{-8} W/cm^2.

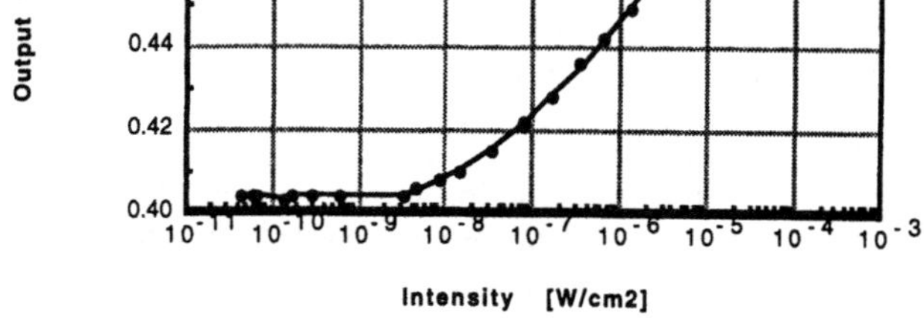

Figure 3 : Output voltage of the image sensor before amplification in function of the light intensity

2.4 Architecture of the Image Sensor

The imager itself contains an array of these pixels, which are selected by address decoders. The architecture is given in figure 4. A row of pixels is selected, by latching in the row co-ordinate with the enable-y signal (E_y). The pixels of this row can be selected with the column multiplexer. This is an address decoder, which selects the column line which is latched in through the enable-x (E_x) signal. The analog signal of the selected pixel goes then through the output amplifier to the analog-digital converter. The ADC has a tristate output port, and it is read out when E_c is low. In this way, the pixel address and the ADC output are placed on the same (bi-directional) bus. This architecture is also used in other random addressable image sensors developed at IMEC [4].

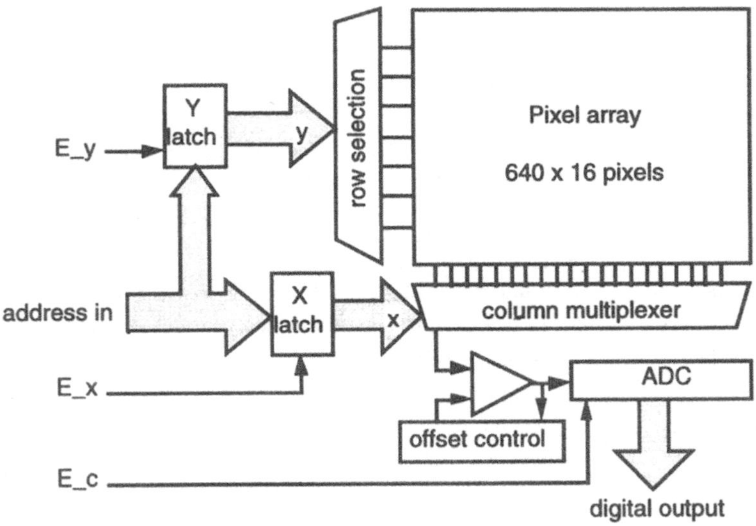

Figure 4 : Architecture of the image sensor

The task of the offset control circuit is to adjust the average output level of the pixels which are read out to the middle of the ADC range. The entire dynamic range of the pixels can be mapped to the ADC input. But, of course, in this case, low accuracy is obtained in a scene with low contrasts. Therefore, the amplification of the output amplifier is increased, but then, several pixels will be clipped to black or white. The offset control circuit adjusts the offset level of the output amplifier so that the average analog output is always in the middle of the ADC range. It is essentially a low-pass filter on the analog output stream.

The analog-to-digital converter is an 8-bit full flash ADC. It converts the input in one clock cycle at a speed of 8 MHz. We prefer one single ADC instead of column-wise parallel ADCs, because we benefit so the most of the random access capabilities of the imager.

2.5 Performance of the Imager

The major specifications of the image sensor are summarised in table 1. The imager can be read out at a speed of 5 MHz in the horizontal (X) direction. Selecting a new row is at a lower speed, depending on the light intensity. At 10^{-6} W/cm^2, the maximal selection speed of the Y-address is 200 kHz. The temporal noise is 1 mV RMS, integrated on a frequency band from 1 Hz to 133 kHz. The spatial or fixed pattern noise is about 12 mV peak-to-peak. An off-chip calibration for correction of this fixed pattern noise is mandatory. This is done by subtracting a reference value for each pixel.

Table 1 : Specifications of the image sensor

Parameter	Specification
Architecture	16 rows of each 640 pixels
Pixel dimensions (X *x* Y)	10 *x* 400 μm^2
Readout system	4 bit Y addressing @ 200 kHz (10 W/cm^2) 10 bit X addressing @ 5 MHz 8 bit digital output
Analog output	5 MHz pixel rate, logarithmic light-to-voltage conversion
Temporal noise	1 mV RMS (integrated from 1 Hz to 133 kHz)
Fixed pattern noise	12 mV peak-to-peak (to be compensated off-chip)
Missing pixels	none (on selected chips)
ADC	on-chip full flash, 10 MHz, tristate outputs
Dissipation (5 MHz)	< 100 mW
Die size	7.5 x 7.5 mm^2
Package	48 pins LCC

3. Stereo Camera

3.1 Description of the Stereo Camera

With the image sensor described in the previous paragraph, a stereo camera was made. It is shown in figure 5. The entire PCB of the camera is 8 by 3 cm^2, with two C-mount lensholders on it.

One of the problems of stereo cameras is the alignment of both image sensors. The 400 µm long pixel shape of our sensor makes it easy to align two sensors on the same PCB. This eliminates the need for a complex calibration procedure used in traditional stereo pairs.

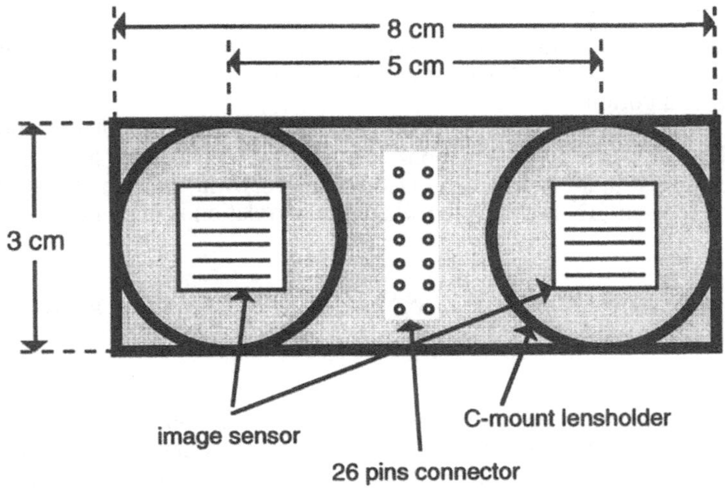

Figure 5 : Stereo camera

3.2 Architecture of the Stereo Camera

The stereo camera has a simple architecture, shown in figure 6. Two sensors are placed in parallel. They share the bi-directional data bus and the E_X and E_Y lines. Both are so addressed in parallel, and the X- and Y-co-ordinates of the pixels which are read out are loaded into the sensors in parallel. Both sensors can then be read out consecutively by pulsing their E_C signals to activate the output buffers of the sensor's ADCs.

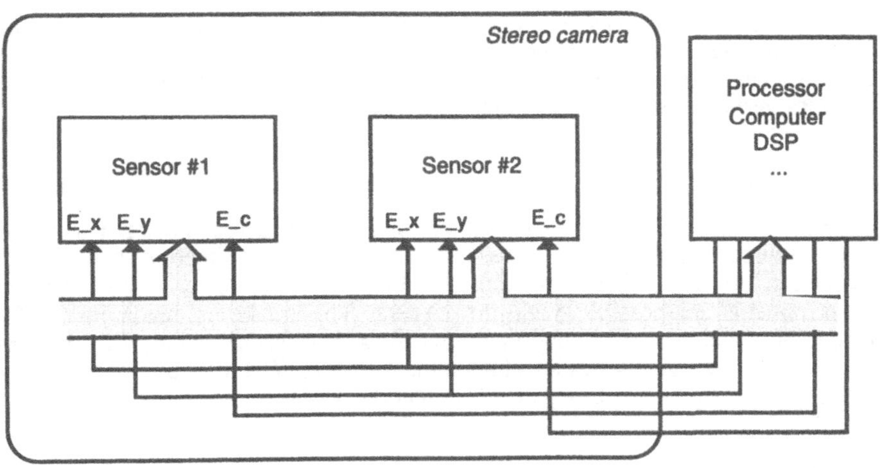

Figure 6 : Architecture of the stereo camera and possible connection to a host computer

3.3 Stereo Matching Algorithms

The algorithm for stereo vision is based on correlations to find a good matching between the two stereo images. We have done experiments with the camera, based on the following algorithm.

The algorithm is based first on the pixel geometry. A Gaussian filtering must be performed in the vertical (y) direction. The first or second derivative of the gaussian filter is used in the x-direction to extract the edges in the imaging scene. Then, the displacement ΔX is calculated where the correlation is maximal.

$$\underset{\Delta x}{\operatorname{argmin}} \sum_{\Delta x} \operatorname{abs}\left[G'_x * G_y * I_1(x,t) - G'_x * G_y * I_2(x+\Delta x,t)\right]$$

In the actual implementation, the gaussian filtering G_y is approximated by the pixel shape. The first derivative of the gaussian G'_x is approximated by a sobel filter. Alternatively, a second derivative of the gaussian (G''_x) can be used, and this can be calculated by a mexican hat filter or a difference of gaussians [5].

4. Applications

There are various applications for the stereo camera. The sensor was developed as a side-development of the ESPRIT II VISTA project. In this project, an optical flow camera was build to predict the time-to-crash [6]. The approach there was with a single image sensor with a special pixel geometry. Stereo vision is an alternative to optical flow for the time-to-crash prediction. Besides that, stereo vision also gives information in stationary scenes, while optical flow needs motion to give an output.

There are a lot of applications for stereo vision in automotive applications : from airbag pretrigger systems to autonomous car park systems, from driving aid to autonomous driving. Other applications are in robotics. The distance measurements provided by the stereo camera can be used for automatic guided vehicles, used in warehouses and fabrication plants.

5. Conclusions

We developed a dedicated image sensor to be used in a stereo camera. The sensor has a sparse sampling with gaussian filtering in one direction and a high resolution in the other direction. In this way, accurate stereo distance measurements are possible with the camera. The sensor is made in CMOS technology, which made it possible to give the sensor features which are necessary in automotive applications, like a high dynamic range. The interface to the sensor is completely digital and based on a true-random access of the pixels.

6. Acknowledgement

The authors want to acknowledge the fruitful collaboration with P. Creanza and N. Ancona from CSATA Technopolis, Bari, Italy. We acknowledge all the partners in the VISTA Esprit II project for their collaboration. Guy Meynants acknowledges the support from the Flemish institute for the promotion of the scientific-technological research in industry (I.W.T.).

7. Bibliography

[1] E. Fossum : "CMOS Image Sensors: Electronic Camera-On-A-Chip", IEEE trans. on Electron Devices, vol. 44, no. 10, October 1997, pp. 1689-1699

[2] N. Ricquier, B. Dierickx, "Pixel structure with logarithmic response for intelligent and flexible imager architectures", Microelectronics Engineering vol. 19, 1992, pp. 631,

[3]N. Ancona, T. Poggio, "Optical Flow from 1D Correlation: Application to a simple Time-To-Crash Detector", proc. Fourth International Conference on Computer Vision, May 1993, pp. 209-214

[4] B. Dierickx, et al., "Random addressable active pixel image sensor", proc. SPIE vol. 2950, December 1996, pp. 2-7

[5] D. Marr, "Vision", 1982

[6] N. Ancona, et al., "A real-time miniaturised optical sensor for motion estimation and time-to-crash detection", proc. SPIE vol. 2950, December 1996, pp. 75-85

Computer Vision Suitable for Vehicle Applications

U. Regensburger, R. Ott, W. Platz, G. Wanielik, N. Appenrodt
Daimler-Benz AG, Research and Technology, 70546 Stuttgart, Germany

Abstract. This paper presents hard- and software components for robust environment acquisition and interpretation for vehicle applications. In particular a high dynamic range CMOS image sensor was designed. For the consecutive processing, analogue image preprocessing units were developed. In addition an image generating radar was realised and evaluated. It is based on the frequency of 77GHz. The above introduced sensors serve as a basis for further environment analysis. Different approaches for object detection and tracking based on radar, grey-scale and colour images are presented. All these methods were developed with a potential analogue implementation in mind. A fusion of video and radar information was performed to increase the reliability of object recognition. The feasibility of the vision sensor suitable for vehicle applications was successfully demonstrated within the application 'truck platooning'.

1. Introduction

A mass market for driver assistance systems is expected in the near future. Those systems are designed to increase the driving comfort or safety. They either warn the driver of dangerous situations or support him during monotonous driving tasks.

The feasibility of driver assistance systems was already demonstrated by the fully autonomous driving testcar VITA II [1] which was developed within the EUREKA project PROMETHEUS. One key-component of such a system is a computer vision platform providing environmental information. State-of-the-art vision systems for vehicle applications lack the necessary compactness and robustness, additionally they are too expensive. Especially CCD cameras, e.g. do not match the temperature range requirements, which can occur inside the car behind the windshield.

The challenge is to close this technology gap. Therefore a BMBF-funded project 'electronic eye' focused on developing computer vision systems and corresponding algorithms suitable for vehicle applications. To adapt the sensor system to the requirements of vehicle applications an early feedback during tests in real traffic scenarios is necessary. This is done within the application „truck platooning".

For safety reasons the driver assistance systems have to detect all relevant objects. Therefore an imaging radar provides additional information about the environment. Furthermore the fusion of vision and radar sensor signals improve the robustness of the 'electronic eye'.

2. Image Sensor and Analogue Image Processing

A camera was designed and constructed to demonstrate the feasibility of an edge detection and image restoration process using recursive analogue networks. The advantage of the analogue technique is that image preprocessing, e.g. edge detection or signal smoothing, is performed simultaneously in the image in real time.

The camera includes three analogue components (fig. 2.1):

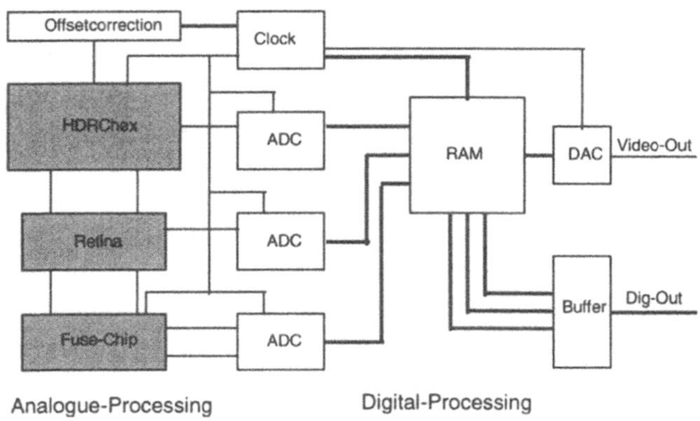

Fig 2.1. Structure of the camera. The digital components generate a standard video and a digital interface.

- **HDRC:** a high dynamic CMOS imager designed by IMS-chips, Stuttgart. The imager has 384*256 pixels arranged in a hexagonal grid. The hexagonal raster was chosen to get a maximum resolution with a minimum of pixels and to support the analogue image processing in the following steps. The logarithmic characteristics allow the operating range of the imager to cover 6 decades. The fixed pattern noise of the imager was reduced by an one point correction.
- **Retina:** the second chip is designed like a 'Mahowald-Retina' and performs a local signal adaption. This step is necessary to match the wide signal range of the imager (up to 1000 noise equivalent signal steps) to the limited input range (about 50 noise equivalent steps) of the following edge extraction chip. The function is achieved in the following way: The output is the difference of the input signal and the local average of the signal formed by a resistive network with a coherence length of a few pixels. It is important to note that the signals are limited by tanh-amplifiers. In this way, the contrast is enhanced, while the information about the absolute lightness is lost. Furthermore, in the input signal the influences of object-reflectivity and spatially uniform illumination are

additively separated by the logarithmic characteristics of the imager. Therefore, the retina output signals depend only on the reflectivity steps at the edges of the objects. Of course, nonuniform illumination (shadows, lights etc.) can lead to additional output signals. The number of pixels of the retina is 144*80. In the demonstrator camera, only a central field of 128*64 pixels was processed to exclude edge effects.

- **Fuse:** in the third chip, edge detection and signal smoothing is performed simultaneously by a resistive grid equipped with fuses [2] [3], i.e. the smoothing operation is interrupted by a switch whenever an edge is sensed by a threshold detection circuit operating on the signal difference of neighbouring pixels. Thus, a mutual stabilisation of the two processes is achieved: around very short edges, the smoothing can reduce the contrast below the detection threshold whereas short missing pieces in long edges can be closed because here the contrast is enhanced by the smoothing which tends to equalise the different signals on each side of the edge.

Fig. 2.2 shows the central field of a HDRC image and the corresponding retina output. Furthermore the edges and the smoothed output of the fuse-chip are presented. The processing was done on the same central field of 128*64 pixels.

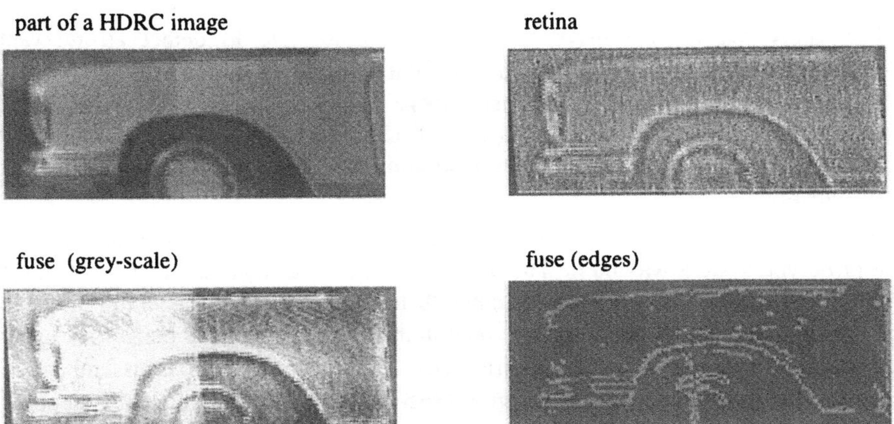

Fig. 2.2. Grey-scale HDRC image and corresponding outputs of retina and fuse. The retina-chip performs a local signal adaption. The contrast of the output image is enhanced, while the information about the absolute lightness is lost. The fuse-chip performs a smoothing operation which is interrupted at edges.

In addition, two fuse chips were combined by a special wiring technique to a single unit of 160*144 pixels. Processed images are shown in fig. 2.3.

Fig. 2.3. Input image (left) and processed images of a combined fuse chip with 144*160 pixels: smoothed image (middle) and edges (right) superimposed with the smoothed image (reduced gray scale)

As a result, the feasibility of an analogue edge extraction process has been demonstrated. To meet the requirements for applications, however, an architecture to cover lager fields must be developed. Furthermore, the fixed pattern noise of the chips, especially of the retina, has to be reduced.

3. Image Generating Radar

For safety reasons the driver assistance systems have to detect all relevant obstacles with a high reliability independent of weather conditions. Decreasing prices for millimeter wave components make radars operating in these frequency ranges a very interesting sensor solution. Due to the short wavelength, even real aperture imaging with high lateral resolution becomes possible with compact systems[4].

Within the 'electronic eye' project an imaging radar approach was investigated and a prototype system was built. The antenna of the radar consists of a focal plane array and a lens to reduce the beamwidth to three degree in elevation and one degree in azimuth. In this prototype version the image scanning is performed using a motor driven mirror for both directions. The radar frontend is assembled in GaAs-MMIC technology. The imaging sensor is realised as a frequency stepped, pulse Doppler radar. With this multi-mode radar a high range resolution of 0.6 meter as well as a Doppler resolution of 1.7 km/h can be achieved.

The measured data set acquired with the imaging radar system represents a three dimensional spatial description of the traffic scene and is called RDC (Radar Data Cube) (fig. 3.1). For further image processing steps the RDC is reduced to a plain image matrix by generating the top projection. This bird view of the underlying traffic scene is transformed to its perspective projection in fig. 3.2. The top projection is best suited for easy scene interpretation. Several adapted signal processing algorithms are used to perform tasks like object segmentation and classification, roadside extraction and object tracking (fig. 3.3)[5].

For the analysis of traffic scenarios the following radar data characteristics can be utilised:

- The discrimination between asphalt and the roadside is possible due to the very high reflectivity of asphalt.
- The width and length of a car can be obtained.
- Even a vehicle occluded by another vehicle can be detected!

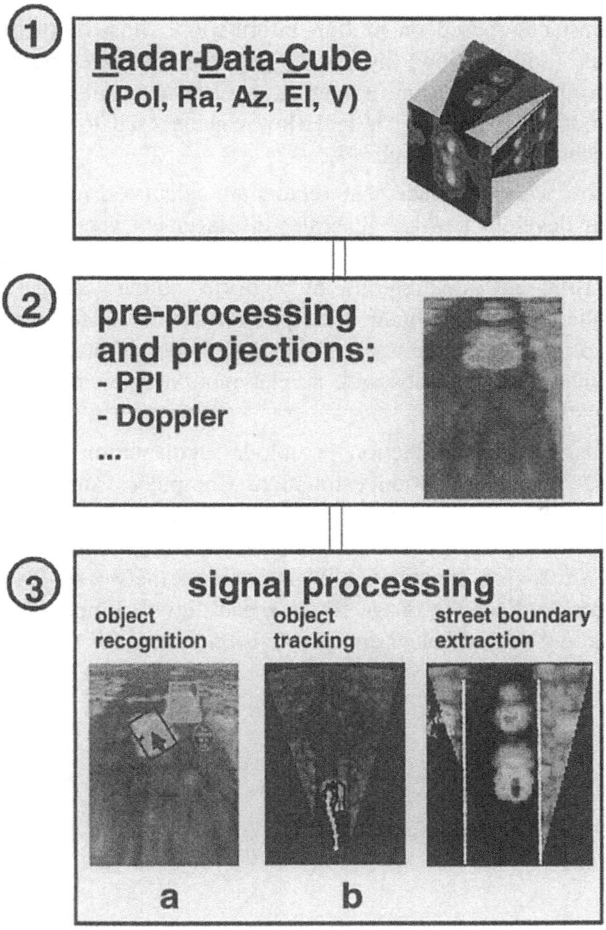

Fig. 3.1. RDC - a 3-D description of a traffic scene

Fig. 3.2. Data reduction to a 2-D image (perspective projection)

Fig. 3.3. Signal processing results of a traffic scene

4. Grey-scale and Colour Based Object Detection and Recognition Approaches

Several approaches have been realised for monocular vision based systems within the 'electronic eye' project.

One group of methods uses a priori knowledge of the obstacle's appearance in the 2-D image plane. They determine the location of certain constellations of image features [6] which are characteristic for given classes of relevant objects (i.e. cars, trucks, motorcycles). These algorithms work well in scenarios where the shapes and the object classes are strongly restricted (i.e. interstate scenario).

Other approaches detect obstacles based on motion information. Most of them estimate the image motion (optical flow) by calculating the spatio-temporal gradient of the image brightness function. Given a calibrated camera and knowledge about the egomotion of the car, optical flow can be used to detect objects on the road independently of their shapes [7].

In contrast to the optical flow where displacement vectors are calculated for each pixel, a method was further developed which generates displacement vectors for colour regions [8]. The first step of this approach is a colour segmentation where each single image is partitioned into regions of uniform colour. Then a connectivity analysis is applied to get a symbolic description of the colour regions. Based on this description colour regions are tracked over a sequence of images to obtain their trajectories. Finally colour blobs with similar motion behaviour are combined into object hypotheses.

The approach performs an early data reduction in colour segmentation which significantly speeds up the following motion estimation. Computer simulation investigations show that a robust segmentation technique can be performed based on relaxation techniques which are computational extensive on digital hardware. These techniques are well suited for implementation in analogue networks. The combination of fast analogue modules for image filtering and digital components for object recognition leads to a very efficient computer vision system for vehicle applications.

Fig. 4.1. Results of object detection. Dark lines represent bounding boxes of the object hypotheses and trajectories of the colour regions.

Some results of the colour based object detection approach are shown in Fig. 4.1. The method successfully detects various types of moving objects (i.e. cars and motorcycles). To detect non-rigid objects we modified our approach and used clustering techniques in the colour segmentation [9]. As shown in Fig. 4.2 moving pedestrians can be detected.

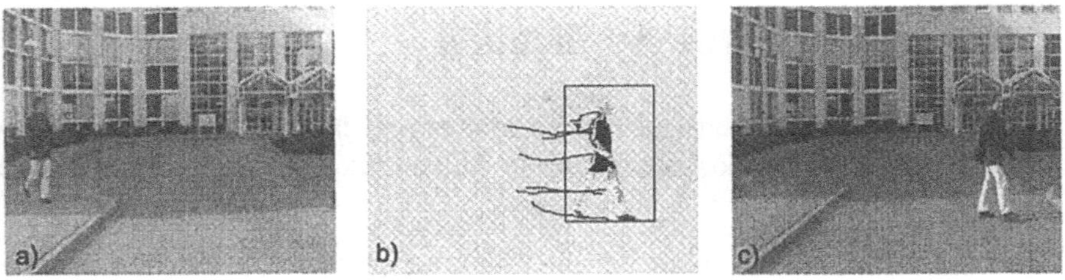

Fig. 4.2. Results of pedestrian detection using clustering techniques. Images (a) and (c) are the first and last image of the original sequence. Image (b) shows the detected pedestrian consisting of several colour clusters.

5. Sensor Fusion Based on Vision and Radar

Fig. 5.2. Complex hull of transformed radar data

A system enhancement with respect to robustness can be realised applying fusion of vision and radar. In safety critical applications a certain degree of redundancy is necessary. By fusing visual and millimeter wave data one can make use of the complementary strengths of these sensors. The vision sensor has a very high spatial resolution, while radar is capable to measure the range and velocity of an object easily. Moving areas in the radar image can be located and transformed into the camera image. These moving areas are represented as groups of voxels in 3-D space (RDC). In the camera image the corresponding convex hull forms a 2-D region, which can be labelled with the affiliated radar features like range and velocity (fig. 5.2). To obtain precise information about the shape and type of the detected moving object, the corresponding area must only be analysed in the video image. So the fusion technique combines the advantages of radar and image processing. A prototype system based on this sensor fusion approach was realised in an simulation environment. The results are very promising.

6. Vehicle Integration and Evaluation of the Vision Sensor within the Application

In order to develop a robust perception system it is not enough to test it in the laboratory. It is of uttermost importance to built a prototype which runs in the real scenario.

The first developed vision component within the project was the high dynamic HDRC vision sensor. A prototype camera with digital output was realised. Due to its high dynamic sensor, no shutter control is necessary. This was extensively tested in high dynamic situation such as tunnel entrances and exits.

Fig. 6.1 shows two grey-scale images of a tunnel entrance. They are obtained from a HDRC and CCD respectively. In both grey-scale images the area inside the tunnel is not visible due to the low dynamic of the print medium itself. In order to illustrate that the high luminance dynamic of the origin scene is still covered by the HDRC image, a simple edge detector is applied on both images. The obtained results are placed on the right column of fig. 6.1. No information inside the tunnel is available in the CCD edge. However, in the HDRC image all relevant edges, e.g. the road lanes, are visible.

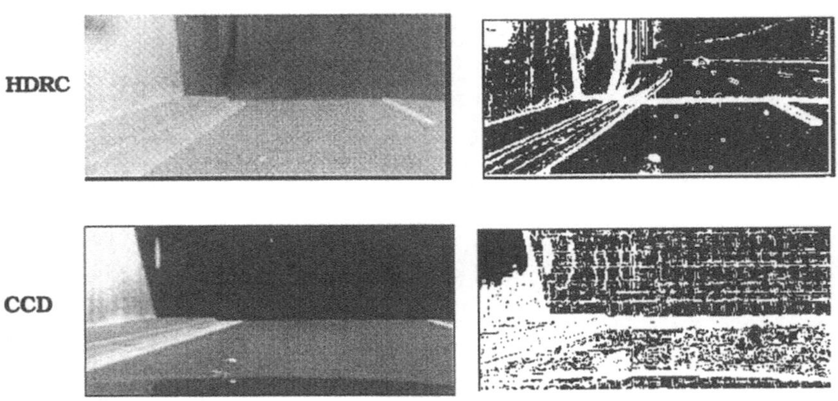

Fig. 6.1. View of a tunnel entrance. The grey-scale images were obtained from HDRC and CCD sensors (left column). The corresponding edge images are shown in the right column.

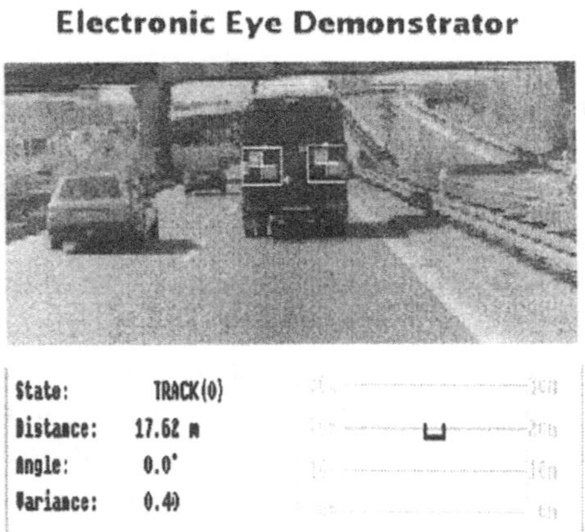

Fig. 6.2. Image taken from the vision system. Ahead the leading truck with marked pattern and its relative position to the ego-vehicle is shown.

The truck platooning application was used for further tests of the camera. This tow bar system, consisting of two trucks was developed in a previous project. The leading truck is equipped with three infrared lights at the rear. They are detected and tracked with a CCD camera system which is installed in the succeeding truck. The idea is to use the HDRC camera and replace the infrared lights with significant chequered pattern, as shown in fig. 6.2. The longitudinal and lateral control is based on the distance and the azimuth angle between both trucks. Their values are calculated from the position of the two pattern centers measured by the image processing. Fig. 6.3 and fig. 6.4 respectively show the calculated distance and the relative velocity to the leading truck. This data was recorded during a real test-drive.

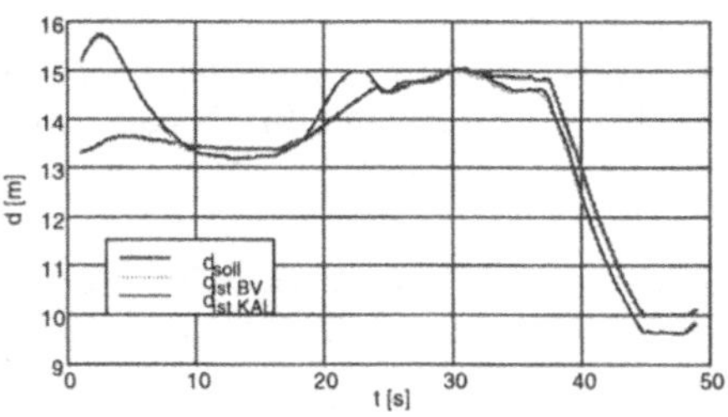

Fig. 6.3. Distance to the leading truck (measured value from the image and reference value).

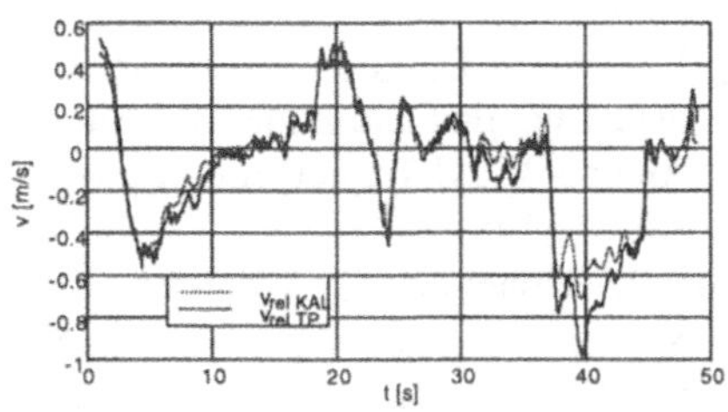

Fig. 6.4. Relative velocity (measured values which are low pass or Kalman filtered).

The vision sensor worked well during various test-drives. It provides image data of the leading truck even under low lightning condition. This allows a precise measurement of the pattern. Nevertheless, there are some problems left:

- Scattered light and inner lens reflections pose a problem. This results in a low signal to noise ratio in dark image areas.
- The resolution of the imager has to be increased.

7. Conclusion

The technical feasibility of both, a high dynamic imager and an analogue preprocessing, was demonstrated. The goal for the future will be to have a compact vision system with analogue processing capabilities. This allows obstacle detection and recognition based on the combination of analogue and digital image processing.

A first step in this direction was done testing the HDRC imager in real traffic scenarios within an application. The results are promising. A CMOS imager matches the temperature range requirements for vehicle application. Furthermore, the CMOS process is a standard technology and the designed chips can in principal be produced all over the world. Due to the fact of falling hardware prices, CMOS vision imager will become cheaper and appear on the market with a high variety. The main task will be to take care off having CMOS imagers not only satisfying the multi media market but also being suitable for vehicle applications.

The imaging radar has to reduce its size and price to be a second perception sensor suitable for vehicle applications.

8. References

[1] B. Ulmer: Vita II - Active collision avoidance in real traffic. IEEE Symposium on Intelligent Vehicles, pp. 1-6, Paris, 1994.

[2] A. Lumsdaine, J. Wyatt and I. Elfadel: Nonlinear analog networks for image smoothing and segmentation. J. of VLSI Signal Processing 3), pp. 53-68, 1990

[3] J. Harris, C. Koch, J. Luo and J. Wyatt: Resistive fuses: Analog hardware for detecting discontinuities in early vision. In: C. Mead and M. Ismail (eds): Analog VLSI Implementation of Neural Systems, pp. 188-202, Norwell, MA, Kluwer 1989.

[4] R. Schneider, G. Wanielik, S. Bhagavathula: A Polarimetric Measurement System for Real Aperture Imaging. Proceedings of the Third International Workshop on Radar Polarimetry, No. 1, pp. 341 - 347, 1995.

[5] R. Schneider, H. Neef, G. Wanielik, N. Appenrodt, J. Wenger: Polarimetric Imaging for Traffic Scene Interpretation. IEEE Symposium on Smart Systems for Transportation, Delft, 1997.

[6] M. Schwarzinger, T. Zielke, D. Noll, M. Brauckmann, W. von Seelen: Vision-Based Car-Following: Detection, Tracking, and Identification. IEEE Symposium on Intelligent Vehicles, pp. 24 - 29, Detroit, 1992.

[7] D. Willersinn, W. Enkelmann: Robust obstacle detection and tracking by motion analysis. IEEE Symposium on Intelligent Transp. Systems, Boston, 1997

[8] B. Heisele, W. Ritter: Obstacle detection based on color blob flow. IEEE Symposium on Intelligent Vehicles, pp. 282 - 286, Detroit, 1995.

[9] B. Heisele: Motion-based object detection and tracking in color image sequences. Submitted to 5th Europ. Conf. on Computer Vision, Frankfurt, 1998.

[10] U. Franke et al.: Truck Platooning in Mixed Traffic. IEEE Symposium on Intelligent Vehicles, pp. 1-6, Detroit, 1995.

Optical Detector System for Water, Sludge and Ice

T. W. Huth-Fehre, F. Kowol and M. Bläsner

Institute for Chemical and Biochemical Sensor Research, Mendelstr. 7,
D-48149 Münster, Germany

1 Introduction

Precise knowledge of the current road conditions concerning water, sludge and ice would improve overall vehicle safety considerably. Furthermore it will improve the performance of several systems as ABS, anti skid (ASC) and vehicle dynamics control (VDC). For adaptive cruise control (ACC) it even might turn out to be crucial.
First attempts to detect water and ice spectroscopically can be traced back for twenty years when a first patent [1] was issued containing the basic idea to spectroscopically measure the shift in vibrational energy between liquid and frozen water. At that time it was far ahead of technological possibilities and public awareness, hence for more than a decade no real attempts were made to really build such a detector.
Seven years ago during the 'Prometheus' project different ways of estimating the friction potential were severely investigated, spectroscopic measurements included[2].
This concept also failed to solve the problem, mainly because of prohibitively high prices for infrared detectors, of a sensitivity of the device against movements of the car and because it could not detect ice under thicker layers of water.
These problems are overcome in the system presented.

2 Basics

When water molecules are irradiated by light of certain wavelengths in the near infrared region of the spectrum, internal vibrations can be excited at the expense of absorbed radiation. This effect can be observed in Figure 1, where water layers on asphalt are irradiated by light from a regular halogen bulb and the amount of backscattered light is plotted versus wavelength. The broad minimum between 950 and 1050 nm, centered at 980 nm, corresponds to the waterspecific absorption. The relation between the thickness of the water layer and absorption strength is given by Beer's law, which for such weak absorbers can be approximated linearly.

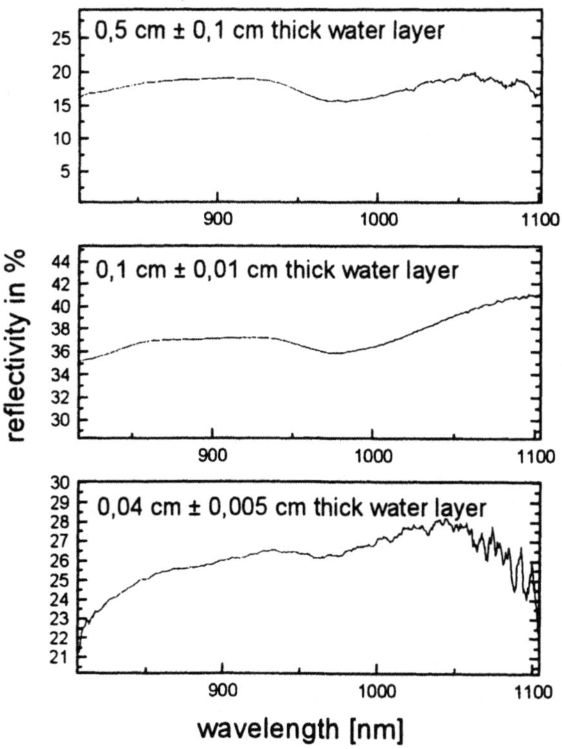

Fig. 1. Spectrally resolved reflectivity of wet asphalt.

When the liquid water forms the ice matrix, the molecules are slightly deformed, resulting in a shift of the resonant wavelength. Figure 2 shows spectra of iced asphalt, where the absorption is shifted to 1030 nm.

All overtones of the stretching vibration of the O-H bonds show a strong difference between ice and liquid water. They mainly differ in their absorption strength, which falls roughly one order of magnitude when going to the next higher harmonic. This also means the penetration depth rises reciprocally at each harmonic. Since only silicon-photodiodes have production prices that justify their use in automotive consumer products, the third harmonic (see figures 1 and 2) for the sensor to work on was chosen. The penetration depth at this wavelength is several centimetres, which means on one hand that ice can still be detected even under a puddle of water, but on the other hand for thin layers the peak absorption is only a few per thousand. Since already a layer of 0.1 mm water reduces friction

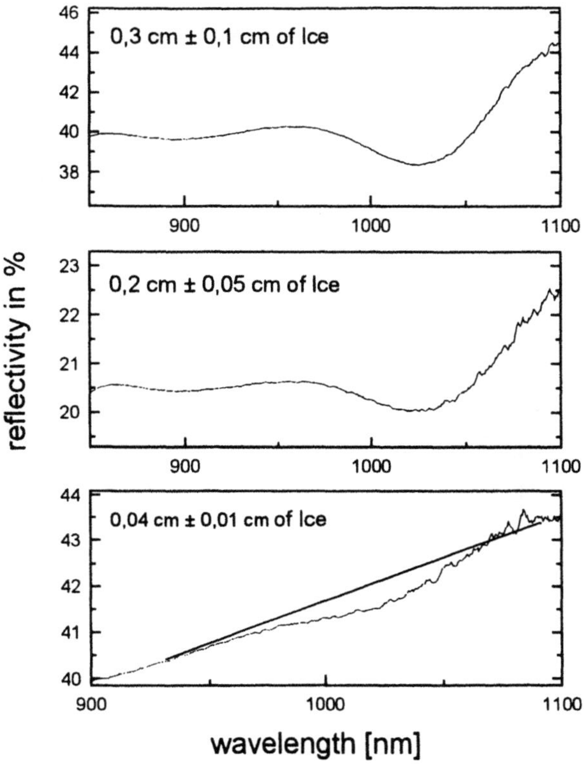

Fig. 2. Spectrally resolved reflectivity of asphalt with ice.

noticeably, the design goal was set to develop a spectrometer system with an SNR of > 1000 : 1 at a measurement time of 10 ms.
Another effect of the penetration depth are features in the spectrum caused by roughness of the underlying asphalt. To compensate for that two further wavelengths outside the resonant bands are used. By this the absorbance of ice and water is linearly corrected.

3 Experimental setup

The road surface is illuminated by a 50 W halogen lamp (Osram) from a 30 cm distance. A BK-7 lens collects the backscattered light into an optical fibre. Those two items are mounted in a water tight housing and can be affixed to the front bumper of a test car. The fibre leads the light into the spectrometer system inside the car. Here it gets equally distributed into four channels in each of which it passes an interference band pass filter and is converted into a photocurrent by solar blind photodiodes. These currents are converted into voltages by gated integrating preamplifiers, again amplified and finally digitised. Figure 3 shows the principle setup:

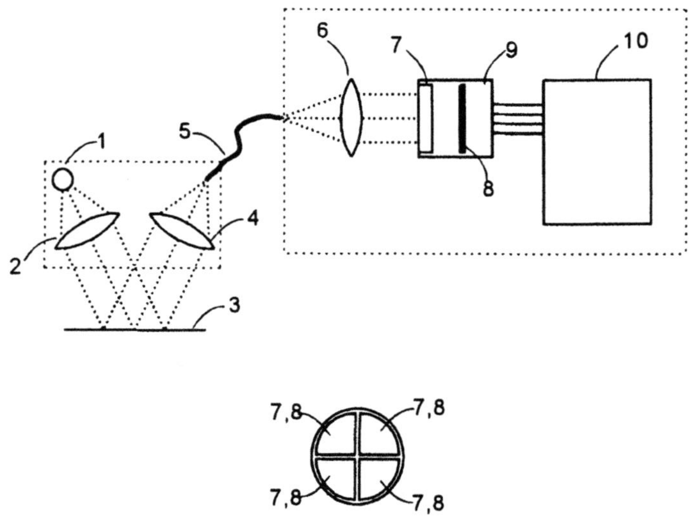

Fig. 3. Principle set-up of the sensor system:
1) lamp; 2) optional condensor lens; 3) road surface; 4) collecting lens; 5) light fibre; 6) beam shaping lens; 7) optional window; 8) filters; 9) four photodiodes (one for ice, one for water, two for linear correction); 10) preamplifiers.

The final conversion of these four raw signals into layer thicknesses of ice and water is performed by software on a standard laptop computer.

4 First test results

Several laboratory set-ups and test drives were done to prove the capacities of the prototype, and to test the sufficient resolution in the 100 μm range.
First a laboratory test sequence was started. Therefore, defined amounts of water were measured by the system to assure the height calibration. Figure 4 shows the water height in mm drawn as a time line (in arbitrary units)
Then the system's capacity of detecting ice and water in snow was demonstrated under laboratory conditions. Several snowflakes, each of different height, were tested. Figure 5 shows a sequence of different flakes. Here again thickness is drawn against an arbitrary time line. The rising portion of molten water in the snow is caused by the test conditions under room temperature.

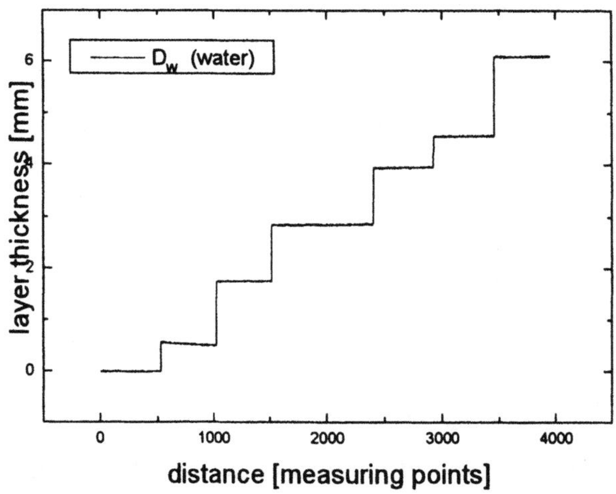

Fig. 4. Laboratory test with water. The system was calibrated by filling certain amounts of water in glass.

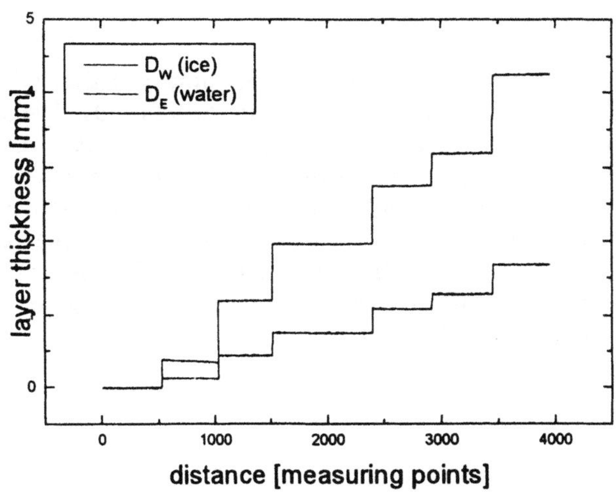

Fig. 5. Laboratory test with snow. Different parts of melting snow were placed below the optical head to test the thickness of frozen and liquid water.

At last a test drive with the prototype fixed to a car was arranged. The optical head was mounted in front of the bumper, while the spectroscopic part with the laptop was put on the front seat for operation. Then a stretch of road was prepared with three ice fields, and the car drove along at 50 km/h while measurements were made. Figure 6 shows the measured sequence, the x-axis represents the length of the test course where about 100 measurements per second were taken. Because of medium external temperatures at that time the sequence shows a 2 mm water film on the ice spots.

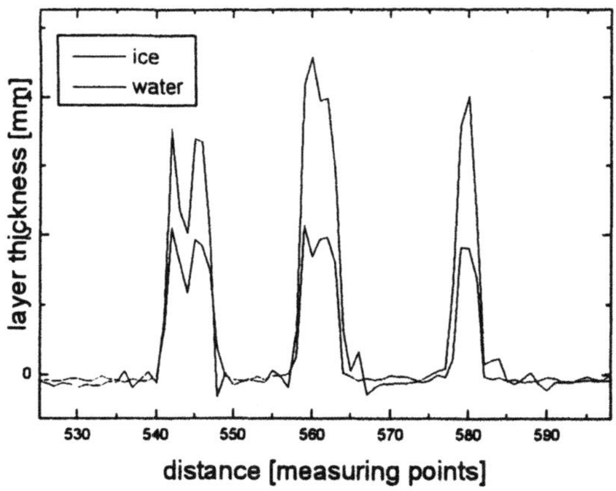

Fig. 6. Test drive with mounted system. Three different ice spots were passed after each other. The x-axis shows arbitrary length units.

5 Conclusion

A fibreoptic filter-spectrometer with a software algorithm can determine the height of ice and water without contact and within 10 milliseconds. The system also allows the conclusion about the roughness of the roadway. By this development it is possible to improve the performance of driving systems like ABS, ACC or ASC. Because of its simple construction and low-cost parts the system is suitable for mass production at a reasonable price. With its inexpensive silicon detectors and interference filters that can be micromachined into an ASIC the development presents a small and compact design which can be easily integrated in today's automotive technology.

6 Literature

1: Peter Decker, German Patent office DE 2712199, 1977

2 F. Holzwarth and U. Eichhorn, Sensors and Actuators A, Vol 37/38, 121-127 (1993)

Engine Management System for Direct Injection Gasoline Engine

Hideyuki Tamura[1]

[1] Manager
Engine Engineering Section No.4
Engine Engineering Department No.2
Powertrain Engineering Division
Nissan Motor Co., Ltd.
6-1, Daikoku-cho, Tsurumi-ku, Yokohama-city, Japan 230

1 Introduction

The need to protect the earth's environment has created a strong trend to improve fuel economy in automobiles. A promising technique to increase engine output while improving fuel economy in gasoline engines is the use of direct injection technology. Due to lean combustion , direct injection engines have less pumping and heat losses. This increases the engine's efficiency and thus improves the fuel economy. The use of direct injection technology also makes it possible to increase the engine's specific power output. This can be achieved, because the low charge temperature improves both volumetric and anti-knocking performance.

This paper describes;

1) the fuel control system needed for direct injection,
2) NTD(Nissan Torque Demand) control system, which maintains torque smoothly and effectively at the required level depending on the engine's operating conditions,
3) the controller technology required to realize both fuel and torque control.

2 System Drawing

Figure 2-1 shows an example control system for a direct injection gasoline engine. This type of engine typically has the following characteristics;

1) High pressure fuel technology
   - a high pressure fuel pump
   - high pressure fuel injectors
   - high pressure fuel pipings
2) Combustion control technology
   - independent intake ports
   - a swirl control valve
   - a high energy ignition coil
3) Air control technology
   - an electronically controlled throttle system
4) The controller technology
   - a *high performance* engine control module

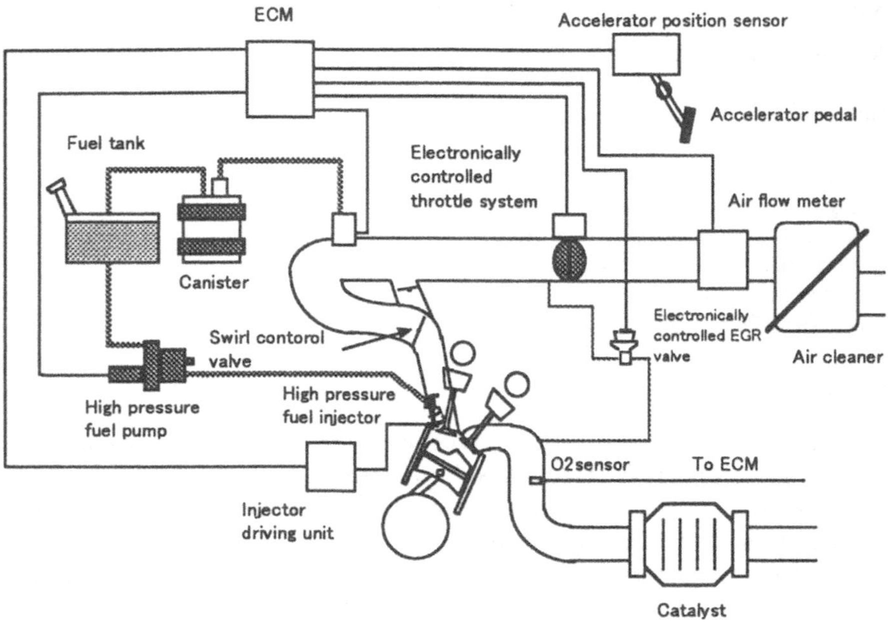

Fig. 2-1. The control system for a direct injection gasoline engine

3 Type of Combustion

The direct injection gasoline engine has two types of combustion.

One is stratified combustion. Under stratified combustion, the injection timing is comparatively close to ignition timing and combustible air-fuel mixture will be formed around the spark plug, though the mean mixture ratio of combustion chamber is lean.

From above mentioned, the energy lost by pumping loss in same work is diminished compared with conventional gasoline engine because more air is inhaled to cylinders and the pressure drop of intake air is decreased. As a result, the fuel consumption rate is reduced. The reduction of the fuel consumption rate is supported by other factors ; the decrease of the rate of the specific heat of working gas and heat loss to cylinder wall. Stratified combustion contributes the decrease of CO_2 emission and other components due to the reduction of the fuel consumption.

The other type of combustion is homogeneous combustion. Under this combustion, the fuel is injected at intake stroke and intake air to cylinders and fuel spray are mixed.

This homogeneous combustion is the same combustion type of conventional gasoline engine. Under this combustion, combustible working gas is filled in the combustion chamber. This means the more quantity of combustible fuel is introduced into the combustion chamber. So that if high power performance is required, this combustion is carried out.

4 Fuel Supply System

To achieve stable stratified combustion at lean air-fuel ratio mixture, following measurements were taken:

1) The fuel injection pressure is between 5 and 9MPa. This high pressure has been achieved by using a high pressure fuel pump.(Fig.4-1)
2) The accurate fuel injection volume and timing were achieved by using a highly responsive fuel injector.(Fig.4-2)

The fuel is pumped from the fuel tank to the engine with an electric fuel pump inside the tank, and the high fuel pressure is obtained and delivered to each cylinder by of a mechanical pump attached to the engine's camshaft. A pressure regulator incorporating a linear solenoid keeps the fuel pressure at the optimum level according to the engine's operating condition.(Fig.4-3.)

Since the Nissan Direct Injection concept implies an engine which is both powerful and environmentally friendly, it was necessary to employ a means of maintaining high power performance during homogeneous combustion while simultaneously utilizing stratified combustion to improve fuel economy. This is accomplished by aerodynamic straight ports and shallow bowl pistons.

Prior stratified combustion observation results have made it clear that , with shallow bowl pistons, it was necessary to use "casting net injectors" to ensure a stable combustion. These "casting net injectors" have biased injection pattern and speed providing a uniform distribution of the fuel mist above the piston.

Fig. 4-1. High Pressure Fuel Pump

Fig. 4-2. High Pressure Fuel Injector

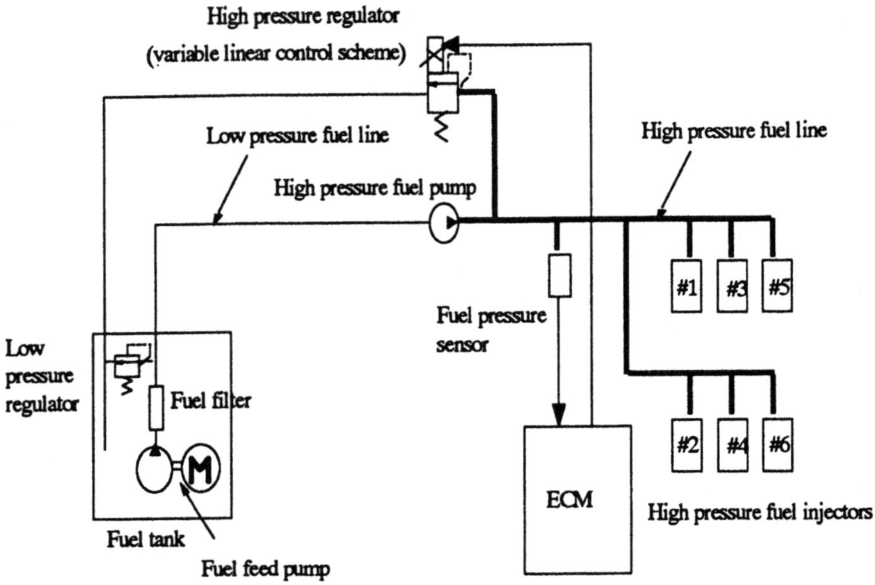

Fig.4-3. Fuel System

5 Variable Fuel Pressure Control

Optimum fuel pressure differs depending upon the engine's operating condition. Furthermore, the allowable deviation of the fuel pressure is low.

This is the reason to apply a variable pressure regulator of the linear type and fuel pressure sensor system.

A fuel pressure sensor is used to provide feedback of the fuel pressure. This is needed to keep the pressure at the required precise level. Thus, the fuel pressure of our direct injection gasoline engine is controlled between 5 and 9MPa though that of conventional port injection engine is only 0.3MPa.

Table 5-1. Direct Injection -Port Injection Comparison

	Port Injection	Direct injection
Injector Location	Intake port	Combustion chamber
Fuel Pressure	0.3MPa	5~9MPa
Fuel particle diameter	200 μ m	25 μ m

6 Fuel Injection Control

The air-fuel ratio is controlled by an O2 sensor with adoptive control strategy. This solution has been chosen instead of a wide-range air /fuel(A/F)sensor.

The system perceives the values corresponding to improper air-fuel ratio during lambda control at homogeneous stoichiometric operation, and uses these values for lean-burn operation.

This makes it possible to control air-fuel ratio both accurately and cost-effectively. The system also includes a fail-safe control, which is activated when the air-fuel ratio deviates widely from the target value, for example as the result of some component failure. This system monitors the engine stability by extracting the specified frequency component of engine speed fluctuations. If the engine stability becomes too low, the system goes to homogeneous stoichiometric operation.

Table 6-1. Combustion Configuration Types

	Homogeneous combustion	Stratified combustion
Fuel injection timing	Intake	Compression
Fuel density inside combustion chamber	Uniform	Rich only near spark plugs
Maximum leanburn A/F ratio	23 ～ 25	40 or more

7 NTD (Nissan Torque Demand) Control

For a direct injection gasoline engine to achieve both high power and environmental friendliness, it must be able to maintain a satisfactory balance between stratified combustion, which achieves high fuel economy, and homogeneous combustion, which achieves high power. To achieve this goal, the system incorporates the NTD control system which enables a "peaceful coexistence" between stratified and homogeneous combustion. (Fig.7-1.)

The scheme enables engine-generated torque to be controlled by setting target torque and controlling the torque to remain within this target torque with a high degree of precision.

The control scheme determines the intention of the driver by monitoring the operation of the accelerator pedal, and then calculates the target engine output torque value using also other variety of information sources. This includes the vehicle speed, the engine speed, and the intake air volume. It then selects the optimum air-fuel ratio needed to achieve the target torque and determines the throttle position. It measures the intake mass air flow induced through throttle valve, and controls the fuel injection amount, fuel injection timing, and ignition timing. (Fig.7-2.)

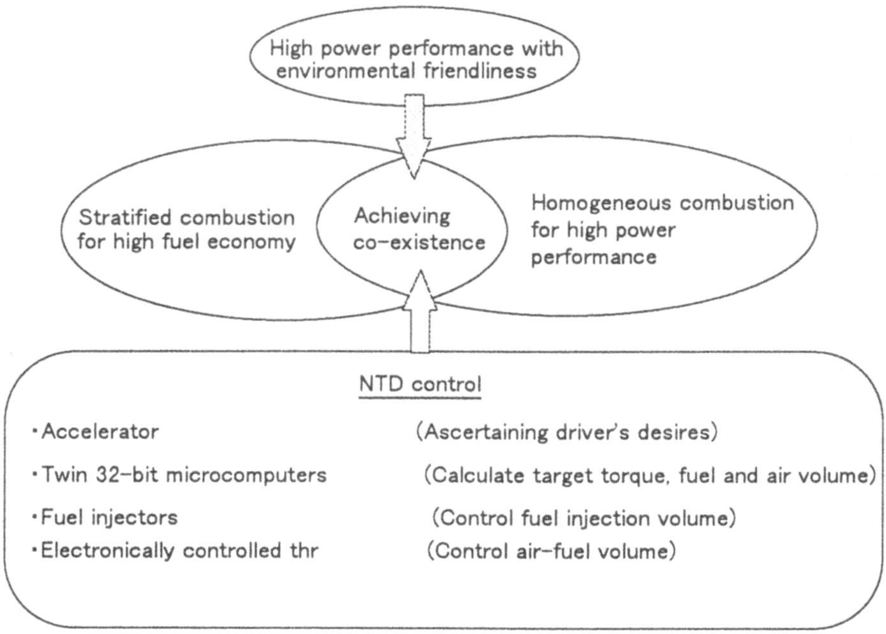

Fig. 7-1. Nissan Torque Demand Control

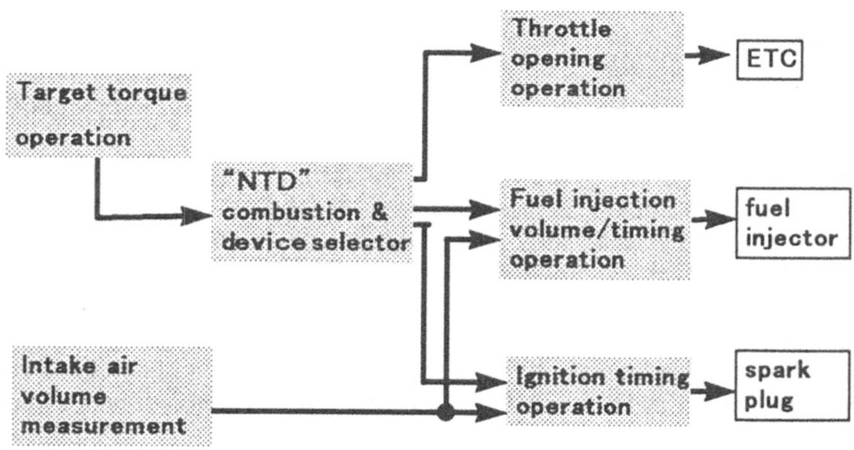

Fig. 7-2. Basic NTD Control Procedure

The operation of this system can be explained concretely in a case where the driver is pushing the accelerator slowly down to the floor. In this case the target torque increase. When the target torque becomes such a level where the combustion has to be switched from stratified to homogeneous, this system can switch the combustion while making sure that there are no discontinuities in the engine torque. Because the air amount of homogeneous combustion is smaller than that of stratified one to generate the same torque , the system close the throttle valve quickly, at the same time the injection timing , the fuel amount and the ignition timing are also controlled precisely. Also for the case of switching from homogeneous to stratified combustion the same strategy is applied. Thus this system can achieve a smooth feeling without torque discontinuities.(Fig.7-3.)

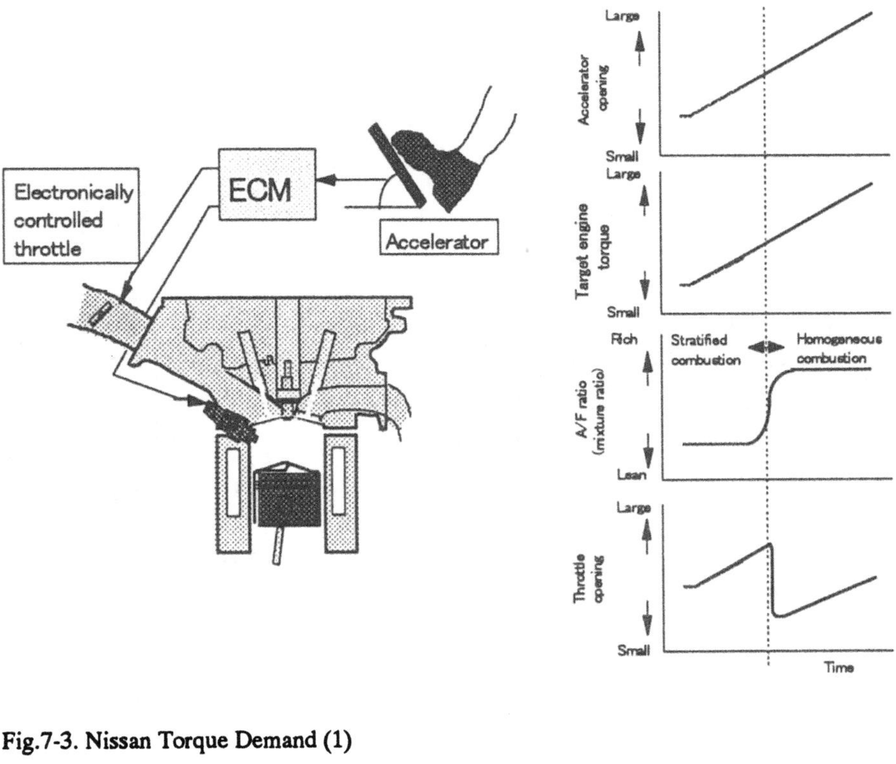

Fig.7-3. Nissan Torque Demand (1)

The results obtained in switching from stratified to homogeneous combustion under idling conditions in an actual vehicle test are shown in Fig.7-4.

Under idling conditions engine revolution fluctuation can occur easily by torque fluctuation. Therefore it is particularly difficult to maintain a steady engine speed at the time of changing from stratified to homogeneous combustion or vice versa .However as the data indicate, in the test there was only a small fluctuation of the engine speed.

Further improvement in fuel economy can be obtained by combining NTD control with CVT. For example, if the accelerator pedal is pressed to achieve fast acceleration, the engine operating condition changes from A to B as shown in the figure. If a controlled CVT is not used , a shift from high-fuel economy stratified combustion to high-torque homogeneous combustion is required.

If, on the other hand, CVT control is used, the operating area changes from A to Ć in the figure, due to the change in CVT transmission ratio. As a result, there is a way to increase the operating area where stratified combustion is possible, thus enabling an improvement in fuel economy(Fig.7-5.)

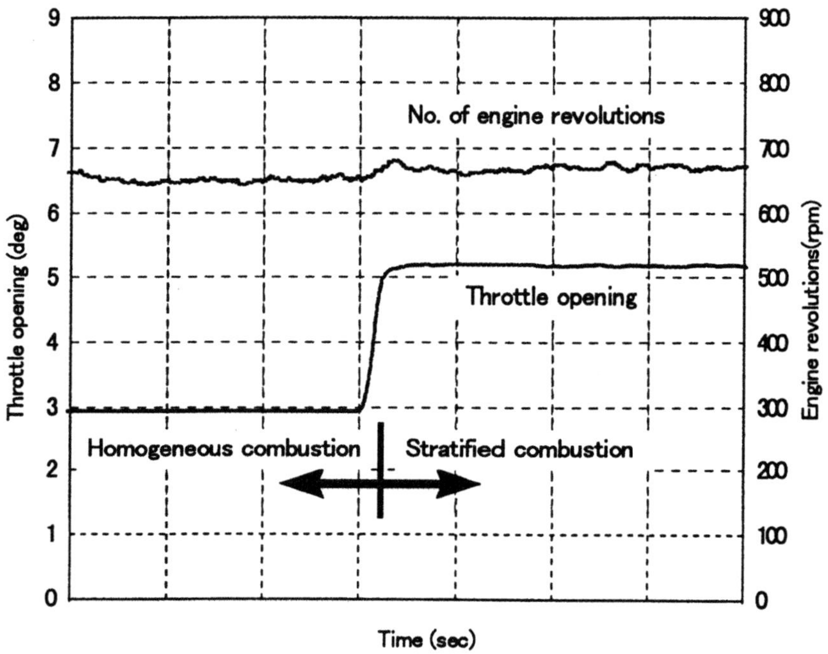

Fig. 7-4 Homogeneous combustion → Stratified combustion switch (idling)

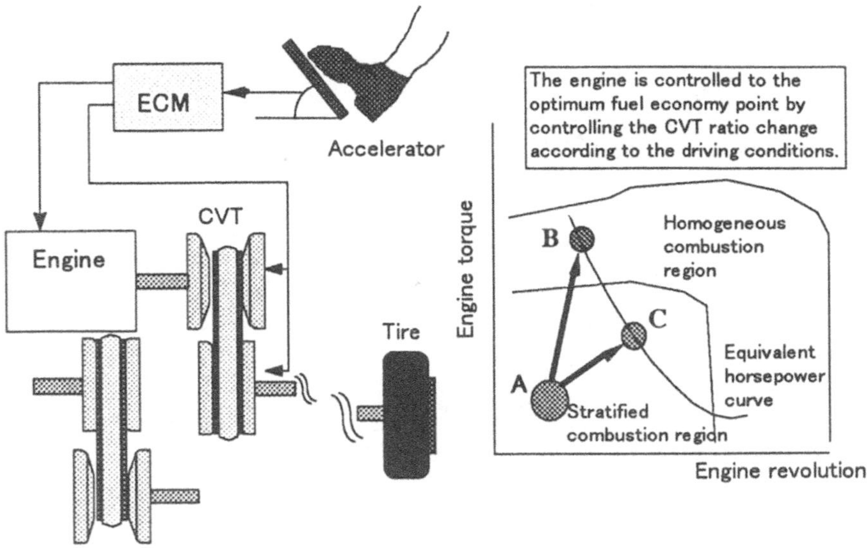

Fig. 7-5 Nissan Torque Demand (2)

8 ETC (Electronically Controlled Throttle Chamber) System

The ETC system consists of the following components (Fig.8-1.), (Fig.8-3.):

- Accelerator position sensor : Linked to accelerator pedal, it detects the driver's desired torque.
- Throttle position sensor : detects the actual degree of throttle-opening
- Throttle motor : a DC motor for driving the throttle valve.
- Throttle valve, throttle valve return spring, gear, etc.
- ETC control logic (ECM)

(The ETC system eliminates the need for the conventional auxiliary air control valve for correcting air volume during idling)

The NTD control system calculated the target throttle opening. The current of the throttle motor is feedback controlled, so that the target throttle opening matches the actual throttle opening.

The ETC system also has to be robust, highly responsive, and have high operational precision (resolution). To fulfill these requirements, the Nissan ETC system employs "Robust Model Matching Control", a modern control logic . This enables both the system's responsiveness and its operational precision to be maintained at a high level, and minimizes the effect on system performance of variations in battery voltage, temperature, and manifold pressure. (Fig.8-2.)

Finally, the system enables both smooth setting off and acceleration performance to be obtained through an additional control scheme which corrects the throttle opening characteristic relative to the accelerator pedal input.

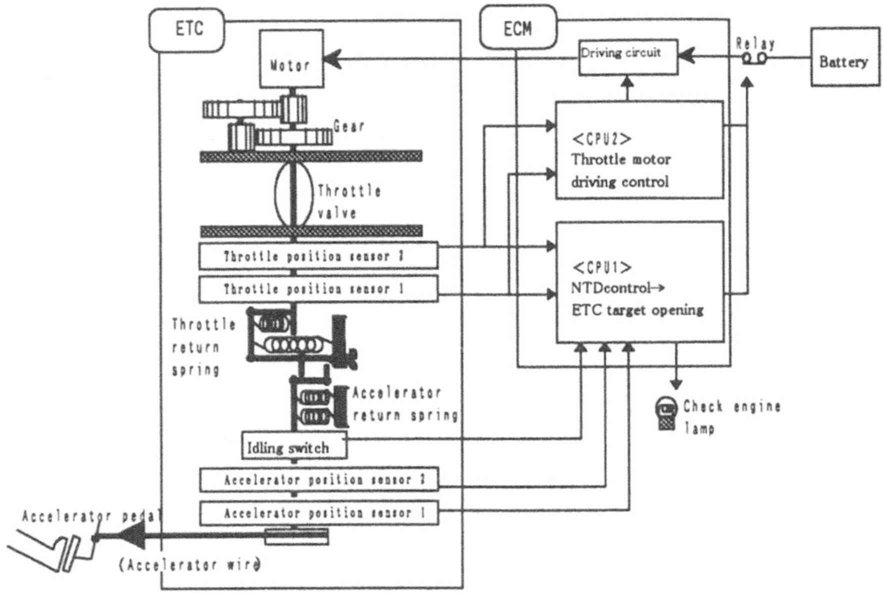

Fig. 8-1. ETC System Configuration

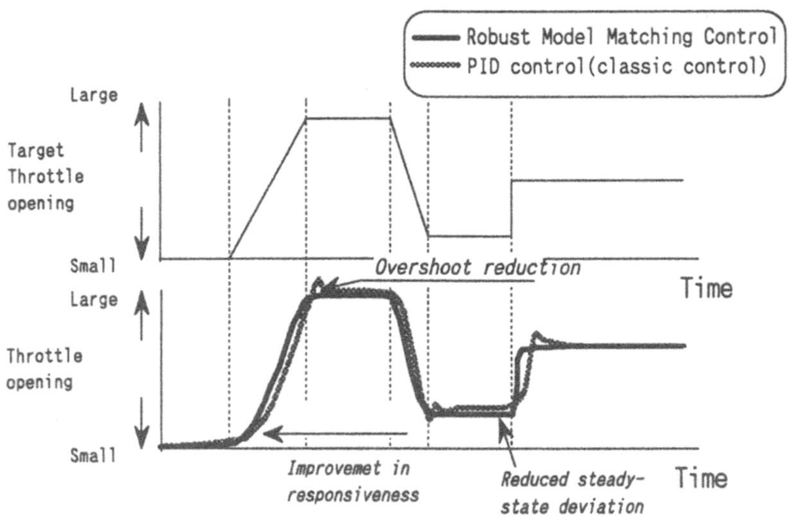

Fig.8-2 Effect of using Robust Model Matching Control

Fig. 8-3 ETC

9 Fail-Save-System

The fail-safe system includes a function to prevent incorrect operating and a function to warn the driver when a failure occurs. The use of the fail-safe system ensures that the same degree of reliability as that of a conventional mechanical throttle system is maintained.

The principal structural features of the system are as follows:

1) Two CPUs monitoring each other and each CPU is monitored by a watchdog timer.
2) Dual input lines from the accelerator position sensors.
3) Dual input lines from the throttle position sensors.
4) A mechanism which monitors the consistency of the target throttle opening with the actual one.

The main functions are as follows:

1) Enabling limp-home function with keeping the electronically controlled throttle as default position by shutting off the throttle motor operation when an abnormality of the electronically controlled throttle is detected.
2) Ensuring the motor operation is shut off through a duplicate safety system: by cutting off the main power relay and the driving circuit power transistor.
3) Limp-home mode being engaged until the key is turned off.
4) Activation of the engine check lamp in the instrument panel to warn the driver in the case of a malfunction.

Table 9-1. Fail Safe

Malfunction component		Abnormality monitoring	Action when abnormality detected
CPU	CPU operation	Monitoring via watchdog timer	1)motor stopped ; limp-home running begins 2)Duplicate motor stopping mechanism ・Turn off driving circuit main power relay ・Turn off driving circuit power transistor 3)Limp-home running continues until key is turned off
		Mutual monitoring of CPU	
Sensor	Accelerator opening sensor	Duplicate mutual monitoring	
Actuator	Throttle opening sensor	Duplicate mutual monitoring	
	Electronically Controlled Throttle System	Monitoring consistency between target throttle opening and actual throttle opening	

10 ECM (Engine Control Module) Construction

A new-generation VP1D-type ECM was developed to obtain the control required for direct-injection engines. Since a greater degree of control will be required for the engines of the future, including direct-injection engines, we used new technologies to achieve significantly improved operating performance and development efficiency in this VP1D-type ECM. (Fig.10-1.)

Nissan adopted many model based controls for direct injection gasoline engines which make, very high precision control possible. Each ECM includes two high-performance 32-bit RISC microcomputers to ensure the required engine and fail-safe control. In addition, C language is now used as the programming language to make the model control and improve the software development efficiency.

A new operating system was also developed. This operating system separates input/output jobs and application jobs. The operating system was designed to be both applicable to program module design and to be capable of preventing timing bugs.

The result of these new developments is an ECM structure that makes it possible to achieve very complex and sophisticated engine control.

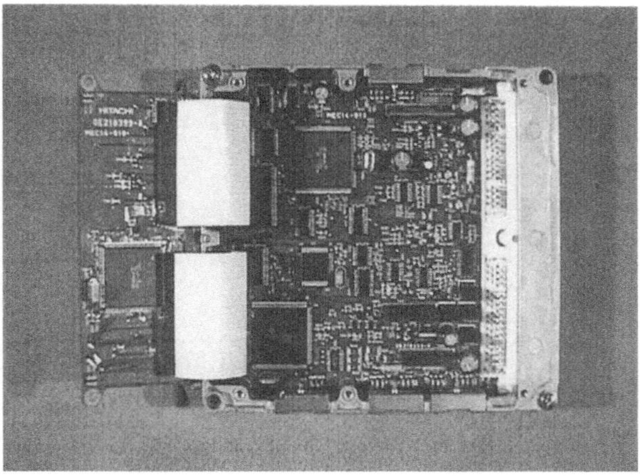

Fig.10-1. ECM

11 Real Time Simulator Outline

When developing the new large-scale control system, one of the key question is the efficiency to develop a control logic for the system. Our target was to improve the efficiency of the development of the control logic. We achieved this by extensively using CAE. The CAE system consisted mainly of a desk-top simulation carried out with an engine simulator.

To do this, we modified existing engine simulators by incorporating the models for the additional system. This enabled us to develop software and test this software on the simulator. The simulator output is equivalent to the output of an actual engine. This means that we did not have to wait until the actual engine was completed.

Through the use of a hardware-in-the-loop system in which the simulator is connected to the actual control unit, we were able, without waiting for the actual engine to be completed, to achieve simultaneous development in which trial production software was operated to achieve results equivalent to those which would be obtained with a connection established to an actual engine.

Results: With the real-time simulator, we were able to verify the simulated operation of the created software under the same conditions as would occur in testing an actual engine. This was very useful in helping to identify timing related bugs caused by the dynamic behavior of the engine. Those bugs could be easily overlooked when using the simple debugging procedure we previously used.
This made it possible to deal with bugs before the actual-engine test. The actual engine testing could be done with software almost free of bugs.

Moreover, the results obtained were very effective in providing information under conditions difficult to reproduce with actual engines, for example fail-safe feature with some broken components. This represents a major improvement in efficiency of the development of control systems as a whole. (Fig.11-1.)

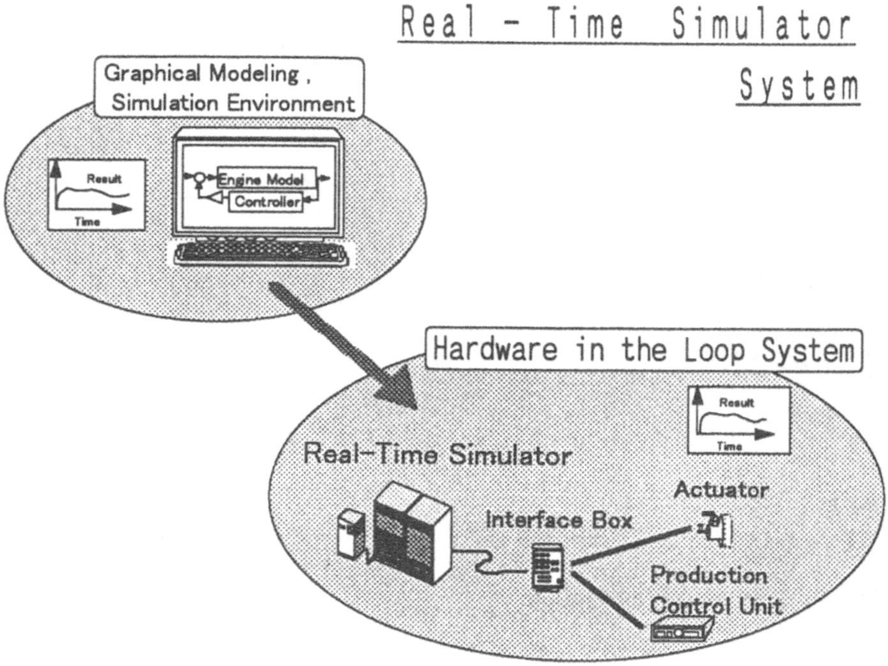

Fig.11-1. Real Time Simulator outline

12 Summary

Gasoline engines with superior fuel economy, power, and drivability can now be developed through the use of a control scheme suited to direct-injection fuel control systems, NTD control, and a new type of controller. However, since direct-injection gasoline engines are still expensive to produce, ways need to be found to reduce the cost of fuel system parts and controllers for these engines. The increasing trend to strengthen exhaust emissions regulations requires that the ways of improving fuel injection control as well as improved catalysts are required.

Integration of Electronic Control Units in Modern Transmission Systems

R. Ingenbleek, G. Birkenmaier, and G. Horsak

ZF Friedrichshafen AG, Corporate Research and Development, Electronics Development Department (TE), 88038 Friedrichshafen, Germany

Abstract. This paper intends to highlight the difficulties and marginal conditions that have to be considered, when electronic control units are to be integrated into a modern transmission system. It discusses the environmental conditions inside an automatic transmission and shows how they, along with other aspects, influence the decision of where to locate a mechatronic module. The paper also illustrates an application of modern assembly and connection technologies for integrating sensors and actuators into a module on a macro level. The implementation of a testing and calibration procedure into the production process is outlined. Finally, the need for new development techniques and for new forms of cooperation is explained.

Keywords. Automatic Transmission, Electronic Control Unit, Integration, System Performance, Sensors, Actuators.

1 Introduction

Increasing demands on automatic transmission systems – with regard to power density, efficiency, safety, reliability, and service life – call for integrated systems. Combining the Electronic Control Unit (ECU) with sensors and actuators and mounting such a module directly onto the hydraulic valve housing of the transmission control, enables complete testing and thus an improvement in system performance. At the same time, costs can be reduced by decreasing the number of components and additional logistical advantages. On the other hand, particular efforts are required to meet the new challenges presented by spatial narrowness and stringent environmental conditions.

Section 2 discusses the problematic nature of integrating standard electronics in modern transmission systems with respect to increased environmental demands. In Section 3, different concepts for integrated systems are described and an outlook on further advantages gained by system calibration is given. Since the development of a mechatronic module cannot be realized by a subsequent design of components, new development techniques have to be applied (Section 4). The Conclusion in section 5 summarizes the main results.

2 Environmental Limits for Standard Electronics

Electronic Control Units (ECU) as specified in today's applications, are located inside the passenger cabin where harmful environmental influences are excluded to a great extent. For a secure operation it is sufficient to specify the maximum ambient temperature. There are no special requirements concerning tightness and resistance against media. This is almost true when the ECU is placed inside the engine compartment.

Today's ZF-transmissions comprise an external stand-alone ECU and an integrated hydraulic control unit with sensors, actuators, and wiring harness. The central transmission connector is the interface between internal and external devices. The ECU, located within the engine compartment, is plugged with the central transmission connector and the position switch separately via the vehicle's external wiring harness (Fig. 2.1).

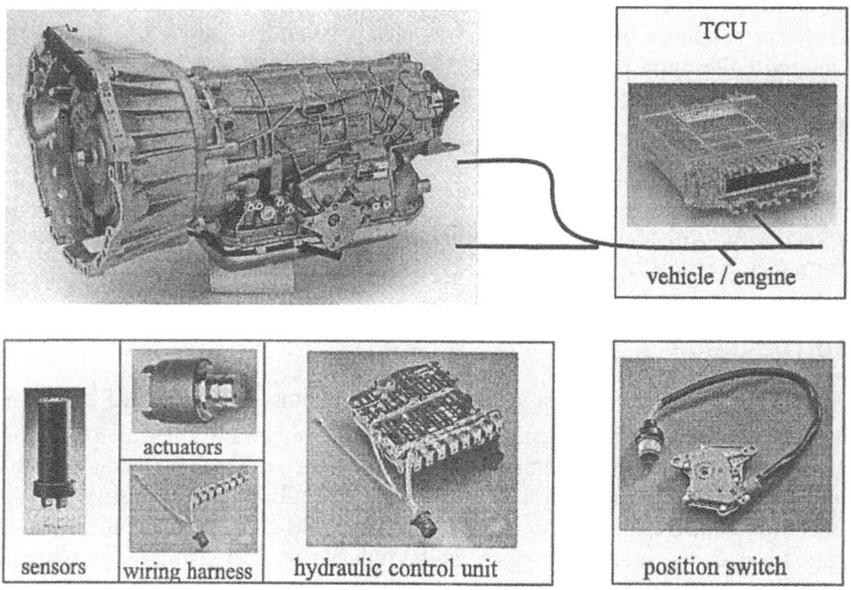

Figure 2.1. Today's Standard Configuration

In contrast to these standard applications, the integration of an electronic module into the transmission is a great challenge both for the system supplier and for the development partners.

Due to extreme temperature peaks and gradients, it is not longer sufficient to specify the technical operation by means of constant temperature levels. Even the worst case cannot simply be characterized by the maximum values. Talking about an operating range of -40°C up to +150°C, standard electronic components are not suitable for a long-term exposure at the extreme values. In view of these difficulties, frequency and duration of the environmental stress has to be taken into account. The conditions of heat transfer become very important and need to be optimized in order to implement standard electronics into the transmission. This leads to a characterization by means of temperature profiles instead of fixed limit values.

To achieve this new quality of specification, temperature curves were recorded at different locations inside the transmission, i. e. the inlet and outlet temperatures of the torque converter and the gearbox oil cooler, and the oil sump temperature. These measurements gave information about the heat distribution inside the transmission. An important result is that there is no specific low temperature area, when different duty cycles are taken into account. High load operating modes of the vehicle like fast up-hill driving, stop-and-go in emergency mode or stall speed yield maximum temperatures in the torque converter. This, in return, affects the temperature level of the oil cooler such that single temperature peaks on its outlet port exceed the oil sump temperature level. It is worth mentioning that this, of course, depends on the duration of the stress since stabilizing temperature may not be reached for each sub-system.

This confirms the fact that constant threshold values are not suitable for specifying the operating range. Furthermore, the duration of the stress modes have to be determined and applied to generalized duty cycles. With respect to the ECU, it is important to know whether low temperature levels with extreme peaks or slow temperature changes with higher average values are preferable.

Besides the problem of heat, it is very important to protect the ECU from the surrounding media. Low viscosity of the transmission fluid together with temperature changes, and thus pressure differences in an encapsulated air volume, require special precautions to ensure the tightness of the module. Passing through the electric terminals from the ECU to peripheral devices is a particularly critical task. While a welded metal housing with glass sealing increases the costs of the packaging, ZF proposes special sealing technology in combination with an ECU plastic housing.

3 System Integration and Calibration

System Integration

Different kinds of integration concepts are shown in Figure 3.1. ZF decided to realize the mechatronic solution, which integrates the ECU, sensors, actuators and wiring technology into a functional unit. This module should be mounted directly onto the hydraulic control unit inside the transmission.

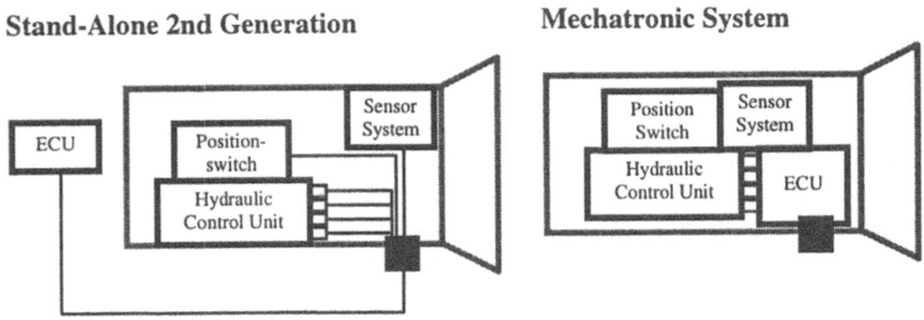

Figure 3.1. Concepts for System Integration

Figure 3.2 shows different levels of modularization. While highly integrated systems offer optimum cost reduction, modular systems present the potential for standardization over different applications.

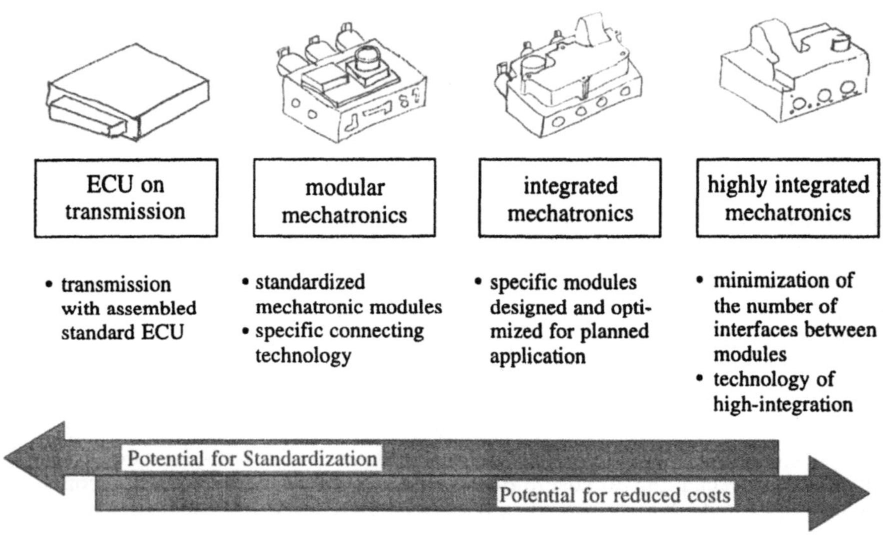

Figure 3.2. Integration Levels for Mechatronic Modules

The decision, as to which level of integration should be realized depends on the production volumes planned.

ZF proposes to place the module above the oil level. This guarantees slow temperature changes without extreme peaks (see section 2). For cooling, the heat transfer between the junction and the environment is optimized by using bare chips, directly bonded onto the microhybrid in combination with a metallic contact to the hydraulic control unit, which itself is placed within the oil sump. This approach ensures a minimum temperature difference of 10K in the entire range of operation. As was previously assumed, measurements have shown that for 80% of the operation, the temperature remains at 90°C. The bare chip technology also contributes to a reduction of installation space.

In order to minimize development risks, ZF intended to use modified standard components for the first application. For integrating sensors and actuators into the module, new connection technologies, as shown in figure 3.3, had to be developed.

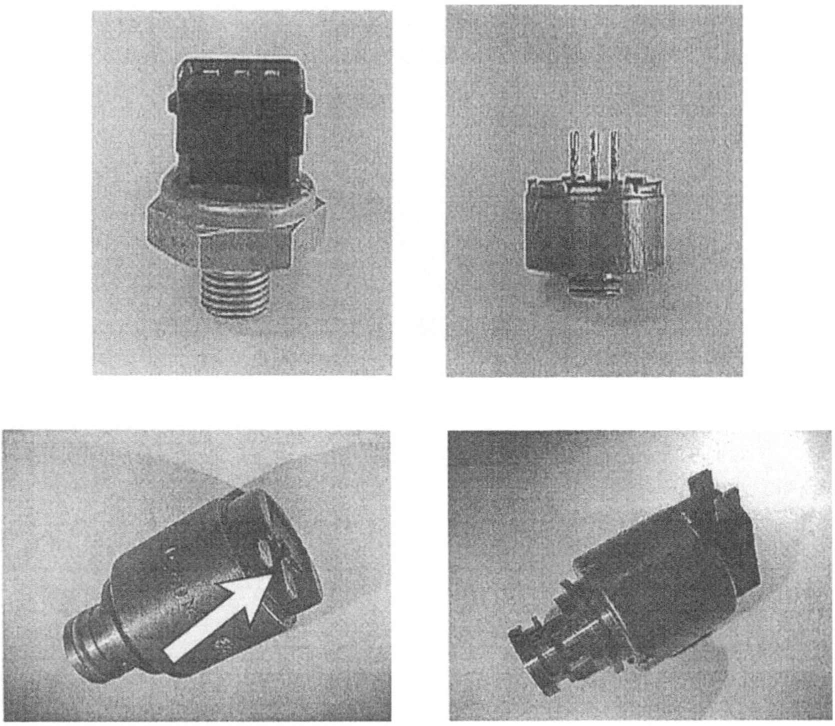

Figure 3.3. Comparison of Conventional and Module-Specific Components

The new terminal design enables direct connection to a circuit board manufactured on the basis of insert mold technology, instead of a wiring harness with single sockets. The advantages for the system supplier result from a simplified assembly and the reduction of separable components. The plastic carrier is also part of the electronics housing.

The cost reduction for the OEM mainly arises from logistical advantages and a reduction of the wiring harness on the vehicle side.

Calibration

Due to logistics, conventional transmission controls are unable to assign the ECU to a specific hydraulic control unit. Thus, system functionality has to be guaranteed for all tolerance conditions since there is no feedback of the system performance in the production process. This restricts tolerance limits of the sub-systems and increases costs since each component has to be as good as possible.

Integrating both electronic and hydraulic sub-systems to a module allows the calibration at the end of the production line. This enables wider tolerance ranges. As a consequence, components need only to be as precise as necessary to ensure system performance. This again, calls for new development techniques because the component itself may not represent a self-contained function unit. New testing facilities in simulation and on the test bench have to be used in order to give proof of a system-oriented specification (Deiss and Krimmel, 1997).

During the calibration process, characteristic curves are determined for each pressure control chain and stored in the ECU memory as shown in figure 3.4.

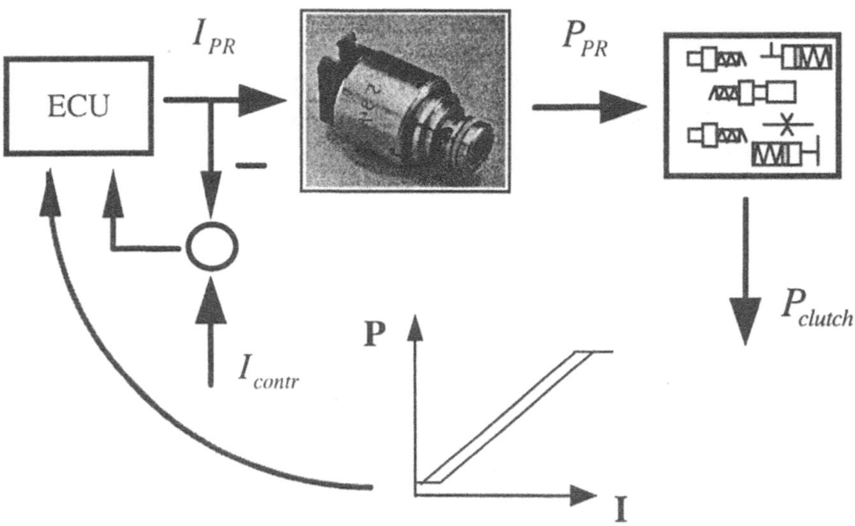

Figure 3.4. System Calibration

The whole procedure must be included in the test cycle at the end of the production line, which is one of the future tasks.

Summarizing the experience ZF has made during the development of the module, it should be mentioned that coping with the challenging demands is only possible when new structures concerning the cooperation of all development partners are used. This is described in more detail in the following section.

4 Application of Modern Development Techniques

The development of mechatronic systems for transmissions requires new forms of interdisciplinary teamwork involving the customer, the system-supplier and the sub-suppliers (Fig. 4.1).

Because of the variety of different system interfaces (mechanical, electric, magnetic, thermal and hydraulic) only a team dedicated to simultaneous engineering is able to create a robust and efficient design with parallel development techniques instead of today's consecutive development steps. The conventional procedure guarantees system functionality by means of optimal performance of all sub-systems, which are specified by their interfaces.

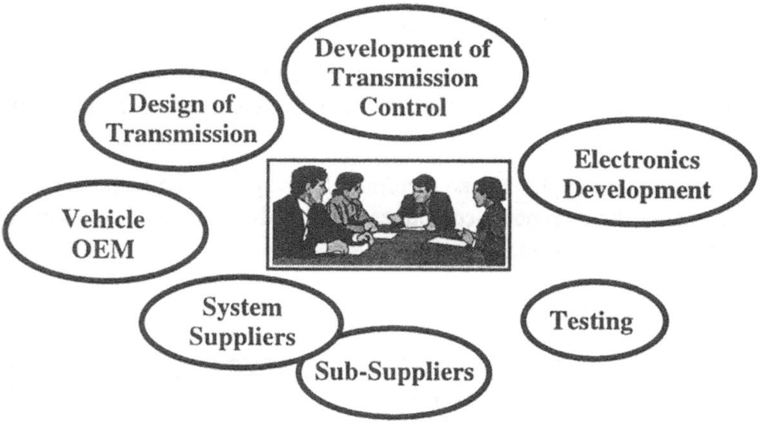

Figure 4.1. Mechatronic Development Requires Simultaneous Engineering right from the Beginning

In contrast to this, parallel development begins with an a priori specification of the system performance as an overall functionality, which is, generally, more than the sum of the sub-functions. Thus, the interplay of the sub-systems is of essential importance in the sense that sub-functions are subordinate (Fig. 4.2).

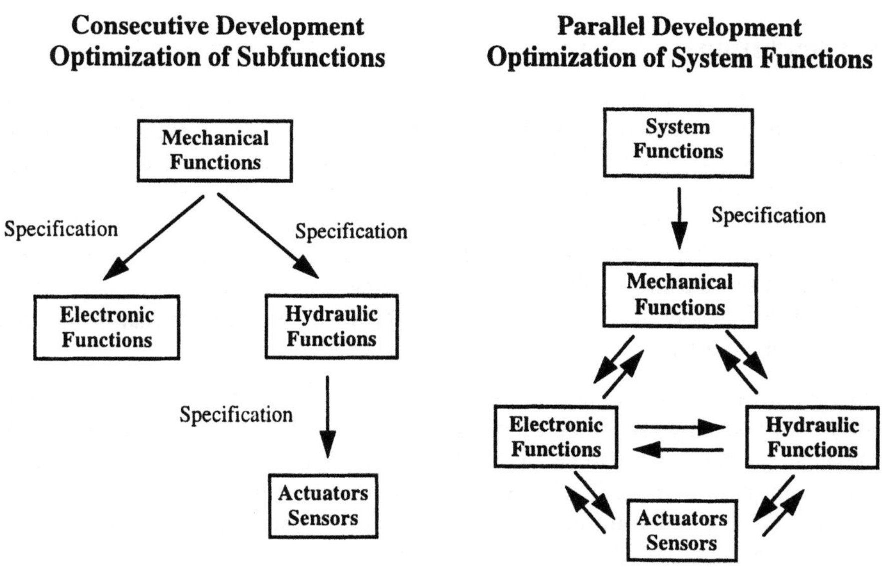

Figure 4.2. *Development Techniques*

Therefore, a sub-function is not necessarily identified by a single sub-system. For example, mechanical functions may be replaced by software; the precision of actuators can be enhanced by sensoring the desired physical quantity in combination with a closed-loop digital control.

Intensive communication within the team also enables early fault detection during the development and prevents expensive and time consuming redesigns in the sample phase.

System suppliers and sub-suppliers need a certain degree of system understanding. This means, that new development techniques require different qualifications of engineers (Fritz and Runge 1997).

Table 4.1. Qualification of Suppliers

	System / Service Concept	Application / Development Tools	Electronic HW SW Sharing Quality	Sensors / Actuators Assembly and Connection Technology
ZF	●●●	●●●	●●●	●●
System Supplier	●	●●●	●●●	●●
Sub-Supplier			●	●●●

Table 4.1 shows how special knowledge is distributed among the development partners. The sub-suppliers, in particular, contribute to a great extent to the system design. ZF, as the transmission supplier, takes on the system leadership.

5 Conclusion

Combining mechanical and electronic devices to a mechatronic system, offers commercial and functional advantages for customers and system suppliers. But on the road to success, all partners are confronted with a variety of new challenging tasks concerning the integration of the sub-systems. In the case of automatic transmissions, stringent environmental conditions and lack of space are the main marginal conditions that prohibit immediate integration of standard control units. Apart from the technical solutions, new organizational structures are required in order to adapt the development process. This makes new demands on the educational qualification of development engineers.

Environmental measurements confirm that constant threshold values are not sufficient to specify the range of operation. Instead, frequency and duration of several stress modes have to be taken into account. A main task for the future is to define appropriate generalized duty cycles for testing the modules. In the ZF solution, the usage of conventional electronic components is possible by ensuring an optimized heat transfer between the electronic junction and the oil sump. Tightness against media is ensured by a plastic housing with an especially developed sealing technology.

Team work and simultaneous engineering are essential for a parallel development, which in return enables the design of a mechatronic system.

The ZF solution, so far, does not exploit the entire potential of system integration especially that offered by micro systems technology. There are several points of departure to optimize costs and functionality. Sensors and actuators, in particular, are still discrete components and not fully integrated. As a partner of the automotive industry, ZF meets the challenges and believes that technical obstacles can be overcome.

6 References

Deiss, H. and **H. Krimmel. 1997.** *Hardware-in-the-Loop Simulation der Spezifikation.* Mechatronik im Maschinen- und Fahrzeugbau. VDI-Berichte 1315, 103-116. Düsseldorf: VDI-Verlag.

Fritz, R. and **W. Runge. 1997.** *Requirement Profile for Development Engineers - Adaptation to new forms of cooperation, to new work processes and to new work content.* Symposium Control Systems for Motor Vehicle Drivelines. September 1997. Berlin.

New High-Temperature Sensors for Innovative Engine Management

G. Krötz[1], W. Wondrak[2], M. Eickhoff[1], V. Lauer[2], E. Obermeier[3] and G. Cavalloni[4]

[1] Daimler Benz AG, Research and Technology, P.O. Box 800465, 81663 München, Germany
[2] Daimler Benz AG, Research and Technology, Goldsteinstraße 235, 60528 Frankfurt, Germany
[3] Technical University of Berlin, Microsensor and Actuator Technology Center, Sekr. TIB 3.1, Gustav-Meyer-Allee 25, 13355 Berlin, Germany
[4] Kistler Instrumente AG, P.O. Box 304, 8408 Winterthur, Switzerland

Abstract. Requirements on future vehicles are reduction of energy consumption and emissions. To fulfil these, an optimization of the combustion process including all relevant parameters is necessary and therefore parameter field based control systems have to be replaced by closed loop regulating circuits. To realize the latter in series applications, new types of sensors designed for high temperatures and harsh environments have to be developed. The present paper describes the development of a combustion pressure sensor based on cubic Silicon Carbide (ß-SiC) layers on Silicon. Results of in-cylinder measurements will be given in comparison to a Kistler reference sensor. SiC technology based on the hexagonal 4H and 6H polytypes we are developing for integrated signal electronics. The latter is needed for smart sensors and independently operating modules, applied in harsh environments. JFET transistors, which are the basis for monolithic circuit design, will be described and I-V-characteristics measured at elevated temperatures up to 400°C will be given.

Keywords: Silicon Carbide, Sensor, Electronics, Harsh Environment, High Temperature

1. Introduction

Silicon based micromachined sensor devices usually cannot withstand harsh environment conditions like high temperature and chemically aggressive media. This considerably limits their use in automotive, aerospace or industrial processing applications, where difficult operation conditions are nothing unusual. More or less expensive measures have to be taken to separate the sensorchip from the surrounding atmosphere. For example pressure sensors for combustion monitoring or exhaust pipe measurements have to be equipped with expensive steel membrans and transmission elements to protect the sensing element against temperature and corrosion. The situation becomes even more complicated, where sensorfunctions require direct contact to media. This is true for mass flow sensors, gas sensors, temperature sensors, petrol and

lubricant quality sensors. One can get out of this dilemma nicely by the combination of Silicon with ß-SiC, a promising semiconductor material for harsh environment sensor applications.
Intelligent automotive systems not only have to be equipped with sensors but also with electronics processing the data gained by the sensors. To get independently working unities, electronics have to be installed directly onto the automotive modules and often has to withstand elevated temperatures. In contrast to Silicon, 4H- and 6H-SiC exhibits very favourable properties for electronic devices operating at high temperatures and is the right choice for the applications mentioned above.

2. Combustion Pressure Sensor

2.1 Sensor Relevant Properties of Cubic SiC

Table 1 shows in comparison to Silicon properties of ß-SiC taken from the literature, which are mainly important for sensor applications. The etch attack of ß-SiC both by acids and alkaline solutions is nearly not measureable, the diffusion constants are very small, the hardness and wear resistance are high. The values of heat conductivity, elastic modulus and hardness are much better than those of Silicon. Micromachined SiC membranes, for example, are very strong and in contrast to silicon are suitable for direct contact with aggressive media and for the operation at high temperatures. The large bandgap of SiC allows for operation temperatures up to 900°C. Piezoresitive sensorelements are principally expected to work up to this temperature. Fig.2.1 shows the gauge factor of mono-crystalline ß-SiC in dependence on the temperature for two different doping concentrations [1]. Sign and size of the piezo resistive effect strongly depends on the crystal orientation and the crystal quality.

Table 2.1 Sensor Relevant Data of ß-SiC

	ß-SiC	Silicon	Quote
Bandgap	2.3 eV	1.1 eV	[2,3]
Maximum Operation Temperature	873°C	300°C	[2,3]
Heat Conductivity	5 W/cm °C	1.5 W/cm °C	[2,3]
Elastic Modulus	3.5 GPa	1.7 GPa	[2,3]
Hardness	3720 kg/mm^2	1120 kg/mm^2	[4]
Etching rate in KOH, (30%, 80°C)	< 0.008 nm/min	1317 nm/min ((100)-Orientation)	
Diffusion Constant of Aluminium at 1100°C	2 10^{-22} cm^2/s	3 10^{-11} - 4 10^{-13} cm^2/s	[5]

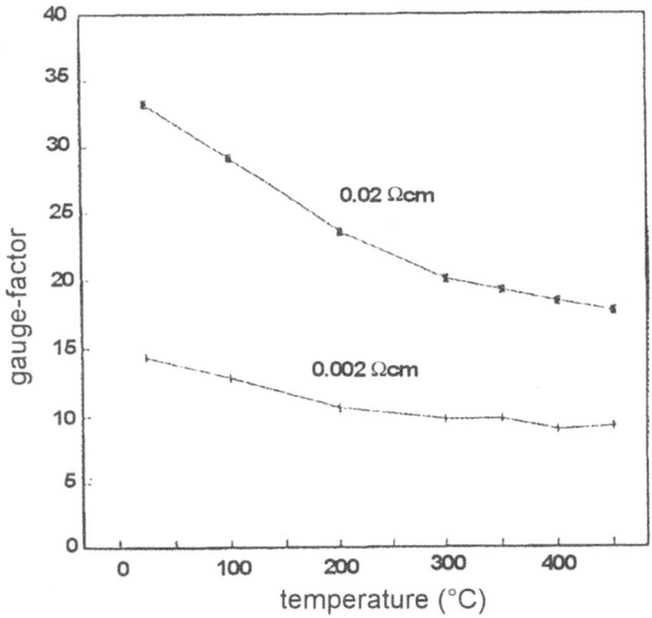

Fig.2.1 Gauge factor in dependence on the temperature [1].

2.2 Sensor and Chip Design

Fig. 2.2a showes a photograph of the sensor chip used in the combustion pressure sensor described in the paper in hand. It was micromachined in Silicon and only the piezoresistive sensorelements were made from cubic SiC. The SiC was epitaxially grown onto SOI substrates. The use of the latter was necessary in order to get a reasonable electric isolation between the SiC sensorelements and the underlying substrate. Fig. 2.2b showes a technical drawing of the housing

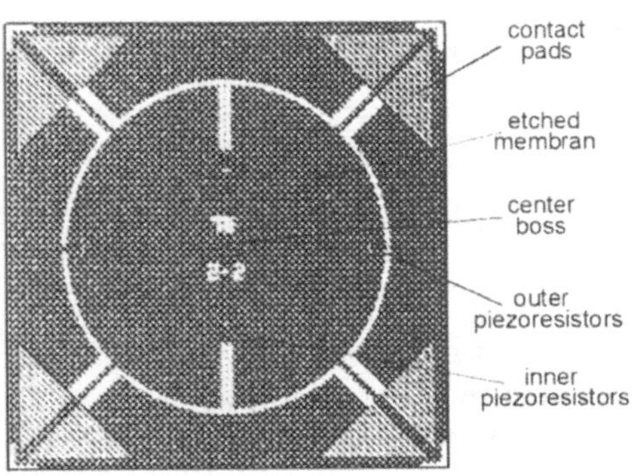

Fig.2.2a: Photograph (top view) of the sensor chip with SiC piezoresistors.

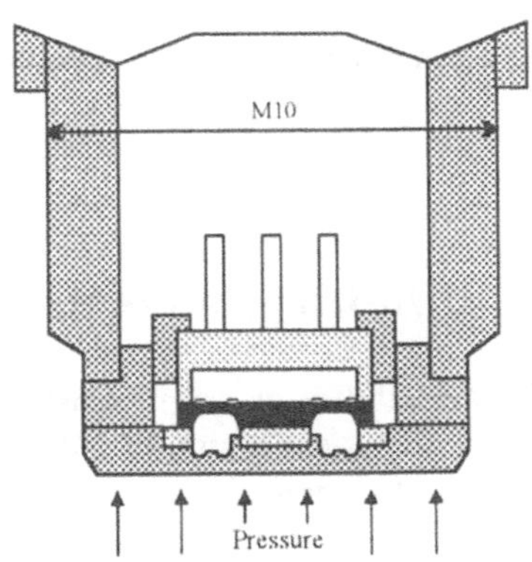

Fig.2.2b: Technical drawing of the housing of the combustion pressure sensor.

holding the sensorchip. In account of the high temperature stability of the sensor chip the housing can be quite simply constructed and no complicated thermal isolation between the sensor chip and the steel membrane is necessary. The whole sensor is described in detail in [6].

2.3 Characterisation

In Fig. 2.3 in cylinder measurements results of the SiC combustion pressure sensor in comparison to a piezoelectric Kistler reference sensor are shown. The measurements were performed in a Ford 1.4 liter gasoline engine. A quite good agreement of the two sensors was found.

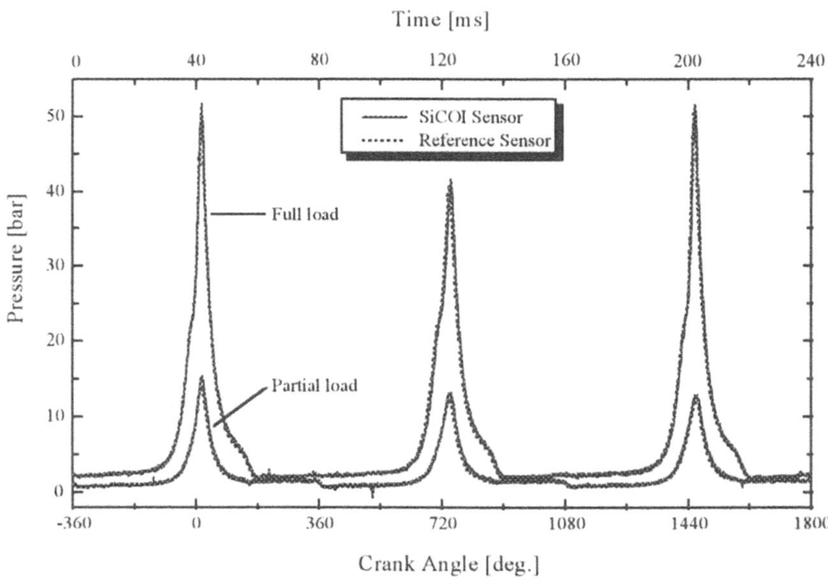

Fig.2.3 In cylinder measurement of the SiC sensor compared to a Kistler reference sensor [6].

3. High-temperature electronics

The availability of smart sensors with integrated electronics will have a big impact on system design in industrial, automotive and aerospace applications. In general, electronic signal conditioning is necessary for signal linerarisation and in order to eliminate temperature drift effects. From a system point of view, sensors with integrated signal conditioning are desired, having standardised output signals or a standardised interface electronics. This leads to a need for high-temperature electronics.

3.1 Properties of SiC relevant for electronics.

SiC has a high potential for high-temperature electronic applications, see table 3.1.

Table 3.1 Data of SiC relevant for electronic applications.

Parameter	Char. Depend.	Si	GaAs	3C-SiC	6H-SiC	4 H-SiC	C
Max.Operat.Temp.°C	E_g	150	350	600	700	750	1100
Leakage Current	$\exp(-E_g/2kT)$	1	10^{-3}	10^{-8}	10^{-16}	10^{-19}	10^{-39}
On-Resistance	$1/\mu E_c^3$	1	0.1	0.02	0.02	0.005	0.003
Min. Thickness	$1/E_c$	1	0.83	0.23	0.16	0.14	0.035
Max. Frequency	$v_s E_c$	1	0.7	10	12	14	73
Power Handling Cap.	λE_c^4	1	0.6	$9x10^2$	$5x10^3$	$8.5x10^3$	$8x10^6$

In the literature, devices operating at more than 600°C have been published. MOSFETs in SiC have been operated even at 650°C [7], and based on NMOS devices, an integrated operational amplifier was presented working at 300°C [8]. Current activities are going towards the development of CMOS technology for integrated circuits in SiC [9]. For high-temperature applications, reliability is a still open question for MOS oxides. As a consequence, we would expect an increasing temperature range beginning from Schottky-based devices over MOS systems to junction devices.

3.2 Prototypic electronic high temperature devices

We have developed lateral diodes and JFETs suited for integrated circuits working at >400°C. The characteristics of lateral diodes is shown in Fig.3.1. They are fabricated by nitrogen implantation into p-type 6H-SiC. We measure a rectifying ratio of 10^9 at room temperature, and still of 10^5 at 400°C. Processing of the normally-on JFETs in SiC has been published before [10]. We used aluminum implants to define p^+-gates over 1µm thin n-doped channel layers. Fig.3.2 shows a photography of a single device. The output characteristics at 400°C is demonstrated in Fig.3.3. These devices exhibit very low leakage currents at high temperatures, and are therefore well suited for the development of integrated circuits operating in harsh environments. The JFETs can be turned off by a negative gate voltage (V_{th} = -4...-8V). The threshold voltage is only weak dependant on temperature, but decreasing carrier mobility leads to a reduction of the current handling capability by a factor of 2.5 going from room temperature to 400°C. These properties have to be considered during the design of integrated SiC circuits. Additional care needs the chip metallisation. At high temperatures, degradation due to migration, corrosion or diffusion can occur. The maximum temperature for Al-based systems will be below 400°C. Gold as a noble metal can go further, but the high diffusivity requires new diffusion barriers above 500°C. The upper temperature limitation of SiC devices is therefore not a question of SiC material alone, but a question of metallization, reliability and packaging as well. Our current activities address these topics.

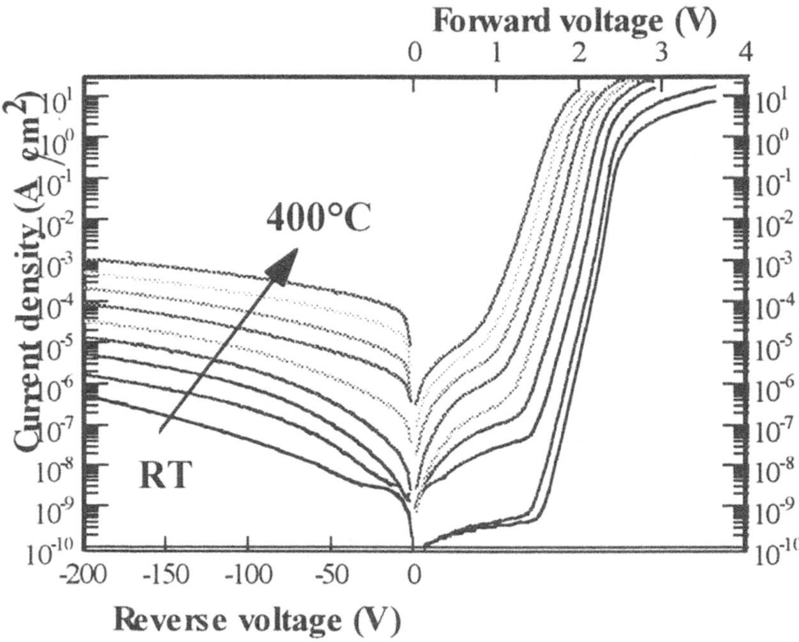

Fig.3.1 Temperature dependence of the characteristics of N-implanted pn-diodes in 6H-SiC

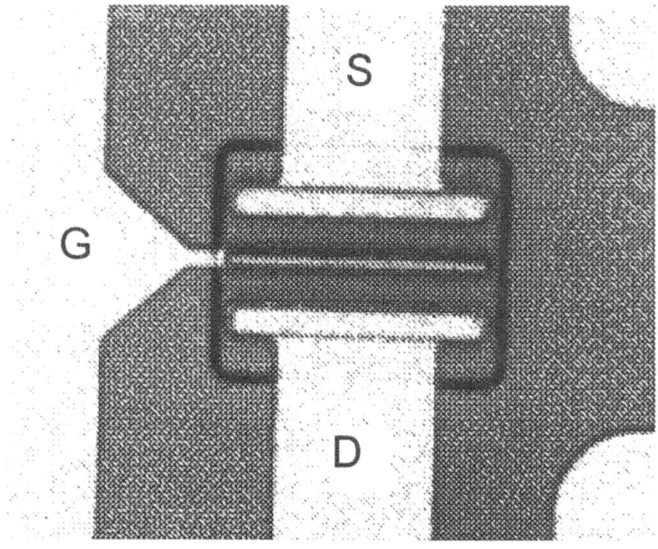

Fig.3.2 Optical photography of a lateral SiC JFET.

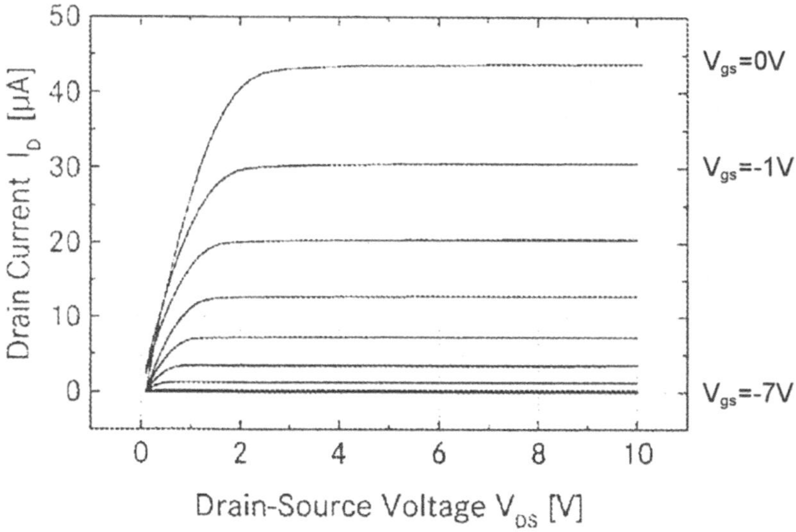

Fig.3.3 Output characteristics of a SiC JFET at 400°C.

4. Production aspects

SiC substrate material development is making rapid progress. In 1999, 3" substrates will be on the market where the average price per cm^2 will be reduced by more than a factor of 10. Furthermore 4" ß-SiC on insulator substrates will be available within a similar period of time. Therefore, mass production of SiC devices seems to be feasible well before the end of this century.

This work has been partially funded by the BMBF

Literature:

[1] J.S. Shor, D. Goldstein, A.D. Kurtz, IEEE Transactions on Electron Devices, Vol. 40, No. 6, 1993, S.1093.

[2] Landolt-Börnstein, Zahlenwerte und Funktionen aus Naturwissenschaften und Technik, Springer-Verlag Berlin, Volume 17, Semiconductors, (1984), S.118.

[3] Data in Science and Technology, O. Madelung (Herausgeber), Springer-Verlag Berlin, Semiconductors, (1991).

[4] G.A. Slack, Silicon Carbide -1973, R.C. Marshall, J.W. Faust jr., C.E. Ryan (Herausgeber), University of South Carolina Press, Columbia South Carolina, (1973), S.527.

[5] Y.A. Vodakov, E.N. Mokhov, Silicon Carbide -1973, R.C. Marshall, J.W. Faust jr., C.E. Ryan (Herausgeber), University of South Carolina Press. Columbia South Carolina, (1973),S.508.

[6] J. von Berg, R. Ziermann, W. Reichert, E. Obermeier, M. Eickhoff, G. Krötz, U. Thoma, Th. Boltshauser, C. Cavalloni, J.P. Nendza, Proceedings ICSC Stockholm, 1997.
[7] J.W.Palmour et al. Appl. Phys. Lett. 51(24), pp 2028 (1987).
[8] D.M.Brown et al. Transactions of the HiTEC 1994, pp XI-17.
[9] D.B. Slater et al. Transactions of the HiTEC 1996, pp XVI 27.
[10] S.Sheppard et al. to be published in Proc. of the ICSCIII-N 1997.

Automotive High-Pressure Sensor

Takeshi Matsui

Denso Corporation, R&D Dept. 1, 1-1 Showa-cho, Kariya-shi Aichi-ken, 448 Japan

Abstract

Recently, a need for detecting high pressure range in electrical automotive systems using oil air and refrigerant media has increased. A high-pressure sensor with simple structure and high reliability is desired.

This report describes the development of simple high- pressure sensor which has new structure fitting up oil in the pressure detecting part and oil sealing approach using resin and plastic instead of metal and glass. In additional, using metal diaphragm in the surface of pressure media this sensor can apply to various pressure media and electrical automotive systems. And integrating a pressure sensing element and electric parts for example amplifier, into one chip semiconductor element, the high reliability was achieved.

The complete manuscript was not available by date of print.

A New Sensor for Continuous Measurement of Fuel Injection Rate and Quantity During Diesel Engine Operation

Oliver Berberig[1], Kay Nottmeyer[1], Takao Iwasaki[2], Hayato Maehara[2], Takashi Kobayashi[2]

[1] ZEXEL Corporation, European R&D Office, Ihmepassage 4, 30449 Hannover, Germany
[2] ZEXEL Corporation, R&D Center, 3-13-26 Yakyu-cho, Higashimatsuyama-shi, Saitama 355, Japan

Keywords. flow sensor, micro turbine, injection system

1 Introduction

For further improvement of emissions and performance of diesel engines, it is necessary to improve the control of the injection system. It is particularly desirable to control injection rate, injection quantity and injection timing during each working cycle of a cylinder. In present conventional fuel injection systems [1] the control of injection quantity and timing depends on the mechanical setting of the fuel injection pump. This method, however, lacks control accuracy, since there is no feedback signal that allows to account for system operation variations due to tolerances and wear of the involved mechanical parts.

There are several injection meters in which the above mentioned parameters can be measured for each injection [2] ~ [4], however, they could not yet be adopted to on-board Diesel engine control. To achieve further engine performance improvement, future systems will have to apply a device that is able to monitor the actual parameters continuously for precise closed-loop control.

A fuel flow meter in the injection pipe encounters severe conditions since the pressure varies rapidly within micro seconds, reaching a maximum pressure far above 100 MPa. Therefore, following requirements have to be satisfied by a sensor for fuel injection rate and quantity measurement:

(1) Miniaturization and cheap mass fabrication of the sensing element
(2) No influence on injection characteristics.
(3) Pressure and shock wave resistance.
(4) Quick response and high resolution.
(5) Insensibility for variations of fuel parameters (density, viscosity).

We have developed a novel fuel quantity measurement device, the Micro Turbine Sensor (MTS), that is designed to measure the injection rate, timing and overall fuel quantity of each injection stroke during engine operation. This device realizes volumetric flow measurement using a tangentially driven turbine as the sensing element which has an outside diameter of 1 mm, and which is located inside the injection line next to the inlet connector of the injection nozzle. The turbine was chosen, since it is basically a volumetric flow meter, which in principle should be insensitive to variations of the thermodynamic properties (density, viscosity) of the fluid.

The feasibility of MTS application to Diesel fuel injection systems is studied, and the actual behavior with respect to fuel viscosity, as well as variations of the turbine holder geometry are investigated.

For that purpose, stationary flow experiments using a real-scale turbine have been conducted. The results are compared to analytical equation calculations that are based on a torque balance around the micro turbine. Finally, MTS measurements under realistic conditions (by installation between an injection pump and nozzle) are compared to measurements of a Bosch type injection rate meter, confirming that a MTS is applicable under such harsh environmental conditions.

2 The Micro Turbine Sensor Concept

2.1 Application Location

Fuel in the injection pipe close to the pump delivery valve flows backward after the end of injection as a consequence of delivery valve downward movement, thus relaxing the remaining pressure in the injection line. Hence, to expose the sensor only to a small amount of backflow, it should be installed in the vicinity of the injection nozzle. For that reason the MTS is positioned in front of the inlet connector, as shown in Fig. 1. The rotational axis of the micro turbine lines up with the pipe wall, thus making use of the Pelton turbine principle.

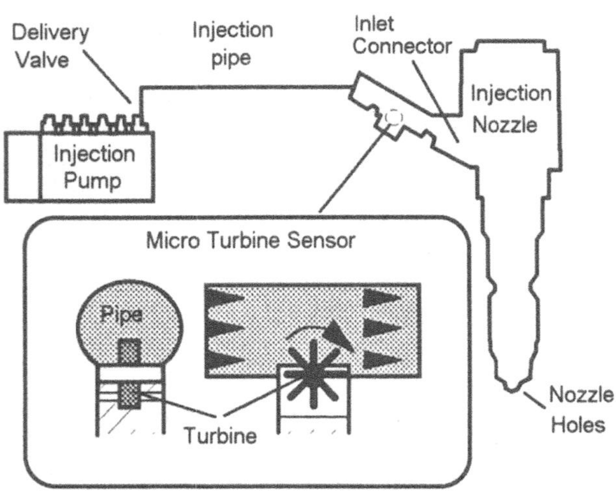

Fig. 1. Concept of MTS application

2.2 Sensor Setup

The turbine's dimension sizes and the different type of holders are shown in Fig. 2. The turbine sizes depend mainly on the application requirements: installation of the sensing element must be possible inside an injection pipe having an inner diameter of 2 mm or less without interference of the nozzle's spray characteristics. In addition to that, small turbine sizes minimize its moment of inertia, thus meeting the demand for quick flow response.

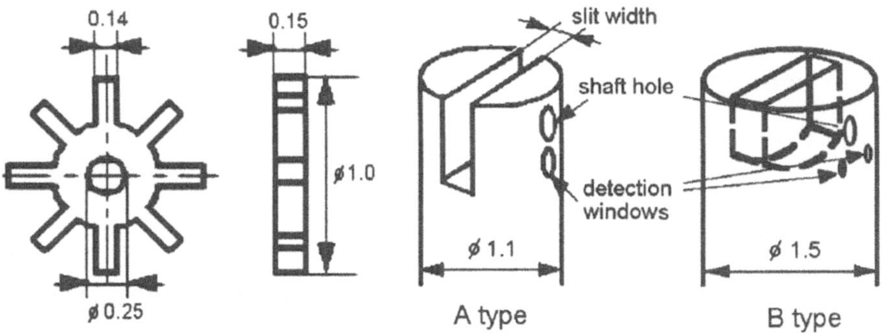

Fig. 2. Sketch of the turbine dimension sizes and holder types

The turbine is fixed at the top of the holder by a shaft. The holder tip contains a slit to install the turbine, and drilled holes. The upper hole holds the turbine shaft, while the lower hole serves as a viewing window for optical detection of the turbine revolution. This is realized by a fiber guided light beam that is chopped by the passing vanes.

Two type of holders are investigated, one of which containing two detection windows for passage of two light beams. Measurement with two light beams not only doubles the sensor resolution, but additionally allows for rotational direction detection, which is important to discriminate backflow.

2.3 Operation Principle

A schematic measurement setup of the MTS system with a B type holder is shown in Fig. 3. The fuel flow acts tangentially on the turbine vanes, thus causing a turbine rotation. Two optical fiber guided light beams pass through the turbine holder via the detection windows. The flow induced turbine rotation chops the beams by the passing vanes, which is detected by two photo sensors located on the opposite side. The photo sensor signals are amplified then, modulated in a subsequent counter, and send to a data acquisition PC, where they can be further processed, recorded and displayed.

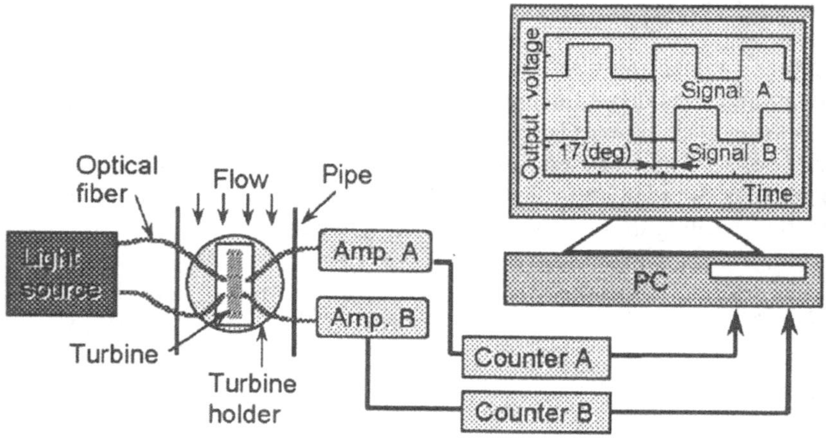

Fig. 3. Schematic arrangement of the sensing system

The injection timing is measured by correlation of the cam shaft position and micro turbine rotation. The injection quantity is derived from the number of measured turbine revolutions, and the injection rate is calculated from the reciprocal of the time between two measured pulses. A phase shift between the two light beam signals indicates a change in rotation direction (caused by backflow during pump valve closure), thus allowing for measurement of only the injected fuel quantity.

2.4 Fabrication

There are a number of technologies, which would be a candidate for fabricating the MTS. The sensing element (turbine) is essentially planar and can be made by various micromachining techniques, which are compared in Tab. 1. All processes are capable to produce 2½ dimensional structures of arbitrary shape in a plane and allow for cheap batch processing. The values given are estimates from information in the open literature [6] and may serve as an overview. Actual values may vary for a particular process or work.

Tab. 1. Comparison of micromachining processes

Method	Material	Thickness [μm]	Tolerance [μm]	Min. bridge [μm]	Min. gap [μm]
surface MM	poly-Si	1~3	< 1	≈1	≈1
resolved wafer	Si	5~10	<1	≈1	≈1
DRIE	Si	20~100	≈1	≈2	≈2
LIGA	Ni	50~1000	≈1	≈5	≈4
thick resist	Ni	10~500	2~5	> 5	> 5

The first two processes, surface micromachining and resolved wafer methods, are principally suitable, but the devices obtained are much thinner than considered here. However, this problem has been overcome by advanced deep dry etching processes, which have become available recently, allowing for thicknesses of up to about 100 µm. The latter two electroplating techniques are both suitable. X-ray lithography (LIGA) gives better accuracy, whereas thick photoresist molds are cheaper, and recent developments of epoxy-based ultra thick resists allow thicknesses in the same range as LIGA.

With a rotor diameter of 1 mm the accuracy requirements are not so demanding and therefore more "conventional" fabrication methods were considered as well.

Tab. 2. Comparison of conventional precision manufacturing techniques

Method	Material	Thickness D [µm]	Resolution [µm]	Min. bridge [µm]	Min. gap [µm]
Wire EDM	metal, Si	100~20000	5	15~20	< 20
Isotropic etching	metals	20~2000	10	0.8 x D	0.8 x D
Laser cutting	steel	50~6000	10	200	100
Water jet cutting	almost any	100~10000	100	200~1000	200~1000

The data are based on manufacturers catalog information and show, that the micro turbine fabrication is well possible. Hence, samples have been made by several precision manufacturing techniques and the accuracy was found to be sufficient. Particularly good results have been achieved by wire electro discharge machining and this method was chosen for prototype production. All data presented here refer to this version of the turbine, which is made from stainless steel and is shown in Fig. 4. The turbine is assembled in the holder made of the same material.

Fig. 4. SEM image of the micro turbine (outer diameter 1 mm, stainless steel)

Wire EDM produces good results and enabled to verify the principle of operation, however, for mass production a micromachining batch process is considered, which moreover holds the promise of integrated fabrication of turbine and holder assembly.

3 Analytical Investigation of the Micro Turbine Operation

In a first attempt to assess the characteristics of the turbine, an analytical study under laminar and turbulent pipe flow conditions was conducted, that is based on a balance of torques around the turbine. To be able to apply analytical equations, some simplifications and assumptions have been necessary:

- Always one turbine wing is fully exposed to the flow, always three turbine wings are fully exposed to the resting fluid on the turbine's lower side (in the holder slit), all other wings are not driven or hindered by the fluid.
- Fluid friction is considered only as shear forces in gaps, i.e. between turbine sides and adjacent holder walls, and between turbine hole and bearing pin. Friction caused by vortices is not taken into account.
- In case of laminar pipe flow (Re < 2300), the flow velocities at a certain pipe radius R are calculated according to the analytical (parabolic) profile.
- In case of turbulent pipe flow (Re > 2300), the velocity profile is assumed to be perfectly rectangular, i.e. $v_{turb}(R)$ = constant for all R (see Fig. 5).

For the analysis, the turbine is investigated at steady state conditions, i.e. the turbine speed at a certain flow condition is neither accelerating, nor decelerating. In that state, an equilibrium of torques is encountered:

$$M_{flow,drive} - M_{fluid,rest} - M_{fluid,frict} = 0 \tag{1}$$

The driving torque M_{drive} is determined by adding up all driving forces which act on the turbine due to differences between local flow velocity and vane speed times the respective turbine radius r:

$$M_{drive,lam/turb} = \int_{r_{root}}^{r_{tip}} F_{flow}(r) \cdot dr \tag{2}$$

The driving forces $F_{flow}(r)$ are a function of velocity differences described above, and with it a function of the assumed flow velocity profile. In the laminar case, a displaced profile according to Fig. 5 is assumed, which is expressed as:

$$v_{lam}(r) = \frac{8\overline{v}_{accel}}{D_d^2} \cdot \left(r \cdot (D_d + 2h_h) - r^2 - D_d h_h - h_h^2\right) \tag{3}$$

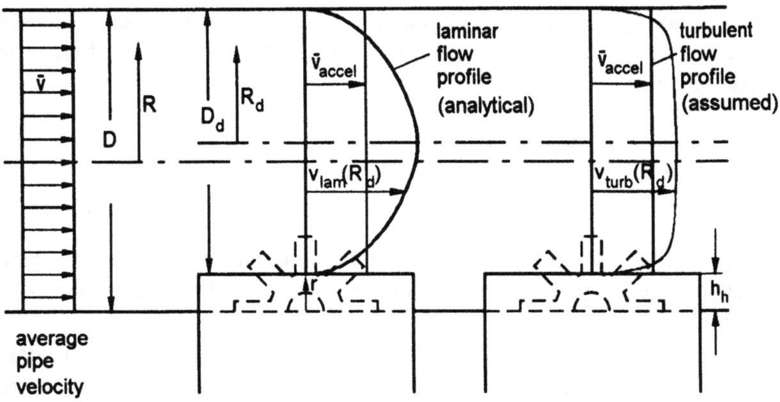

Fig. 5. Flow conditions in the injection pipe around the MTS

In the turbulent case a rectangular profile is assumed, with

$$v_{turb}(r) = \bar{v}_{accel} \tag{4}$$

Thus, the driving torque can be expressed as

$$M_{drive,turb} = c_d w_t \frac{\rho}{2} \cdot \left[\frac{\bar{v}_{ac}^2}{2} \left(r_{tip}^2 - r_{root}^2 \right) - \frac{2}{3} v_{av} \omega \left(r_{tip}^3 - r_{root}^3 \right) + \frac{\omega^2}{4} \left(r_{tip}^4 - r_{root}^4 \right) \right] \tag{5}$$

with c_d = drag coefficient
w_t = turbine width
ρ = liquid density
ω = angular velocity of the turbine

The expression for the laminar profile is more complex due to the parabolic flow velocity function (eq. (3)), and is not presented here. Also a retarding torque has to be taken into account for where the vane velocity exceeds the local flow speed.

The losses produced by the lower vanes running in the slit are expressed as

$$M_{rest} = 3 \cdot \int_{r_{root}}^{r_{tip}} F_{rest}(r) \cdot dr\,, \tag{6}$$

where the flow velocity in the slit is assumed to be zero. Three vanes are taken into account for. Flow losses due to vortices are neglected. Hence, the loss term by drag forces becomes

$$M_{rest} = \frac{3}{8} c_d \, w_t \, \rho \omega^2 \cdot \left(r_{tip}^4 - r_{root}^4 \right) \tag{7}$$

Furthermore, there are shear forces in the gaps around the micro turbine, which can be divided into the gaps between the turbine sides and the holder slit walls, and the journal bearing gap.

For calculation of the further, the turbine is considered to be a full disk. However, since about one quarter of the turbine extends into the flow, only three quarters of each side are taken into account:

$$M_{frict,\,side} = 2 \cdot \frac{3}{4} \cdot \int_{r_{hole}}^{r_{tip}} F_{side}(r) \cdot dr \tag{8}$$

In analogy, the decelerating torque due to the bearing shear forces is

$$M_{frict,\,bear} = F_{bear} \cdot r_{bear} \tag{9}$$

With summation of expressions (8) and (9), the loss term by shear forces becomes

$$M_{fluid,\,frict} = \pi \, \mu_{fluid} \, \omega \cdot \left[\frac{3}{4 \cdot h_{gap,\,side}} \cdot \left(r_{tip}^4 - r_{hole}^4 \right) + \frac{2 \, w_{bear}}{h_{gap,\,bear}} \cdot r_{bear}^3 \right] \tag{10}$$

with μ_{fluid} = dynamic viscosity of the fluid
$h_{gap,side}$ = clearance between turbine side and holder wall
$h_{gap,bear}$ = clearance of the journal bearing
w_{bear} = bearing width
r_{bear} = bearing shaft radius

After inserting all torques in eq. (1) and some transformations, the turbine speed ω is obtained in a second-order equation that can be solved explicitly, or in the laminar case iteratively. This expression then allows to carry out parameter variation calculations in order to investigate the influence of single parameters on the turbine speed. The results for the actual turbine and holder dimensions are displayed in comparison with stationary experiments in Fig. 8.

4 Experimental Investigation under Stationary Flow Conditions

4.1 Experimental Setup

Fig. 6 shows the block diagram of the experimental setup to investigate the turbine response under steady flow conditions. The fuel in the constant-temperature tank is fed to the surge tank by a pump. The fuel flow smoothed in the surge tank is supplied to the MTS and the reference sensor through the injection pipe having 2 mm inside diameter. The fuel is returned to in the constant-temperature tank then. Output data from the MTS and the reference sensor is modulated by the counter, digitized by an A/D converter and accumulated in the computer with a sampling frequency of 100 kHz. The reference sensor has a specified 1% linearity in the range of 0.3 to 9 *l*/min. The counter can process a frequency range up to 100 kHz. As a substitute for fuel, light oil has been used. Temperature fluctuations of the constant-temperature tank have been within ±1 K for the set value.

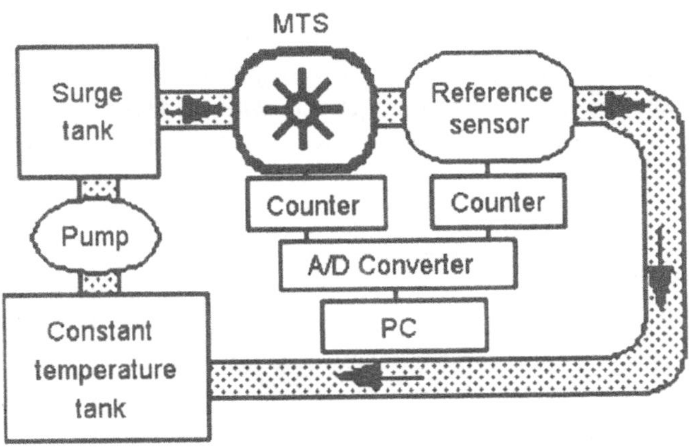

Fig. 6. Block diagram of the experimental setup for stationary flow investigations

4.2 Influence of Temperature Variations

In general, the dynamic viscosity of liquids decreases with increasing temperature due to the temperature dependence of the intermolecular adhesion forces. This also applies to light oil that has been used during investigation of temperature variation influence on the turbine speed.

Table 3. Miscellaneous fluid data of light oil at different temperatures

Temperature [°C]	Density [kg/m³]	Dynamic viscosity [Pa • s]	Kinematic viscosity [m²/s]
30	833	2.33×10^{-3}	2.8×10^{-6}
40	826	1.98×10^{-3}	2.4×10^{-6}
50	819	1.64×10^{-3}	2.0×10^{-6}

Since the decelerating torques due to liquid shear forces acting on the turbine decrease with rising liquid temperature, the turbine speed should increase. This theory based conclusion has been confirmed experimentally, as it can be seen in Fig. 7. At a flow rate of 1.8 l/min for example, the turbine speed is about 1065/1110/1160 rotations per second at 30/40/50°C, respectively.

This result already indicates that viscosity variations as a function of temperature (and pressure) must be taken into account for MTS applications, depending on the required measurement accuracy. Otherwise, an approximately linear relation is found in the region above 0.9 l/min.

The turbine speed graphs as a function of flow rate in Fig. 7 also reveal another phenomenon: the slope of the turbine speed functions varies below 0.9 l/min, with significant changes between 0.5 and 0.9 l/min. This correlates with the laminar/turbulent transition of the pipe flow. The viscosity dependence is clearly seen. Hence, for accurate measurement of fuel quantity with a MTS displaying such a characteristic, a case distinction has to be made on the signal processing side.

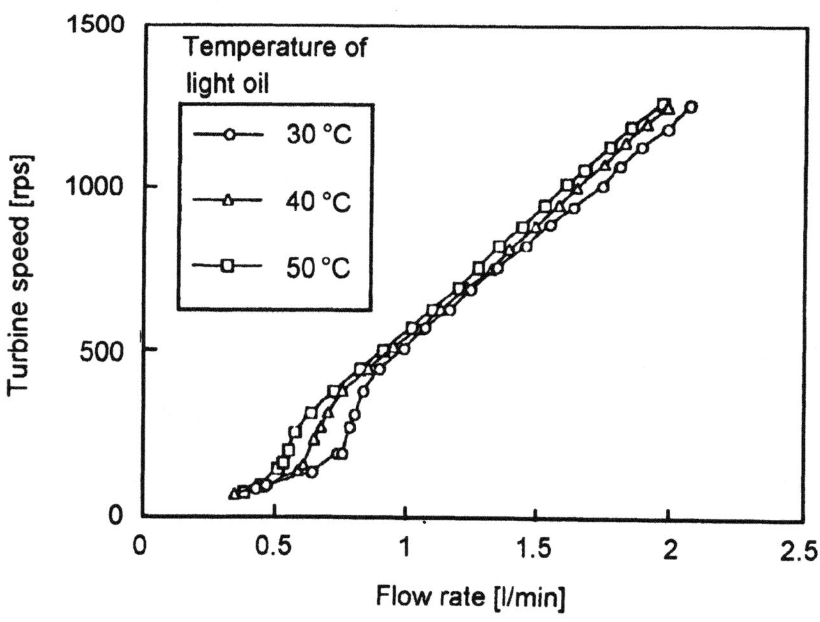

Fig. 7. Turbine speed graphs as a function of oil temperature

4.3 Influence of Holder Slit Geometry Variations

Stationary flow experiments also included the investigation of two different type of turbine holders (see Fig. 2). Contrary to the slit that has been cut completely through the holder tip of holder type A, the B type holder slit has been discharge machined in axial rod direction, leaving an unbroken rim of material around the turbine. Additionally, it contains two viewing windows for direction detection, but this feature does not influence the turbine's operation characteristics.

The turbine speed as a function of flow rate has been measured for the two types of holders at constant temperature (see Fig. 8). The results compare reasonably well with calculations using the theory discussed in section 3.

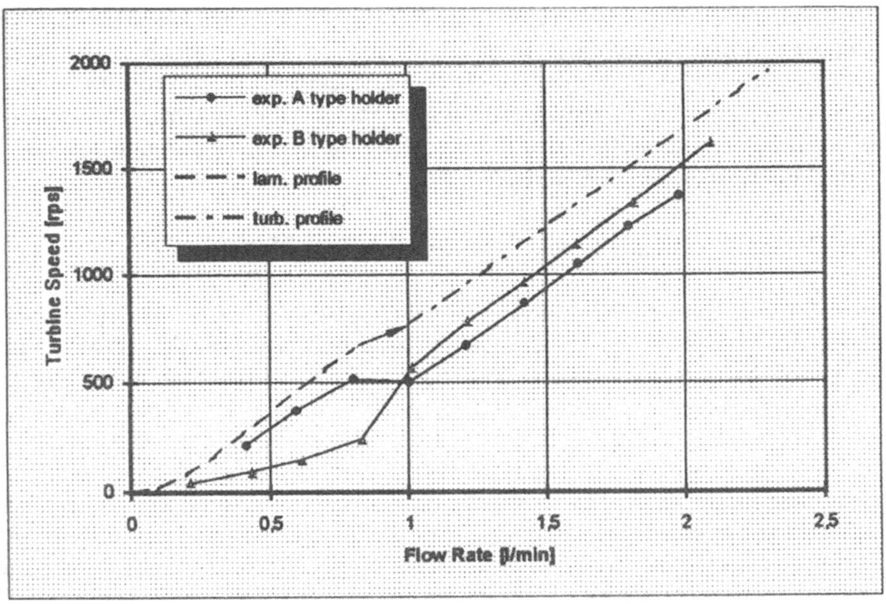

Fig. 8. Comparison of experimental and calculated results

In case of turbulent flow (above 1 l/min), both holder types show the same slope, however with a slight offset. This offset has been clarified by flow visualization experiments using acrylic large-scale models: under turbulent flow conditions the slit of holder type A allows for a counterflow underneath the turbine against the rotation direction, which does not occur in type B due to the closed rim of the holder tip.

In case of laminar flow (below 0.8 l/min), the slopes of type A and B differ considerably, which could not yet be explained. At least, for both cases the turbine speed function is piecewise linear, indicating the possibility to approximate the MTS behavior by two simple linear equations.

In order to match the slope of the experimental graphs in the turbulent region, the flow velocities in the calculations were reduced by a factor of 0.8. This corresponds to an additional loss, which is attributed to vortex induced losses that could not be taken into account for analytically. In the laminar region, the slope of the calculated graph is even steeper than the A type holder curve, indicating that the flow conditions around the turbine in the laminar case differ from the assumptions made in section 3.

5 Experimental Investigations under Transient Flow Conditions

5.1 Experimental Setup

As illustrated in Fig. 1, the MTS has been located between a jerk type injection pump and injection nozzle in front of the inlet connector, which is the actual arrangement of the application the MTS is developed for. In order to achieve maximum measurement resolution, a B type holder with two detection windows, 0.2 mm holder slit width and a turbine whose vanes extend 0.5 mm into the 2 mm diameter injection line has been applied.

5.2 Comparison with a Bosch Type Injection Rate Meter

In order to assess the capabilities of the MTS system, the injection nozzle holes have been connected to a Bosch type injection rate meter [4] that serves as the reference measurement system. The measurement results that are presented subsequently have been obtained with a five-hole nozzle, a single-spring nozzle holder, a constant injection quantity of 20 mm^3/stroke, and a light oil temperature of 30°C. The variable parameter has been the rotational speed of the injection pump (see figure 9.a-c). To simplify the comparison of the three measurements at different pump speed, the same time scale is selected, however, the scale of the ordinate, representing the injection rate in liter per minute, has been adopted to the maximum occurring rate.

In all three figures, the shape of the injection rate functions achieved with the MTS agree well with the measured results of the reference system. For higher pump speeds, however, the MTS signal slightly leads the Bosch type meter signal. One possible explanation is the different location of measurement in combination with liquid compressibility effects.

The steep rise of the MTS curves, representing very quick response to the pipe flow, proves the suitability of the MTS for injection system application. The resolution of the actual injection rate curve (presently, an average value of 2 mm^3/pulse is achieved), however, must be further improved.

The graph of Fig. 8 showed a characteristic variation depending on the flow conditions, with a maximum change in the laminar/turbulent transition area. The beginning of this area is indicated in Fig. 9.a-c as a horizontal line. It can be seen, that the influence of the transition has a different significance, depending on the operating point of the injection system.

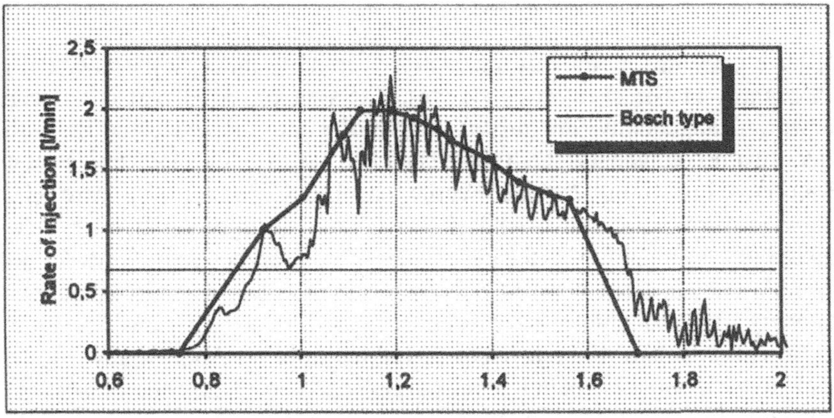

Fig. 9 a. Comparison of the MTS and the Bosch type meter at 500 rpm pump speed

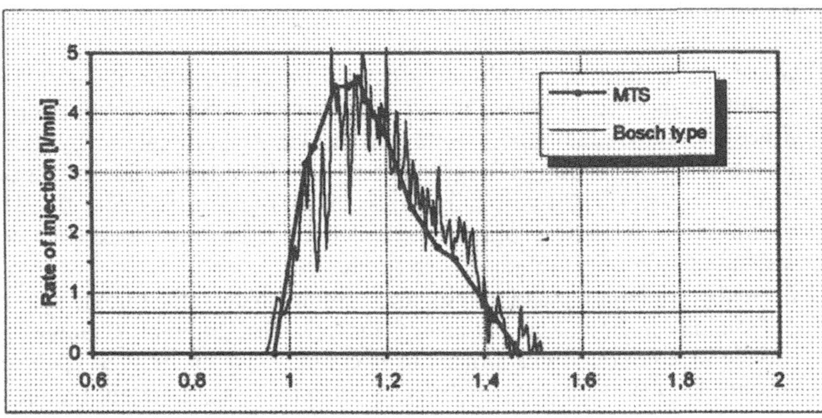

Fig. 9 b. Comparison at 900 rpm pump speed

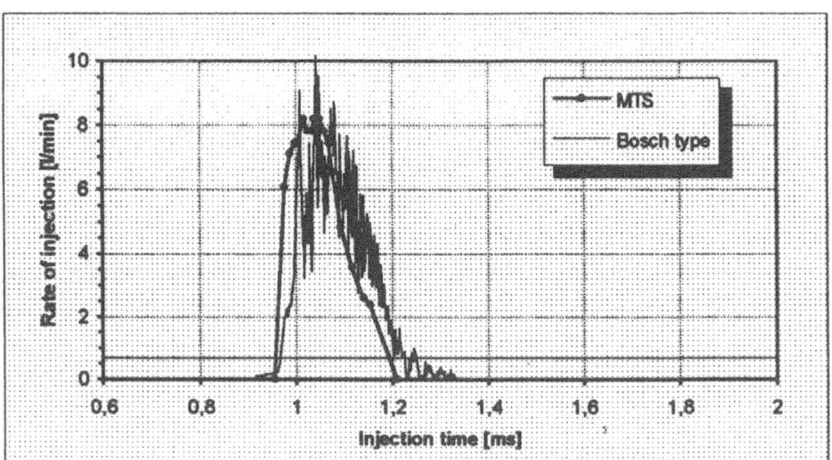

Fig. 9 c. Comparison at 1500 rpm pump speed

6 Summary and Conclusions

We have presented a new instrumentation method to measure the injection rate and quantity of Diesel injection systems. The system's novel part is its sensing element, a tangentially flow driven turbine of 1 mm diameter, and the turbine holder that accommodates the turbine shaft and two windows for turbine rotation detection. This *M*icro *T*urbine *S*ensor has been fabricated by wire electro discharge machining, and extends 0.5 mm into the injection line, thus being exposed to maximum flow velocities of about 40 m/s and pressure loads far above 100 MPa. Calculations according to an analytical study of MTS operations at stationary flow conditions have been compared to steady flow experiments, carried out with two different types of turbine holders. Additionally, the MTS behavior as a function of oil temperature has been examined. Finally, MTS application under realistic conditions has been investigated, and its capabilities have been compared to a Bosch type injection rate meter.

From our studies it can be concluded:

(1) Since the turbine speed slightly depends on liquid viscosity, this fluid parameter must be monitored if high measurement accuracy is required.

(2) Since the flow conditions around the turbine are difficult to predict or unknown for different MTS geometries, and flow losses due to vortices cannot be described analytically, it is *not* possible to describe the MTS operation by analytical equation calculations. However, tendencies (e.g. response to viscosity variations) can be predicted.

(3) Presently, for both types of investigated turbine holders, the turbine speed as a function of flow rate varies significantly for different (laminar and turbulent) flow conditions. Since this operation behavior must be taken into account for the evaluation (in order to correctly relate a certain number of counted pulses to the injected fuel quantity), the next stage of MTS development will include design variations in order to linearize the turbine speed over the full flow rate range.

(4) MTS experiments under realistic conditions prove the feasibility of the new sensor concept: the shape of the injection rate functions obtained by the MTS are in good agreement with the measurements of the reference system. Hence, important requirements for a sensor of fuel injection rate and quantity measurement, like pressure/shock wave resistance and quick response, have successfully been verified.

Future work will concentrate on the characteristics of the device with respect to durability as well as cost effective batch fabrication by micromachining techniques.

References

[1] H. Ishiwata, et al.: 'Recent Progress in Rate Shaping Technology for Diesel In-Line Pumps', SAE paper No. 940194 (1994)

[2] W. Bosch: 'The Fuel Rate Indicator: A New Measuring Instrument for Display of the Characteristics of Individual Injections', SAE paper No. 660749 (1966)

[3] A. Takamura, et al.: 'Development of a New Measurement Tool for Fuel Injection Rate in Diesel Engines', SAE paper No. 890317 (1989)

[4] G. R. Bower, D. E. Foster: 'A Comparison of the Bosch and Zeuch type Injection Rate Meters', SAE paper No. 910724 (1991)

[5] T. Iwasaki, et al.: 'Study of a Sensor for Fuel Injection Quantity', SAE paper No. 970533 (1997)

[6] S. Büttgenbach: 'Mikromechanik', Teubner Verlag, Stuttgart, 1991

Measurement of Physical Parameters of Car Tires Using Passive SAW Sensors

Alfred Pohl [1] and Leonhard Reindl [2]

[1] University of Technology Vienna, E3592, Applied Electronics Laboratory, A-1040 Vienna, Austria
[2] Siemens AG München, ZFE T KM 1, D-81739 München, Germany

Abstract. A new type of sensors based on surface acoustic wave devices are introduced. The sensor elements are totally passive and are only connected to an antenna. The readout is performed wirelessly via a radio link. The sensors are characterized by high thermal, mechanical and electromagnetic loadability and nearly unlimited lifetime. Affecting the sensors with the measurand a lot of physical parameters can be measured without wired connection. The application of these new elements for measurements in cars focused to the tires are discussed. The hardware required and examples for application in automotive industry are presented.

Keywords. Wireless remote sensing, Surface Acoustic Wave (SAW) sensors, Automotive applications of SAW sensors, Tire monitoring

1 Introduction

Electronic control plays an important role in today's systems. Looking at car control, for a long period of time most circuits were directed by the man or the woman at the steering wheel, the sensing was made with the sensory perception of him or her respectively. For a safe operation, nowadays the driver is excluded from most of the control loops. Automatic brake aid systems, dynamic acceleration and skid avoidance circuits help to make driving much more secure, electronic motor management and catalytic converters yield reduced fuel consumption and further lower emissions.
To measure the parameters of the process to control, the electronic systems use a lot of sensors converting the physical measurands to electric signals.
For some applications, a wired connection between the sensor and the system cannot be established since the measurand have to be collected on an rotating or fast moving machinery part. Then remote measurement with wireless data transmission have to be implemented. Therefore a lot of systems have been developed. Most of them are active circuits and contain semiconductors and batteries.

Here a type of passive sensors based on surface acoustic wave (SAW) devices for wirelessly readout is introduced.
In Chapter 2 the fundamentals of these SAW sensors are described. Then, in chapter 3, the required hardware and the signal processing required for wireless readout is discussed. In chapter 4 an overview of possible and experimentally verified applications for vehicular technology especially for car tires is given. Finally, a brief summary conclude the properties and benefits of this new type of sensor elements.

2 Surface Acoustic Wave Devices for Wireless Sensing

More than thirty years ago, surface acoustic wave devices were invented with the invention of the interdigital transducer (IDT), a metallic comb structure on the surface of a plain polished piezoelectric substrate (Lithiumniobate, Quartz, etc.) [1]. An electric radio frequency (RF) signal into this IDT yields a tangential electric field strength distribution on the surface (fig. 1) and generates an acoustic wave propagating on the surface.

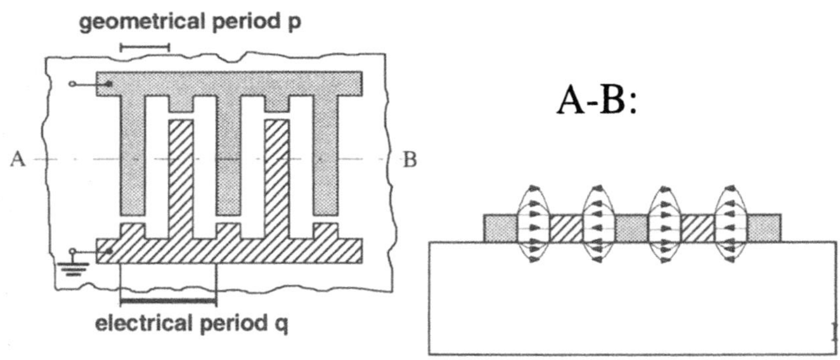

Fig. 1. Interdigital transducer for excitation of surface acoustic waves

Figure 2 shows the cross-section of a crystal substrate, with the SAW-propagation occurring in the z-direction.

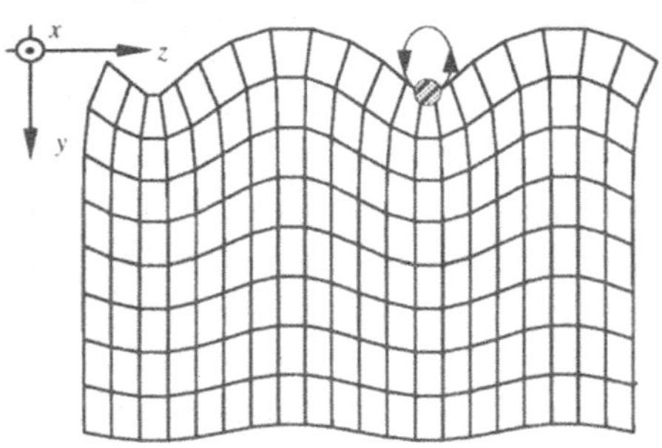

Fig. 2. Drawing of an acoustic wave propagating on a crystal's surface

After passing a propagation distance, another IDT can convert the SAW back to an electric signal. Designing the IDT's impulse response, a wide variety of electronic bandpass filters and resonators for television sets, mobile communication systems etc. can be built. Today several hundreds of million SAW devices are manufactured.

A few years ago, the application of SAW devices for remote passive sensor purposes were proposed [2, 3, 4]. Fig 3. shows the steps from an IDT exciting SAWs to a coded reflecting delay line. If the electric port is connected to an antenna, an RF interrogation signal can be fed via a radio link. In the IDT this signal is converted to an SAW, propagating on the surface and being partially reflected by each of the reflectors arranged. After a delay time of some microseconds all direct electromagnetic reflections of the interrogation signal have been vanished. Then the sensor responds on the antenna port with a train of pictures of the interrogation signal. If a burst was transmitted, after a delay corresponding to the ratio of doubled propagation length to the reflectors and the SAW's velocity, single bursts for each reflector on the surface are retransmitted by the sensor's antenna.

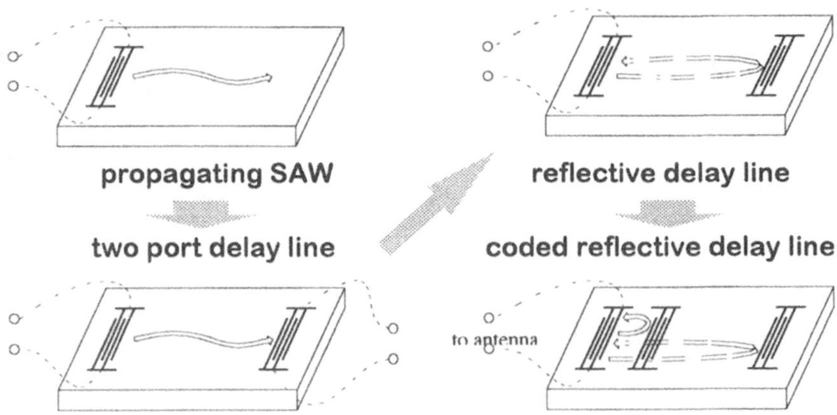

Fig. 3. Passive wirelessly interrogable SAW device

This burst train can be interpreted to be a serial data word, carrying an identification information. Fig. 4 shows the envelope of the interrogation burst, the environmental echoes fading away and the acoustically delayed response bits of a SAW ID sensor with code "10000111100011". Since this kind of SAW devices was implemented first for identification purposes, this sensor type is called ID tag.

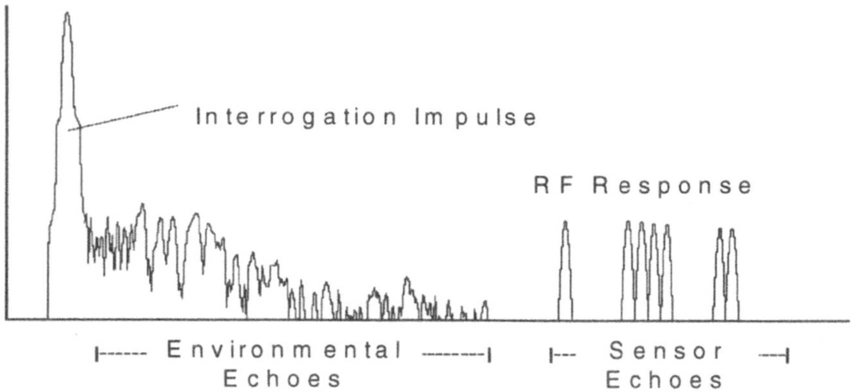

Fig. 4. Burst response of a coded reflective SAW delay line

To measure physical quantities, the sensor can be affected by temperature, mechanical stress or strain and mass loading. This affects the substrate's dimensions and the elastic crystal constants, determining the SAW's propagation velocity. The response of the sensor is scaled by a factor $s=1+\varepsilon$. Evaluating the delay between the response bits, the measurand can be calculated from s and ε respectively. Utilizing differential methods, influence of distance, movement and unwanted effects (e.g. temperature for mechanic sensors) can be compensated.

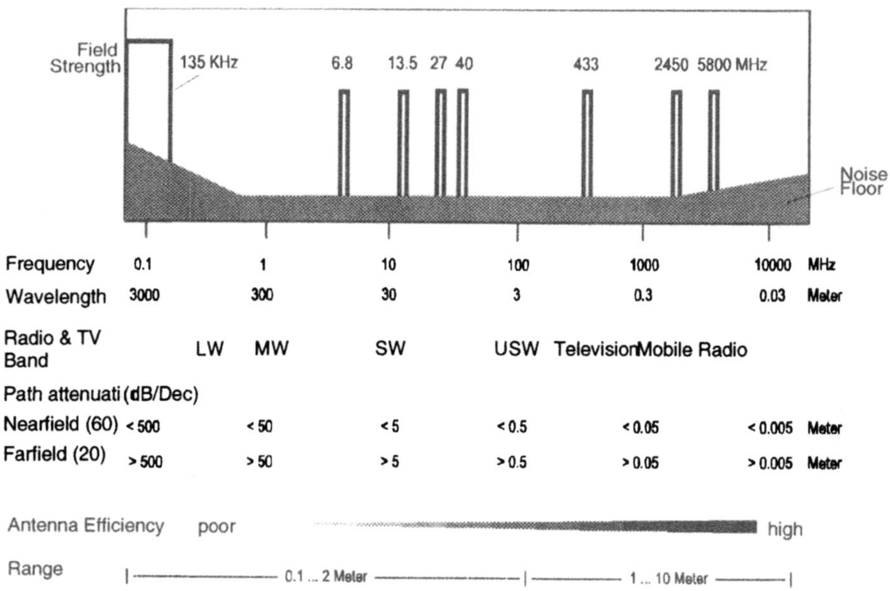

Fig. 5. The allowed frequency bands and the associated achievable ranges for radio-readable identification systems.

SAW sensors are bandpass devices and can be designed to operate in RF bands between 30 and 2450 MHz. For wireless ID-tagging, only certain frequency bands are allowed. Figure 5 shows different frequency bands, with their respective attenuations and achievable path lengths.
Due to the technological and antenna limitations only the 434 MHz and 2.45 GHz ranges are practical for SAW-based systems.
Particular attention must be paid to the transmission path between the interrogation device and the ID-tag. For ID-tags without any power supply the attenuation is doubled, as the signal must travel from the base to the receiving unit and back again without amplification along the way. Additional losses also occur within the antenna and associated hardware of the ID-tag. Thus, the answer signal is several orders of magnitude smaller than the interrogation signal.
With today's SAW sensors, interrogation distances of 0.1 to a few meters in the 2.45 GHz band and between 1 to 20 meters at 433 MHz are feasible.
The most advantageous of SAW sensors is that they contain neither battery nor semiconductors or capacitors etc. Since they only consist of a substrate with metallic structures on its surface and a metallic antenna, they withstand high thermal load. SAW sensors were published to operate up to 1000 degree centigrades [5]. For the same reason, SAW sensors are capable to withstand hard radiation and high electromagnetic interferences.

3 Wirelessly Readout of Passive SAW Sensors

The interrogation unit have to transmit the interrogation signal, to receive the sensor's response after a delay, to evaluate the scaling s and to calculate the measurand and the identification information of the sensor. Figure 6 shows a block diagram of a system.
SAW ID-tags only store the energy from the incoming interrogation impulse, and thus require no power supply. Because the reading takes only a few microseconds, over 100 000 interrogations per second are possible. This allows the interrogation of even fast moving objects. Since SAW ID-tags work completely linear no frequency mixing occurs and therefore they answer with a defined phase. Thus also a coherent system can be built and consequently, like done in radar systems, many RF responses can be summed up to enhance the signal-to-noise ratio, leading to an improvement of the maximum read-out distance. Therefore a quadrature demodulator is implemented in the receiver unit before sampling and digitizing. Like in radar applications an enlargement of the signal-to-noise ratio by 12 dB doubles the read-out distance.

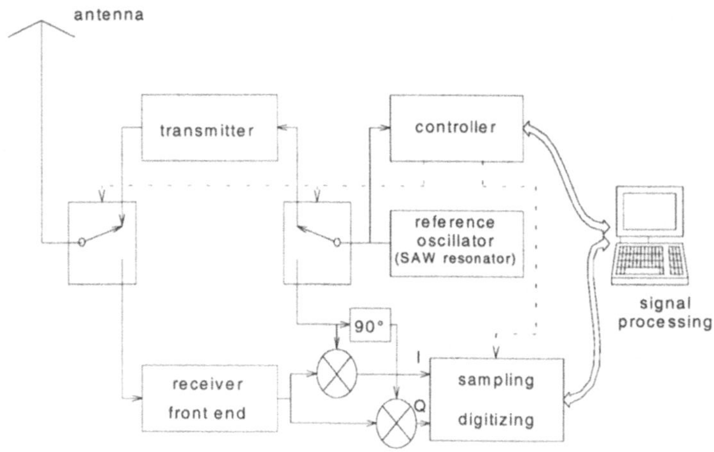

Fig. 6. Interrogation system

Figure 7 shows a photograph of a 2.45 GHz interrogation unit for industrial applications.

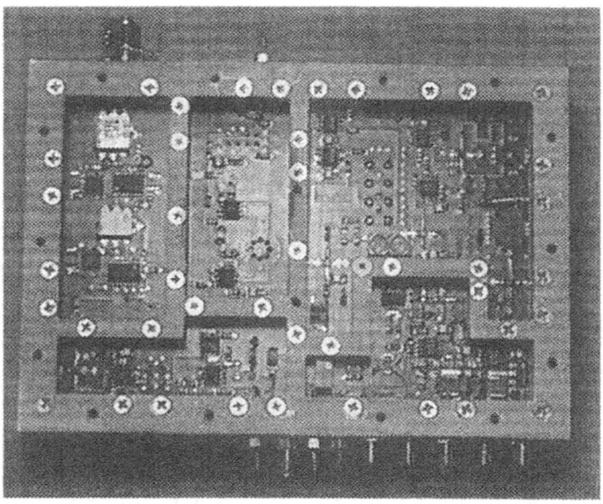

Fig. 7. Interrogation unit for 2.45 GHz SAW sensors

Figure 8 points out a measured sensor response with inphase and quadrature phase components and magnitude calculated out of these.

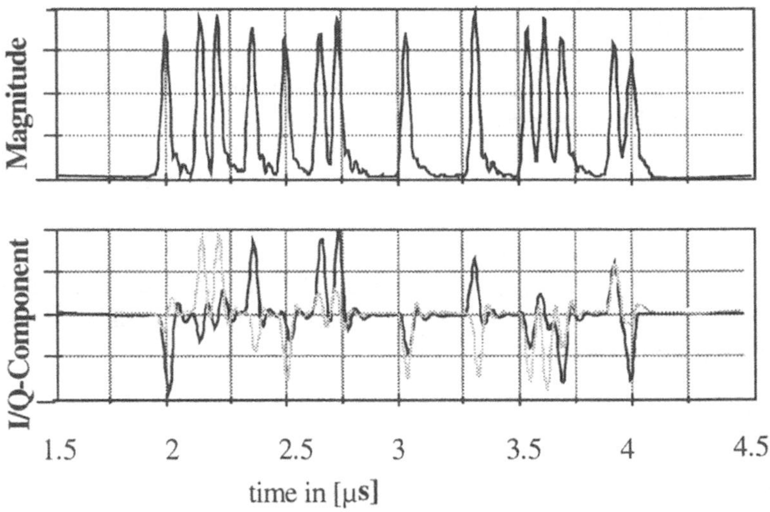

Fig. 8. Sensor response

The principle of signal processing is shown in fig. 9. The inphase and quadraturephase components are sampled and digitized. The coarse delay between the response bursts can be detected by envelope tracking. Fine measurements with enhancement of the resolution to a total value of a few ppm can be performed by phase difference measurements.

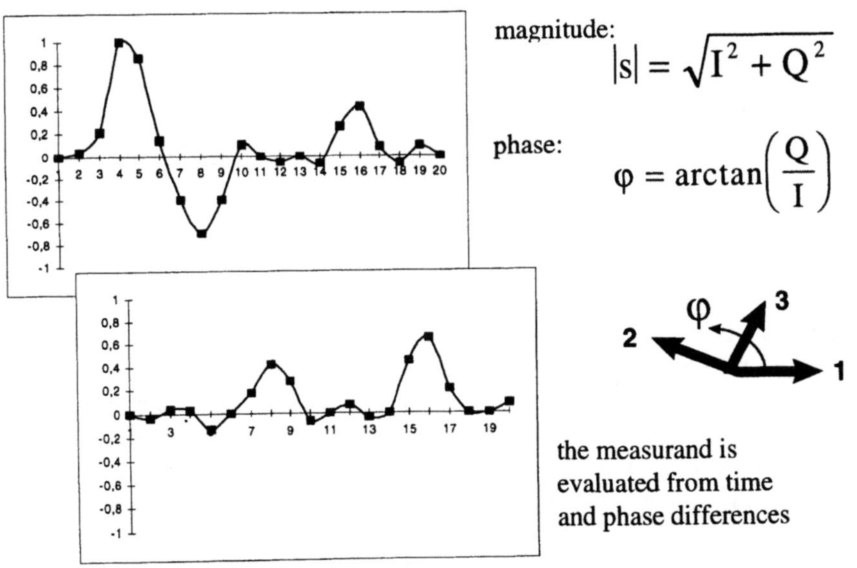

Fig. 9. Principle of signal processing for interrogation

4 Vehicular Applications

In this chapter an overview to vehicular application is presented in examples. Then it is focused to the measurements in car tires.

The first implemented vehicular application of SAW sensors is an ID System used for road pricing in Norway. There, the cars to identify have SAW ID tags with a 32 bit code behind their windscreen and are tolled automatically.
Nowadays systems for vehicle identification with SAW ID tags also are installed at the subway in Munich, Germany. Figure 10 shows a photograph of the ID tag with antenna fixed to a wagon.

Fig. 10. SAW ID tag on a subway wagon

If only the difference in phase between the answer impulses is evaluated, the sensor signal is independent of distance. Through the evaluation of many phase differences, the sensor temperature is unequivocally defined.
The accuracy of the temperature measurement is within ±0.1K, but can rise for particular measurements. Measuring of remote temperatures can also be made from moving objects, as occurs in the case of a rotating wave. To avoid dangerous loss of brake retardation due to overheated brake disks, permanent monitoring can be performed using SAW sensors fixed directly to the disks. Figure 11 shows the change in temperature of the breaks on a train as it slows down to enter a station.

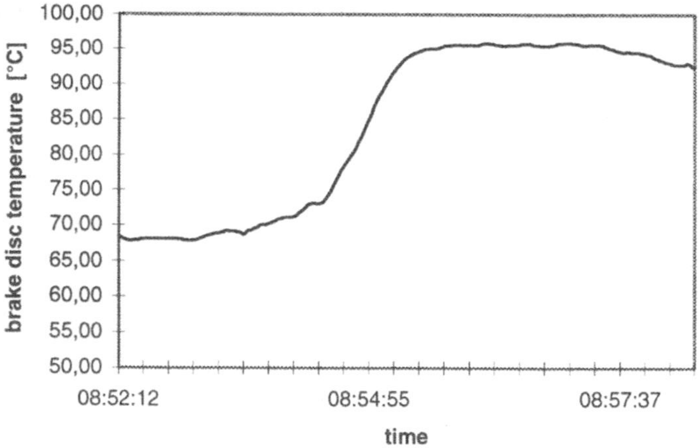

Fig. 11. Temperature of a brake disk of a train during stop in a railway station

Other demands in vehicular technology are measurement of revolutions per time, torque and, the product of these, the transmitted power. The revolutions per time and the angle speed can be measured evaluating the presence of SAW sensors at the circumference of the transmission shaft. rotation angle is measured evaluating the phase difference of the response signals of two different sensors.

For torque measurement, two sensors are fixed right angled on the shaft. If torque is applied, one will be stretched, the other will be compressed. The torque can be evaluated without errors due to temperature etc. from the difference of the sensors scaling. Figure 12 sketches the principle of wireless torque measurement with SAW sensors.

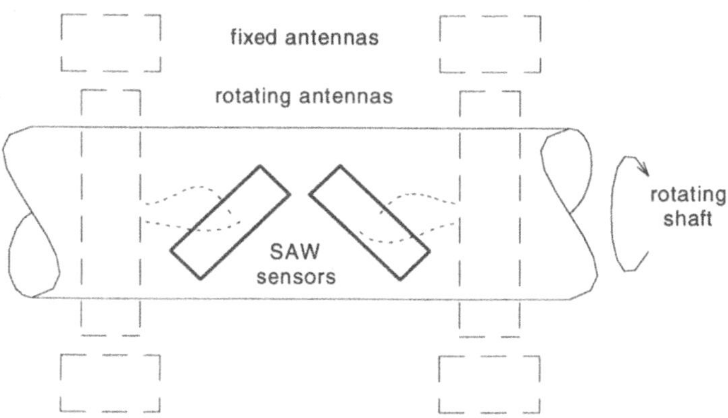

Fig. 12. Wireless torque measurement with passive SAW sensors

The sensor's sensitivity is shown in fig. 13. The phase variation versus the relative extension of one SAW sensor.

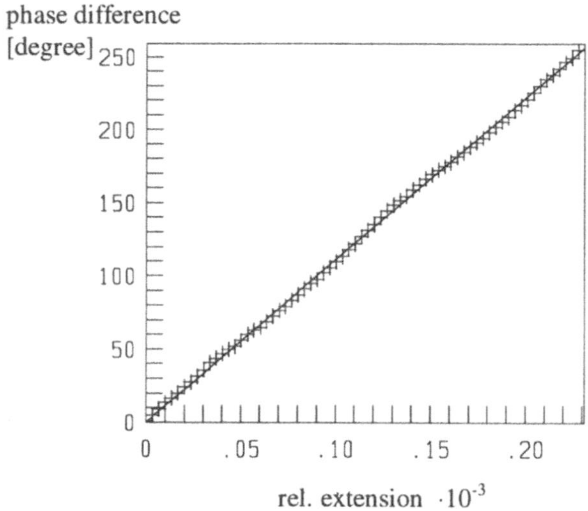

Fig. 13. Sensitivity of a SAW torque sensor

If it is assumed, that phase differences can be measured with an accuracy of only 3 degrees, a resolution of less than 1 percent is achieved.
For mechanical sensing, the SAW sensor is loaded to be bent, again the differential phase shift between reflectors is evaluated. Figure 14 shows the phase shift between two reflectors with an unscaled spacing of 3 μs versus displacement of one edge of the sensor fixed at the opposite edge.

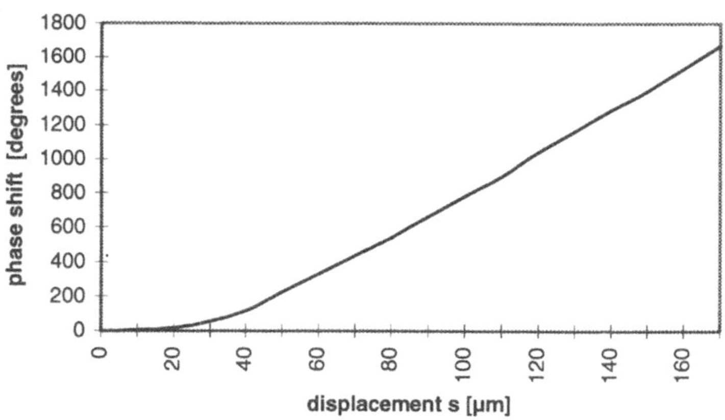

Fig. 14. Phase shift versus bending the SAW sensor

The tires of cars nowadays are not monitored by electronic systems although they are responsible for the save contact between car and the road and so they contribute a lot to a safe operation of the vehicles. Overheating tires due to intensified fulling because of wrong tire pressure are said to be responsible for many terrible accidents. The vibrations and the unbalance of a tire will affect the comfort of driving and the wear of the whole car. Further it is the main topic in making cars quieter.
So, the most important parameters to measure in car tires are the temperature within and at the running surface and the air pressure inside. Further dynamic parameters like vibrations of the flanks and the running surface and some deformations should be measured to supervise the tires comprehensively.

Monitoring of temperature is easy with SAW sensors discussed above. The sensors have to be fixed directly to the location of interest. The sensor of interest can be selected by antenna selection (only one sensor in the range of the interrogation system) or by coding and correlative signal processing [6].

Measurement of tire pressure becomes an important topic in today's cars. Some active systems have been presented and will be implemented in the next years. One disadvantage of these systems is their mass. Driving with 100 km/h, the radial acceleration in the tire becomes more than a few hundreds of g. Further, the today's systems are powered by batteries and semiconductors. Therefore the lifetime is limited, the permissible sensor's environment conditions are limited in temperature and electromagnetic interference. Here passive SAW could be a reliable substitute without narrow limits and with a mass of only a few (0.3 - 3) grams. For pressure measurements, an integrated SAW pressure sensor with identification and temperature measurement facilities was developed [7] (fig. 15).

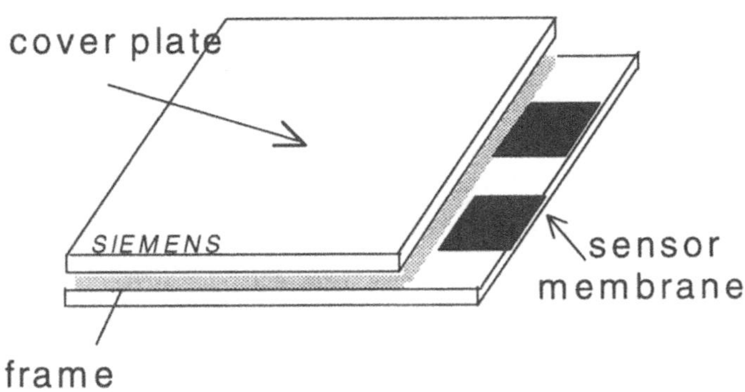

Fig. 15. Integrated SAW pressure sensor

The sensor consists of a quartz membrane with the SAW exciting structures on its surface. Around this surface, a frame keeps a constant distance to a cover plate, surrounding a cavity at reference pressure. The electrical port is to be connected to a small antenna. If the pressure outside is larger than that inside, the membranes will be vaulted towards each other, the surface with the SAW's propagation path will be enlarged. The scaling of the sensor response is a measure of the pressure and the temperature. Evaluating the phase shift of different bits, the temperature and the pressure can be measured clearly. This sensor was experimentally implemented into car tires [8]. The metallic shaft of the rubber made valve was used as antenna.
Figure 16 shows a photograph of an experimentally setup, figure 17 points out the measured tire pressure during passing a two track grade crossing with adjacent water channel. The resolution achieved with SAW sensors for pressure measurements was better than 0.1 Bar.

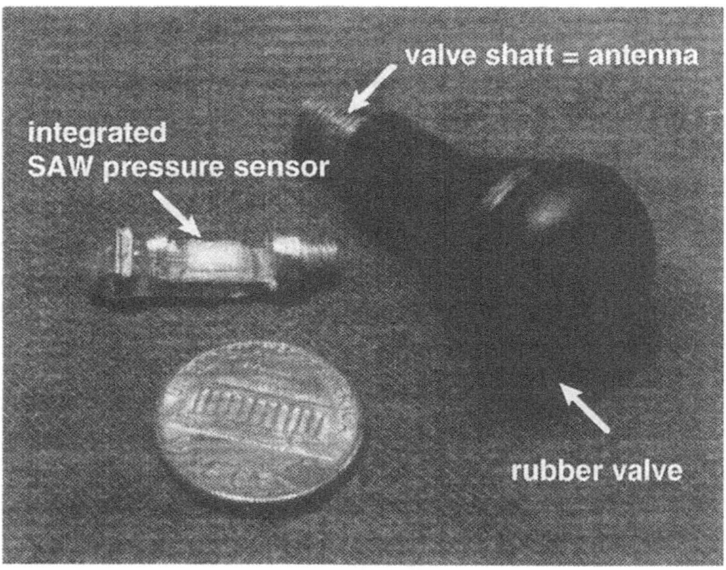

Fig. 16. Experimentally setup for monitoring the tire pressure with SAW sensors

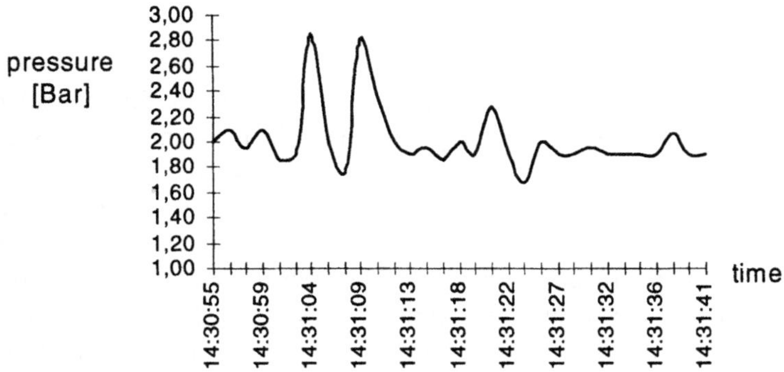

Fig. 17. Tire pressure passing a grade crossing

The measurement of acceleration and vibrations can be performed using SAW sensors to be bent. According to figure 14 a vibration yields a periodic displacement with periodic changing phase differences. Since the sensor can be interrogated with up to a few hundreds of thousand times per second, a mechanic vibration with a frequency of up to several tenths of kHz can be monitored. A sensor assembly consists of the SAW sensor loaded by a seismic mass, like depicted in fig. 18.

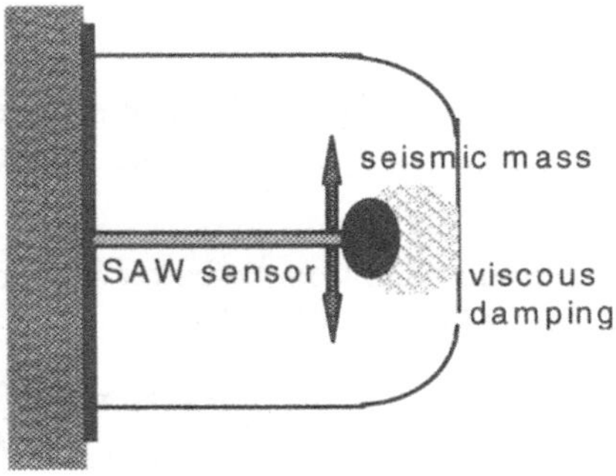

Fig. 18. Passive wirelessly interrogable SAW vibration sensor

Further application can monitor some deformations of the tire and enable extensive measurement and better results during tire design. On the other hand the deformations could be observed and an alert can be given if a dangerous case of operation occurs.

The interrogation antenna at the car body can be fixed behind the car wing's cover made of synthetic material.

5 Conclusion

Passive SAW sensors have been shown to be reliable devices for vehicular applications and well suited for monitoring car tires. Their advantages are that they contain no semiconductors and no charge storage devices. The limitations due to environmental conditions are much less rigid than for active systems, they withstand heat, dust, strong EMI, etc. Advantageous for applications on fast moving or rotating systems is the low mass of SAW sensor devices. Application for identification, measurement of temperature and mechanical parameters like bending, vibration, pressure, etc. were discussed, experimentally results have been shown.

References

[1] D.P. Morgan, *Surface Wave Devices for Signal Processing*, Elsevier, Amsterdam, 1985.

[2] P.A. Nysen, H. Skeie and D. Armstrong, System for interrogating a passive transponder carrying phase-encoded information, *US Patent Nos. 4 725 841; 4 625 207; 4 625 208* (1983-1986).

[3] L. Reindl, F. Müller, C. Ruppel, W.E. Bulst and F. Seifert, Passive surface wave sensors which can be wirelessly interrogated, *International Patent Appl. WO 93/13495* (1992).

[4] F. Seifert, W.E. Bulst and C. Ruppel, Mechanical sensors based on surface acoustic waves, *Sensors and Actuators, A44* (1994), pp. 231-239.

[5] L. Reindl, G. Scholl, T. Ostertag, F. Schmidt und A Pohl, Funksensorik und Identifikation mit OFW-Sensoren, Tagungsband der Sensortagung Bad Nauheim, Germany, 1998, in print.

[6] A. Pohl, G. Ostermayer, L. Reindl and F. Seifert, Spread Spectrum Techniques for Wirelessly Interrogable Passive SAW Sensors, *Proc. IEEE International Symposium on Spread Spectrum Techniques & Applications 1996*, pp. 730-734.

[7] H. Scherr, G. Scholl, F. Seifert, R. Weigel " Quartz pressure sensor based on SAW reflective delay lines", Proc. IEEE Ultrasonics Symp. 1996.

[8] A. Pohl, L. Reindl, H. Scherr, Wirelessly Interrogable Passive SAW Sensors, Application for Permanent Monitoring of Tire Pressure, Tagungsband der 6. VDI Fachtagung Reifen, Fahrwerk, Fahrbahn, Hannover, Germany, 1997.

A Rotary Position Sensor System for Automotive Applications

Dieter Schödlbauer

ruf electronics GmbH, Bahnhofstraße 26-28, 85635 Höhenkirchen

Abstract. Electronic components have been playing an important role in motor vehicle applications for already more than two decades, with growing demands on system components like high resolution position sensors over the last several years. This paper will give an example from a sensor supplier's point of view, depicting a non-contacting rotary position sensor concept which will replace some of the standard, resistor based potentiometric sensors in future applications. Starting with a brief look at the conventional potentiometer, the paper will bridge the gap to a modern sensor system. It is based on a magnetoresistive microsensor front-end, combined with a special two-dimensional signal evaluation. This non-contacting sensor is destined for motor vehicle applications, designed to meet improved specifications in terms of performance , reliability and cost.

1 Introduction

Electronic systems can be found in a great variety of vehicle applications, including more and more sophisticated sensor components. Pedal position sensors (PPS) and throttle position sensors (TPS) are well-known engine management components in the effort to optimize power output while keeping fuel consumption and noxious exhaust fumes as low as possible. Beside these applications, there is an ongoing market for position sensors in the field of passenger safety and comfort, e.g. headlight control, variable suspension and various (position) memory functions.

On the one hand, there is an increasing demand on the performance and the reliability of motor vehicle systems, and thus on the durability of the sensor functions involved (Fig. 1.1. and 1.2.). On the other hand, there is a tremendous pricing pressure on the sensor and system suppliers to reduce costs, resulting in a selling price for some special design that may be around 40 % of that in 1980.

This altogether is leading to changes in sensor design as well as in present system architectures. Regarding the motor management example, improving nowadays TPS and PPS function with respect to reliability means that non-contacting principles will supersede the present potentiometer based on resistor tracks and brush wipers.

On the face of it, this seems to increase the sensor costs and thus the overall system costs. The answer may be treating the sensor as a smart (sub-)system, capable of self-diagnosis and fitted with low cost, mass production microelectronics in order not to inflate the system costs.

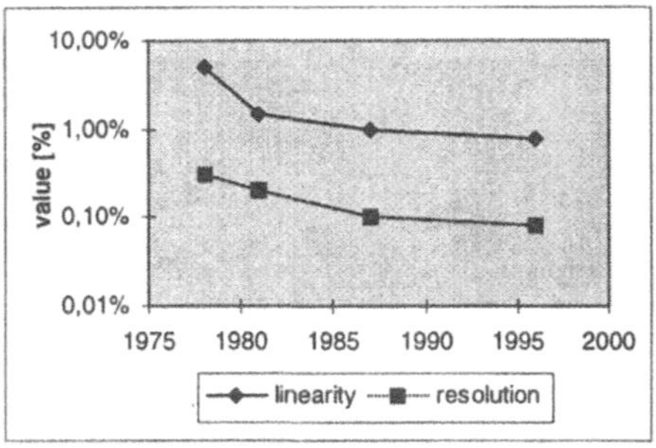

Figure 1.1.
TPS linearity and resolution demands

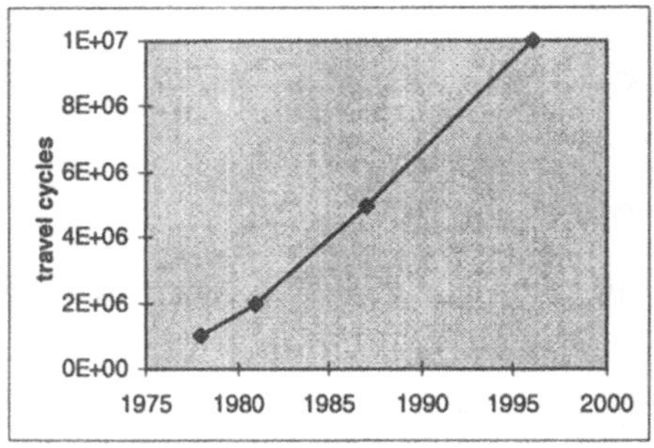

Figure 1.2.
Number of TPS travel cycles

1.1. Conventional Potentiometric Sensors

The most commonly used present technology for low cost analog position sensors is the precision potentiometer, based on a multiwire brush wiper sliding on a thickfilm conductive plastic. Usually, there is a low ohmic „collector" track parallel to the resistor track, with the wiper forming the desired electrical „short circuit" in between (Fig. 1.3.). The materials of both the track and the brush wires are carefully chosen in order to cope with frictional wear, allowing an altogether travel of more than 400 km within lifetime for a current TPS. A typical design for a standard add-on TPS is given in Fig. 1.4.

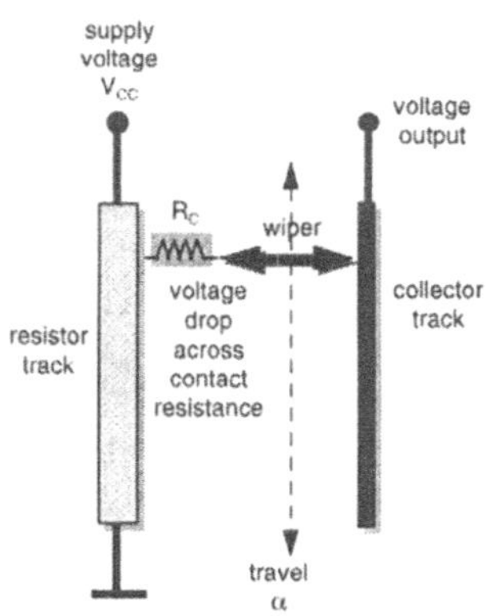

Figure 1.3.
Electrical schematics of a standard TPS potentiometer

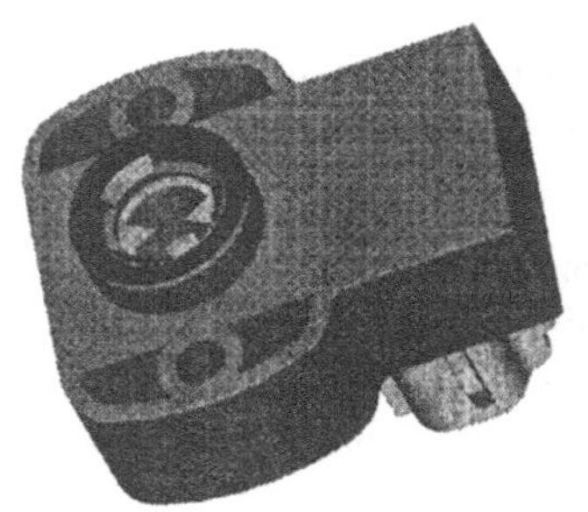

Figure 1.4.
Add-on throttle position sensor

By means of modern screen printing techniques without further trimming of the resistor track, it is standard to achieve linearity of better than +/- 3 %, with relative gradient variation (RGV 0.1°) in the range of +/- 40 %. But during lifetime, exposed to an underhood environment, a serious deterioration may arise. Bearing wear may result in faulty wiper contact force, grease or other chemical residues may deposit on the track or on the wiper wire ends. In any case, the voltage output of the potentiometer will suffer from increased contact resistance between the wiper and the track, as indicated in Figs. 1.3. and 1.5. This problem is to overcome only by a non-contacting sensor principle.

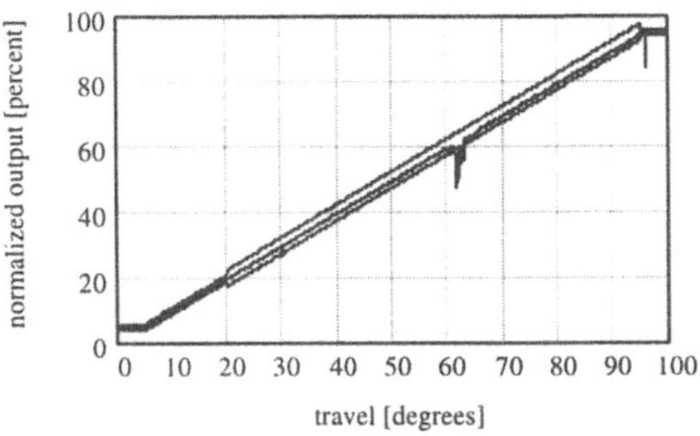

Figure 1.5.
Excessive contact resistance causes voltage output to violate tolerance band

2 Design Objectives for a Contactless Angle Sensor

Before starting a discussion on different physical principles suitable for non-contacting rotary position sensors, it may help to set up a list of some requirements from the practical side of view, especially for TPS applications.

- Almost any position sensor for the automotive range, no matter whether non-contacting or not, must be low cost, volume production. It will not be a competitor to high-end sensors in industrial applications. Any sophisticated (and thus expensive) production steps, high precision mechanics, etc. should be avoided as far as possible. There will be an advantage for a sensor design qualified for add-on solutions as well as for direct integrating into the application package, e.g. the throttle body. Consequently, the sensor principle should be tolerant of non-ideal system configurations like rotational axis displacement and related effects as far as possible.
- Special care should be taken with respect to the usual automotive temperature range, e.g. -40°C to +140°C or even higher for TPS. Even if it is common use to compensate for a mean temperature coefficient, e.g. for the sensor sensitivity, there might emerge a problem related to temperature gradients. This may cause a system failure, at least for a short time duration.
- Compared with conventional potentiometers, any non-contacting sensor will show higher complexity, regardless of the underlying principle. In order to

meet system cost requirements, signal evaluation should allow for simple calibration, diagnosis and data interfacing.

3 The Non-Contacting Rotary Position Sensor

In order to meet the design aims stated above, ruf propose a concept that is based on an anisotrope magnetoresistive (AMR) sensor chip, stimulated by a rotating permanent magnet and completed with special signal evaluation electronics to be integrated into an ASIC.

3.1 The Magnetoresistive Twin Wheatstone Bridge Microsensor

Thin film permalloy stripes show electrical resistivity being sensitive to an external magnetic field. The magnitude of this effect is determined by the angle between the internal magnetization and the current flow in the stripes (Fig. 3.1.). A comprehensive selection of publications on AMR physics and applications can be found in Ref. [1-5].

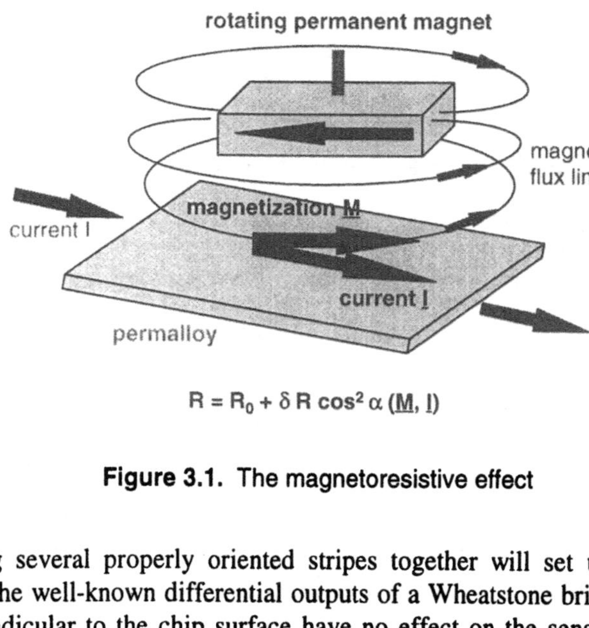

Figure 3.1. The magnetoresistive effect

Connecting several properly oriented stripes together will set up an angular sensor with the well-known differential outputs of a Wheatstone bridge. Magnetic fields perpendicular to the chip surface have no effect on the sensor. A rotating field in the sensor plane however will give rise to a sinusoidal output, provided the field is strong enough to keep the permalloy saturated [5-7]. In this case, the internal magnetization vector is following the direction of the external field. The square law of the AMR effect causes the phase angle of the signals to be twice that of the travel, and as a consequence, the rotational period of the sensor turns out to be 180 degrees. An second sinusoidal signal is generated by means of a supplementary bridge mounted on the same chip, aligned with the first one by an angle of 45 degrees [5-7]. Therefore, both a sine U_0 and a cosine U_1 are provided for evaluating the phase angle, as shown in Fig. 3.2.

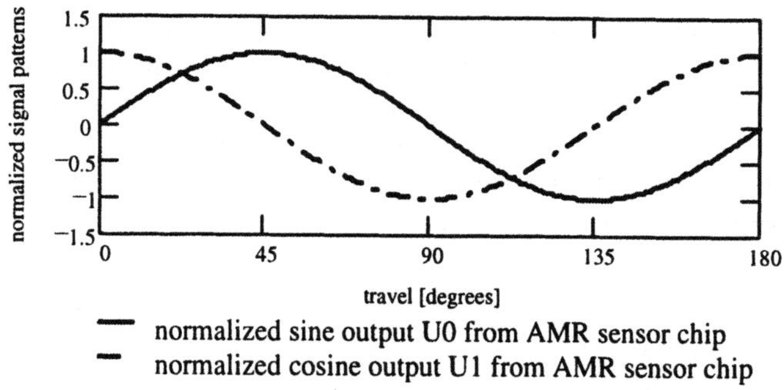

Figure 3.2.

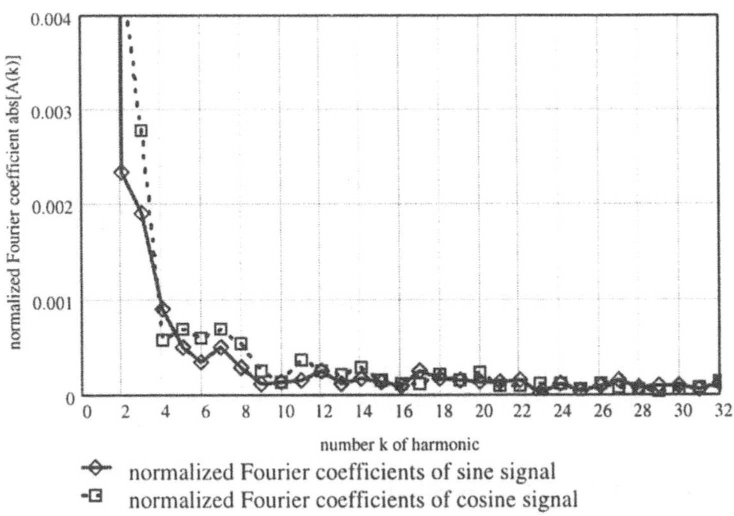

Figure 3.3.
Philips KMZ41 with magnet Sm2Co17, d = 0.2 mm (see text)

This type of angle sensor device already matches the first of the system design goals given in section 2. Provided that the driving magnetic field is not too weak, i.e. $H \cong 100$ kA/m for the Philips KMZ41 [6], the output signals are affected only by the direction of the magnetic flux lines and not by field strength. Therefore, the sensor arrangement shows the desired, restricted sensitivity for mechanical tolerances. This does not only apply to rotational axis eccentricity, but of course also to varying space between the sensor chip and the driver magnet. This is illustrated in Figs. 3.3. and 3.4., showing an FFT analysis of KMZ41 output signals at two different values for the distance d between the magnet and the front of the sensor. In this example, the magnet is rare earth Sm2Co17 with edges 8 x 3 x 7.5 mm, the magnetization being parallel to the latter dimension. The magnitude of all the higher harmonics is (far) below one percent of the fundamental. Therefore, the quality of the sinusoidal curves remains almost unchanged in an axial tolerance range of 1 mm and even more. Combined with the fact that even a simply shaped permanent magnet can be used, this set-up gives way to modular sensor concepts as described later on.

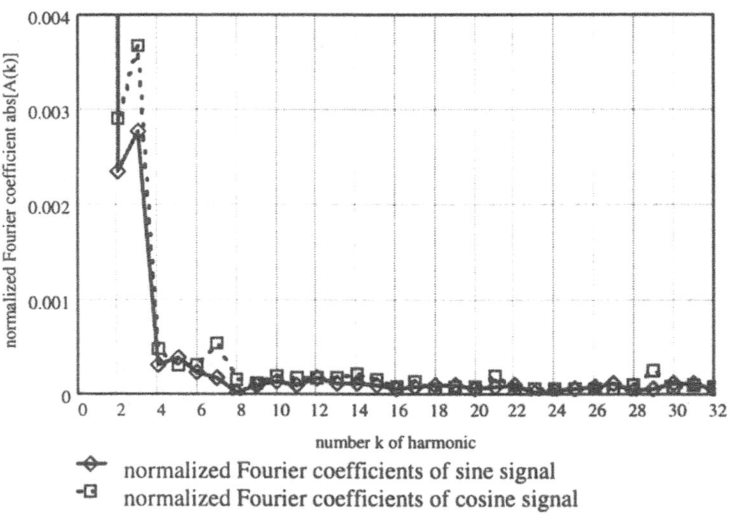

Figure 3.4.
Philips KMZ41 with magnet Sm2Co17, d = 0.8 mm (see text)

3.2 The Sine-Cosine Signal Evaluation

Taking advantage of the orthogonality relation, the phase angle α can be readily calculated, avoiding the influence of varying signal amplitudes. From a mathematical point of view, it is straight forward to divide the sensor signals

$$U_0(\alpha) = A \cdot \sin(\alpha) \text{ and } U_1(\alpha) = A \cdot \cos(\alpha) \quad (1)$$

by each other to get the tangent or the cotangent of α, respectively. With the same magnitude A for both the signals, it is obvious that α does not depend on A, and the phase angle can be found by simply utilizing the reverse functions. Fig. 3.5. shows the result of such a calculation.

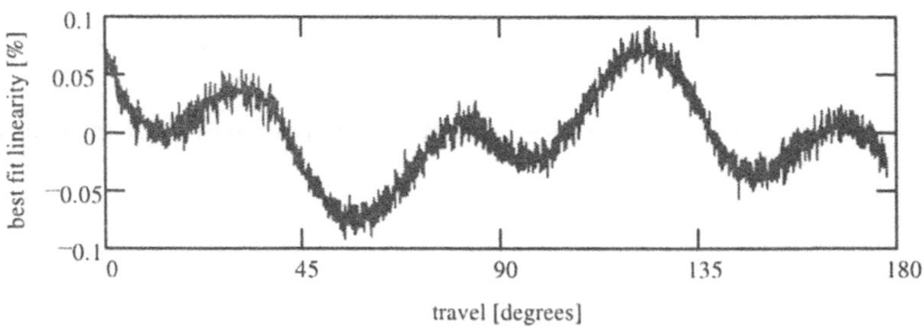

Figure 3.5.
Calculated linearity plot of typical Philips KMZ41 output data

From the statements above it is clear that the sine-cosine evaluation in principle can avoid the influence of signal drifting caused by temperature, aging etc. It turns out to be a promising approach to catch the second design goal given in section 2, provided that there is an appropriate implementation available.

Regardless of other competitive proposals published in succession [8, 9], ruf had started investigations into this field besides current conventional potentiometer development in 1994. This section will give only one example out of a range of prototypes which have been launched so far.

When calculating the tangent of α by dividing the AMR signals given in eq. (1), singularities must be avoided, while it is favourable to evaluate the tangent only in the range between -1 and +1, with its limited non-linearity. Both of these features can be satisfied by introducing four subsequent quadrants, establishing an alternate evaluation of the tangent and the cotangent.

Based on the two additional signal patterns

$$U_2(\alpha) = U_1(\alpha) + U_0(\alpha) = A \cdot [\cos(\alpha) + \sin(\alpha)] \quad (2)$$

$$U_3(\alpha) = U_1(\alpha) - U_0(\alpha) = A \cdot [\cos(\alpha) - \sin(\alpha)] \quad (3)$$

and on the definition of a function f(α),

$$f(\alpha) = \frac{U_3(\alpha)}{U_2(\alpha)}, \quad \text{valid in first and third quadrant,} \quad (4)$$

$$f(\alpha) = \frac{U_2(\alpha)}{U_3(\alpha)}, \quad \text{valid in second and fourth quadrant,} \quad (5)$$

it can easily be shown that

$$U_1(\alpha) - \frac{1}{2} \cdot [f(\alpha) + 1] \cdot U_2(\alpha) = 0 \quad \text{in first and third quadrant, and} \quad (6)$$

$$U_1(\alpha) - \frac{1}{2} \cdot [f(\alpha) + 1] \cdot U_3(\alpha) = 0 \quad \text{in second and fourth quadrant.} \tag{7}$$

$f(\alpha)$ turns out to be a tangent function with the desired behaviour. With regard to the abbreviation $G(\alpha)$,

$$G(\alpha) = \frac{1}{2} \cdot [f(\alpha) + 1], \tag{8}$$

equations (6) and (7) may be rewritten as

$$U_1(\alpha) - G(\alpha) \cdot U_2(\alpha) = 0 \quad \text{in first and third quadrant, and} \tag{9}$$

$$U_1(\alpha) - G(\alpha) \cdot U_3(\alpha) = 0 \quad \text{in second and fourth quadrant.} \tag{10}$$

Based on the fact that $0 \le G(\alpha) \le 1$, equations (9) and (10) can be evaluated in a mixed-signal circuit, with a multiplying digital-to-analog converter (DAC) being the core function. With alternating U_2 and U_3 fed into the DAC's analog reference input, the conversion cycle concludes with the DAC register representing $G(\alpha)$. This pattern D_{n-1} ... D_0 is used as the address input for an angular decoding table with n Bit resolution. The microcontroller (µC) allows for the individual manipulation of the output data, which may be accessed via a serial data interface. The scheme of the described circuit is shown in Fig. 3.6.

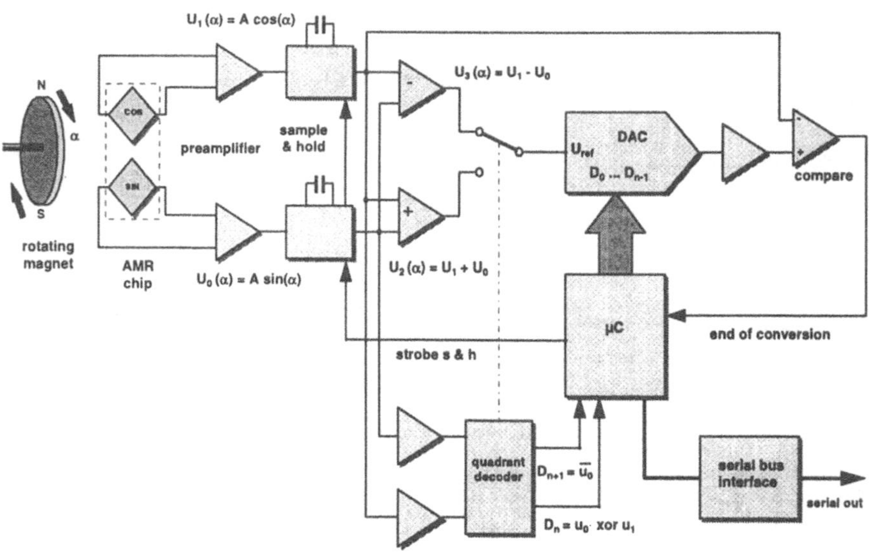

Figure 3.6.
Angular sensor schematic circuit

The valid quadrant is derived from U_0 and U_1 by means of a 2-Bit single-step code, finally given as $D_{n+1}D_n$. Consequently, the final resolution is n+2 Bit for a maximum angular stroke of 180 degrees. Figs. 3.7. and 3.8. show the result for a system with 12 bit resolution. The measurement was carried out at ambient temperature.

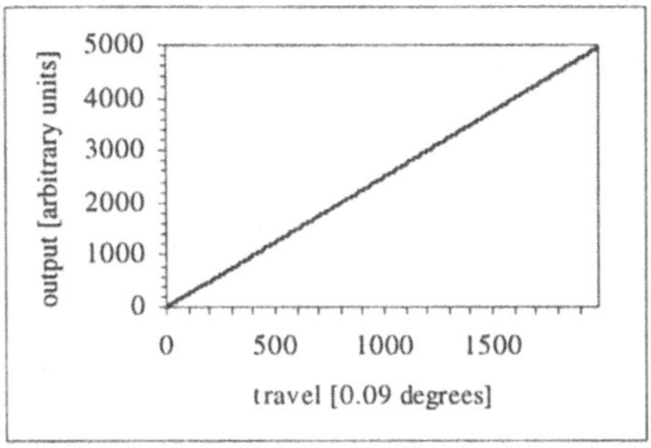

Figure 3.7.
Final output curve

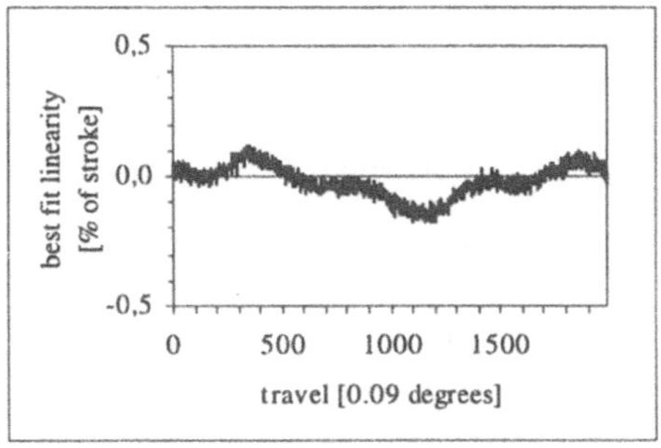

Figure 3.8.
Linearity plot for 12 Bit resolution, normalized to the total stroke of 180°

4 Outlook

The system described in section 3 can be used for various rotational sensor set-ups. However, it is qualified in particular for non-contacting TPS applications (NTPS), based on features matching the design objectives noted in section 2. Fig. 3.9. shows the exploded view of an NTPS design, tracing its origin to a standard TPS. Fig. 3.10. gives an outlook to a more advanced NTPS. It is directly adapted to the throttle shaft, in order to avoid any additional bearing.

Figure 3.9.
Add-on NTPS design

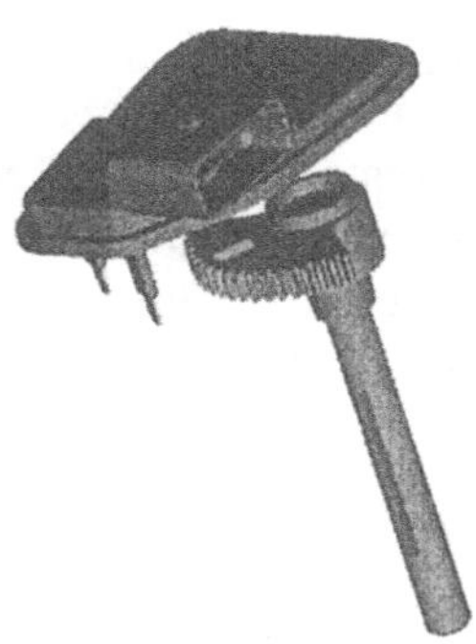

Figure 3.10.
Advanced NTPS design

Acknowledgment

The author would like to thank Jürgen Schmidt, Hans-Christian Steeg, Ralph Steige and Helmut Wagatha from ruf electronics GmbH for their help during the preparation of this paper.

References

[1] T. R. McGuire and R. I. Potter, „Anisotropic Magnetoresistance in Ferromagnetic 3d Alloys", IEEE Transactions on Magnetics, Vol. Mag-11, No. 4, pp. 1018-1038, July 1975

[2] David A. Thompson, Lubomyr T. Romankiw and A. F. Mayadas, „Thin Film Magnetoresistors in Memory, Storage, and Related Applications", IEEE Transactions on Magnetics, Vol. Mag-11, No. 4, pp. 1039-1050, July 1975

[3] U. Dibbern, „Magnetic Field Sensors using the Magnetoresistive effect", Sensors and Actuators, 10 (1986), pp. 127-140

[4] J. P. J. Groenland, C. J. M. Eijkel, J. H. Fluitman and R. M. de Ridder, „Permalloy Thin-Film Magnetic Sensors", Sensors and Actuators A, 30 (1992) pp. 89-100

[5] Philips Data Sheet „General, Angular Measurement", File under Discrete Semiconductors, SC17, 1996

[6] Preliminary specification „Magnetic Field Sensor KMZ41", Philips Semiconductor Sensors Data Handbook SC17, pp. 110-113, 1997

[7] Preliminary Data „Magnetic Field Sensors, Angle Sensor KMT 31", HL-Planartechnik GmbH, Hauert 13, Dortmund, Germany, 1997

[8] C. J. M. Eijkel and J. H. J. Fluitman, MESA Institute, University of Twente, Enschede, The Netherlands, H. Leeuwis and D. J. M. van Mierlo, 3T Twente Technology Transfer BV, Enschede, The Netherlands, „Contactless Angle Detector for Control Applications", presented at the 3rd Symposium on Magnetoresistive Sensors, Wetzlar, Germany, April 1995

[9] Objective specification „Sensor Conditioning Electronics UZZ9000", Philips Semiconductor Sensors Data Handbook SC17, pp. 224 ff, 1997

Saw and IDC Devices as Oil Quality Sensors

Anton Leidl, Bernhard Mader, Stephan Drost
Fraunhofer Institute for Solid State Technology,
Hansastr. 27d, 80686 München, Germany,
e-Mail: leidl@ift.fhg.de

Mechanical machinery has to be lubricated by oil. Lubricating oil loses its capability of protecting the metal surface during its application. This will lead to failures like wear, stiction, seizure, erosion and corrosion. To maintain for example a combustion engine of a car in good condition an oil change is necessary after a given number of kilometers. But maybe at that time the lubricating oil is still able to protect the engine and a change just means an additional economical and ecological load. Or even worse it is too late for the oil change and the engine is damaged seriously. To avoid both cases you have to perform oil analysis.

Different on-site analysis systems are available [1]. These systems determine viscosity, water content, total base number, total acid number, solid contents, oxidation levels and other physical properties of lubricating oil. These systems are very large and expensive. For these reasons they are not suitable for installation in a car or truck.

To fulfill the demands on modern sensor systems we have engineered a small, robust and cheap sensor system to determine viscosity, dielectric properties and temperature of lubricating oil on-bord (figure 3). Thus the system monitors most of the relevant physical properties of the lube oil during its entire use. With an appropriate signal processing this sensor system will determine individually the change intervals of lube oil.

The sensor system bases on a surface acoustic wave (SAW) device and a interdigital capacitor (IDC) [2]. The SAW device operates approximately at a frequency of 70 MHz. Therefore the sensor measures the viscosity at a shear rate of $\omega = 4.4 \cdot 10^8 s^{-1}$. In the same way the engine stresses the lube oil. Except of SAW devices there are no other viscometers that measure the viscosity at such a high shear rate especially without heating up the oil. As a consequence SAW devices fit very well for this purpose. The IDC is a planar capacitor and so measures the dielectric properties of the lube oil. To determine the oil temperature, we apply a PT100 or a PT1000 temperature sensor.

Figure 2 shows the sensor signal of the SAW-device in dependence on the kinematic viscosity ν of a mineral oil between 20°C and 60°C. The damping of the SAW is proportional to the square root of the kinematic viscosity. This viscosity was measured with a capillary viscometer at a low shear rate. Mineral oils without additives react in measurements with capillary viscometers as a newtonian fluid. It still has to be investigated whether mineral oils react equally in measurements with SAW-devices.

We manufacture the SAW device and the IDC in thin film technology in our fab on one single crystal quartz substrate. The Interdigital Transducer (IDT) and the IDC consist of gold metallization (figure 1). Silicon carbide (SiC) covers the whole sensor device and protects it against mechanical and chemical impact.

Applying an AC to one of the IDT it generates a SAW. This signal is received at the second IDT. The quartz substrate and the viscoelastic properties of an adjacent liquid, for example lube oil, determine the velocity and attenuation of the SAW. As a consequence a change of the viscoelastic properties of the lube oil changes the velocity and attenuation of the SAW. An oscillator circuit converts it into a shift of frequency and voltage. To track the voltage is sufficient normally. An IQ-demodulator transforms the complex impedance of the IDC into two voltages, too.

To sum up with four voltages (1 SAW, 2 IDC, 1 temperature), perhaps further parameters of the system (kind of the oil, type of motor, ...) and an appropriate signal processing this sensor system will determine individually the change intervals of lube oil. Further work is essential for investigating the long term stability of the system and an appropriate signal processing.

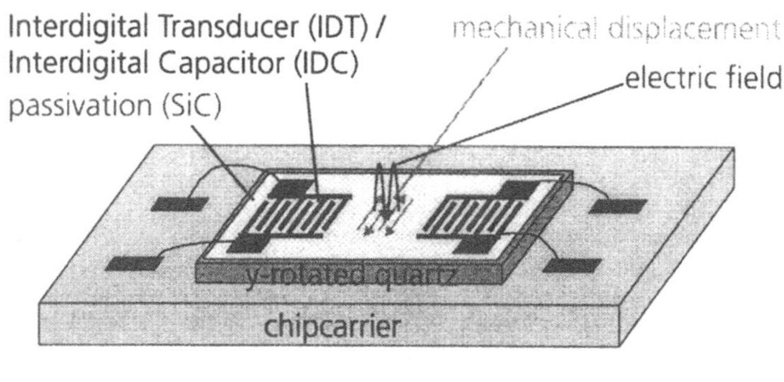

Figure 1: Principle of SAW devices. One of the IDTs acts as an IDC.

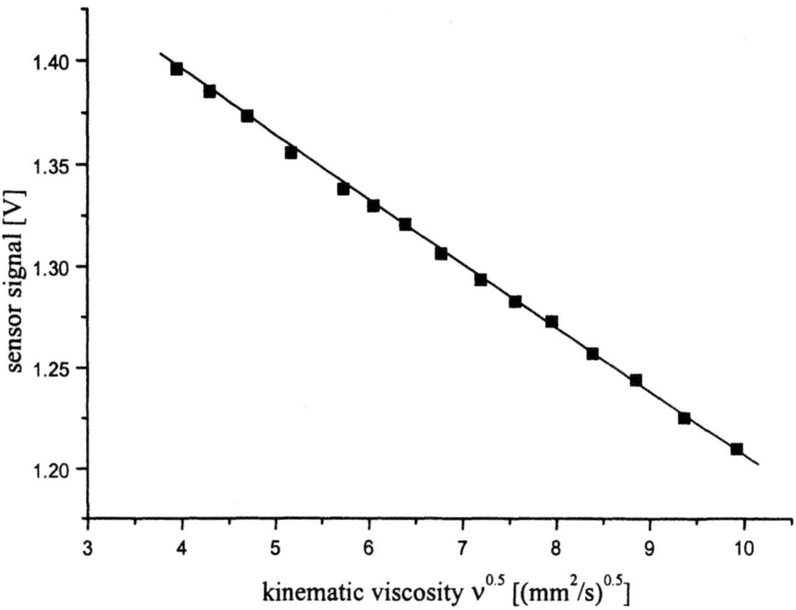

Figure 2: Sensor signal in dependence on the kinematic viscosity of a mineral oil

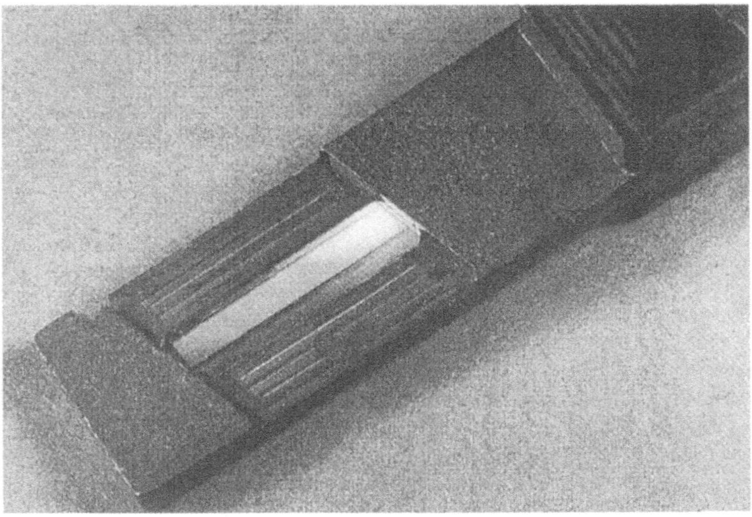

Figure 3: Photography of the sensor

[1] DIAGNETICS, 5410 S. 94TH E. AVE., TULSA, OK 74145

[2] A. Leidl, I. Oberlack, U. Schaber, B. Mader and S. Drost, Surface acosutic wave devices and applications in liquid sensing, Smart Mater. Struct. 6 (1997) 1-9

A New Silicon Rate Gyroscope with Decoupled Oscillation Modes

W. Geiger, B. Folkmer, M. Kieninger, J. Merz, H. Förstermann, H. Sandmaier, and W. Lang

Hahn-Schickard-Gesellschaft,
Institute of Micromachining and Information Technology, HSG-IMIT
Wilhelm-Schickard-Str. 10, D-78052 Villingen-Schwenningen

Abstract

HSG-IMIT is developing a new silicon rate gyroscope of very small size, low cost, and high performance. The device is called MARS-RR, which means Micromachined Angular Rate Sensor with two Rotary oscillation modes. First prototypes, MARS-RR1 yielded random walk and bias stability as low as 0.27 deg/√h and 65 deg/h, respectively. The noise equivalent rate (3 σ) corresponds to a resolution of 0.096 deg/s in a 50 Hz bandwidth.

1. Introduction

The main sources of errors of comb driven gyroscopes are mechanical and electromechanical coupling effects and the small deflections which are to be measured. The last mentioned problem has to be solved by using a detection capacitance as large as possible. The coupling effects include purely mechanical crosstalk and at least two electromechanical effects. First of all, due to the substrate supporting the comb-drives the electric field is unsymmetric and electrostatic forces arise, which pull the mechanical structure upwards. This electromechanical coupling effect is called levitation [1]. Secondly, if the comb drives are not decoupled from the secondary oscillation, the overlap of a pair of combs changes in dependence of the input rate. Thus the driving forces change and a nonlinear behaviour results. MARS-RR was designed in a way that a large detection capacitance allows very sensitive measurement and that the mentioned coupling effects are reduced to a great extend.

2. Theory of Operation

Figure 1 shows a schematic drawing of MARS-RR. The mechanical sensor element consists of comb drives, which build the spokes of the inner wheel, and an outer rectangular structure, which is called secondary oscillator. The entire movable structure is electrostatically driven to a rotary oscillation around the z-axis by four comb drives (primary mode). The remaining four comb capacitors are used to detect the primary oscillation. When the device is turned around its sensitive axis, the x-axis, Coriolis forces arise, which cause a rotary oscillation around the y-axis (secondary mode). In this direction the high stiffness of the beam suspension suppresses an oscillation of the inner wheel. Only the rectangular structure can follow the Coriolis forces, because it is decoupled from the inner wheel by torsional springs. The oscillation of the secondary oscillator around the y-axis is capacitively

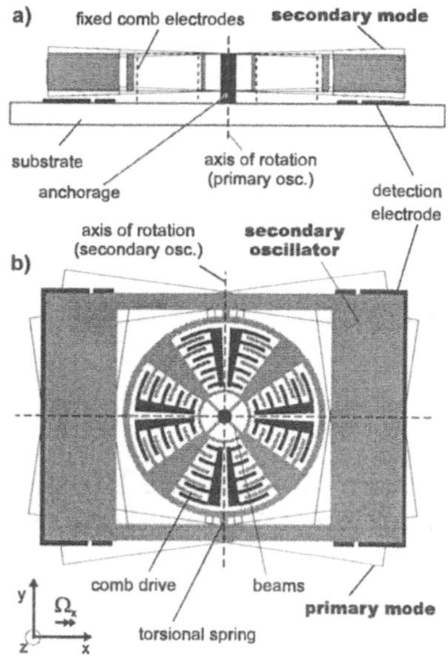

Fig. 1 a) cross-section and b) top view of MARS-RR.

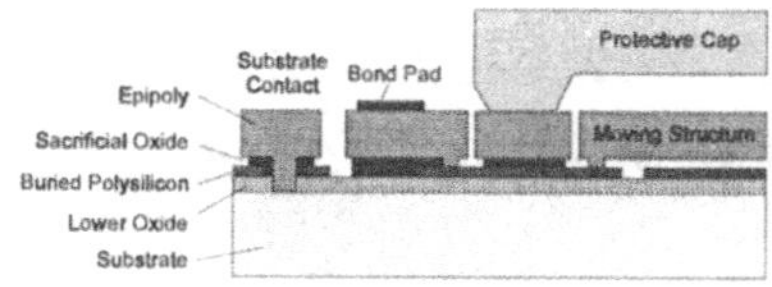

Fig. 2 Layer structure of the Bosch Foundry Process [3].

Fig. 3 SEM of MARS-RR1.

detected by substrate electrodes. The suppression of an out of plane motion of the inner wheel prevents any changes of the overlap of the combs in z-direction. Thus undesired changes of the driving force are suppressed, which would yield a nonlinear output characteristics. As pointed out in [2] the use of eight pairs of combs and their proper, highly symmetrical arrangement compensates the levitation forces up to second order. Terms of higher order are effectively suppressed by the decoupling and the high stiffness of the suspension beams in z-direction.

3. Fabrication

MARS-RR1 was produced within the Bosch Foundry process [3], which is a process similar to conventional surface micromachining with an additional protective silicon cap (figure 2). The process features a polycrystalline silicon layer with a thickness of 10.3 µm for the freestanding structure. The large thickness is achieved by using an epitaxial deposition of polysilicon ('epipoly'). A micrograph of the realized sensor is shown in figure 3.

4. Test Results

The readout of the primary and the secondary oscillation is accomplished by using an amplitude modulation technique. For driving the primary oscillation complementary 3 V_{pp} signals with an offset of 1.5 V are applied to the two pairs of comb drives. The measurements are made without any frequency matching. Tests were carried through with the sensor placed in a vacuum chamber at a pressure of 10^{-2} mbar. The electrical signals are transmitted by slip rings.

Figure 4 shows the sensor noise at zero rate input measured using a low pass filter first order with a corner frequency of 50 Hz. The standard deviation (1 σ) amounts

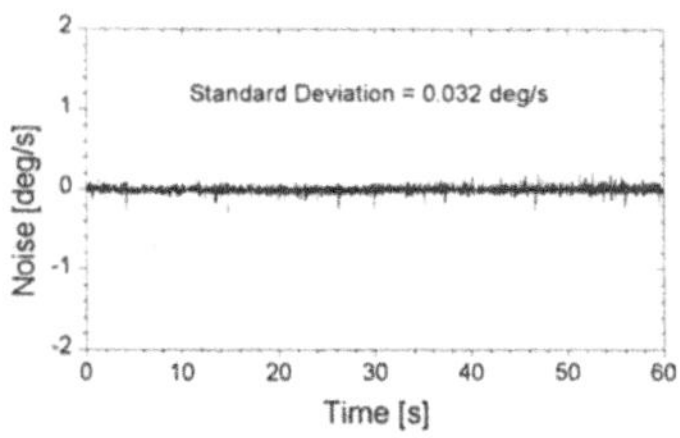

Fig. 4 Excellent short term noise.

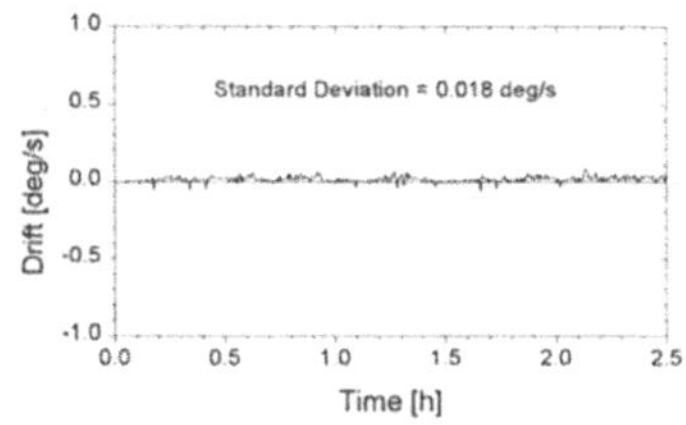

Fig. 5 Offset drift of MARS-RR1.

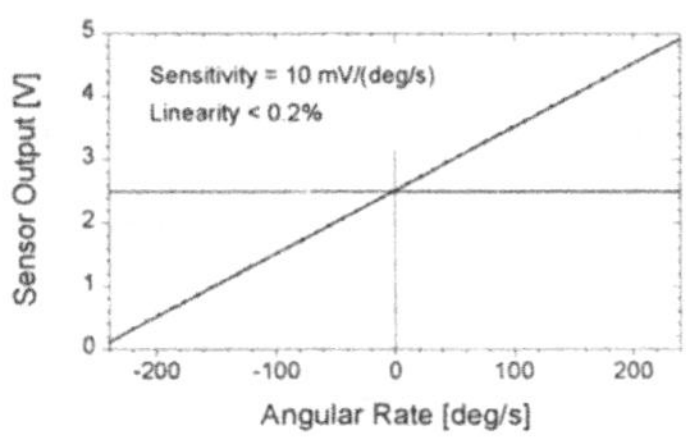

Fig. 6 Sensor output versus rate input.

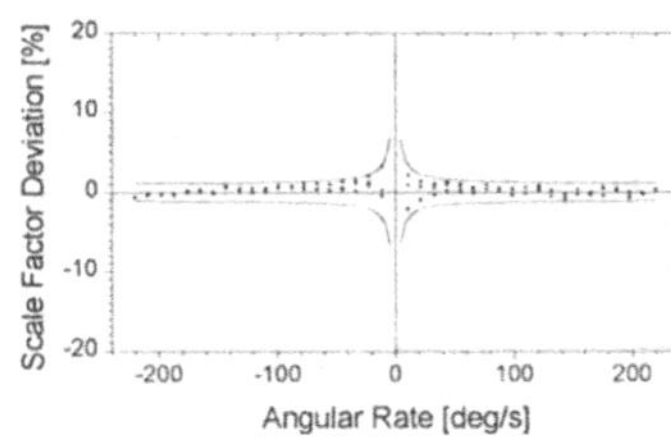

Fig. 7 Scale factor deviation.

to 0.032 deg/s which is equivalent to 0.27 deg/√h or to a resolution of 0.096 deg/s defined as the 3 σ noise equivalent rate.

The sensor drift is shown in Figure 5. For the displayed time of 2.5 h the standard deviation of the drift amounts to 0.018 deg/s or 65 deg/h.

Three measurements of the sensor output versus the rate input are shown in figure 6. The sensitivity is adjusted to 10 mV/(deg/s) with a sensor output of 2.5 V at zero rate input. Calculating the linearity based on the end point straight line methode measured at the midrange input rate yields a linearity smaller than 0.2% full scale span. This method does not give the "worst case" error. For example deviations of full scale input with different sense of rotation do not necessarily exacerbate the calculated linearity.

A much stronger criterion of the linearity is the scale factor deviation. At this time the scale factor reveals in particular imperfections of the measurement setup, especially deviations of the turn table rate, which is calibrated with a tolerance of 1% only and a noise caused by the slip rings and the electronic motor signals. The dots in figure 7 represent the measured scale factor deviation, whereby at each input rate three measurements are taken. The solid lines indicate the maximum theoretical deviation caused by the rate tolerance and by the exacerbated resolution of 0.25 deg/s. The measured points are inside of the tolerance band and the scale factor deviation is smaller than 2% for rates of 20 deg/s to 220 deg/s.

Improvements of the calibration and of the noise caused by the turn table motor should greatly improve the scale factor deviation. The theoretical deviation caused by the nonlinear behaviour of the capacitance is smaller than 160 ppm and the

Table 1 Technical Data of MARS-RR1.

TECHNICAL DATA		MARS-RR1	
Performance			
Bias Stability	(1 σ)	0.018 deg/s	= 65 deg/h
Noise	(1 σ)	0.0045 deg/s/√Hz	= 0.27 deg/√h
Sensitivity		10 mV/(deg/s)	
Linearity	(end point straight line)	< 0.2%	
Scale Factor Linearity	(20 - 220 °/sec)	< 2%	
Power Supply (temporary)			
Supply Voltage		15 V	
Current		20 mA	
Environment (FEM-Model)			
Shock Survival	(1 ms, 1/2 sine)	1000 g's	

deviation caused by a sensor noise of 0.27 deg/√h is smaller and than 0.5% for rates of 20 deg/s to 220 deg/s.

Finite element modelling of a 1000 g's shock (1 ms, 1/2 sine) yield a maximum stress of 25.4 MPa which is clearly below the fracture stress of silicon (approx. 160-700 MPa [4]). First drop tests have confirmed the simulations.

The measured and simulated parameters are summarized in table 1.

5. Conclusion

MARS-RR1 was designed to overcome the main drawbacks of micromachined comb driven rate gyroscopes. The outstanding performance of the device proves the conceptual approach. Our further work on MARS-RR1 will concentrate on packaging and the readout electronics to develop a sensor ready for production.

6. Acknowledgements

We sincerely appreciate that the device is manufactured by Robert Bosch GmbH within the Bosch Foundry process and we wish to express our special thanks to M. Illing for his permanent assistance. The finite element modelling done by N. Hey and S. Messner has contributed a great deal to the success of the project.

References

[1] W. C. Tang, et al., *Electrostatic Comb Drive Levitation and Control Method*, Journal of Microelectromechanical Systems, Vol. 1, No. 4, December 1992

[2] W. Geiger, et al, *New Designs of Micromachined Vibrating Rate Gyroscopes with Decoupled Oscillation Modes*, Transducers '97, pp 1129-1132

[3] M. Illing, *Micromachining Foundry Designrules, Version 1.0*, Bosch Mikroelektronik

[4] M. Paulsen, *Wacker-Silizium Wafer Produktinformation*, Wacker-Chemitronic GmbH, D-82633 Burghausen, April 92

Miniaturized Crash Test System

Gerhard Haas[1]

[1] Kayser-Threde GmbH, Perchtinger Str. 3, Munich, Germany

1 General

Kayser-Threde has been manufacturing crash test data acquisition systems since 1976. The demands on such systems have steadily risen over the years, especially on the number of channels available. The Kayser-Threde engineers have continuously implemented state-of-the-art technology to increase, initially, the number of channels for measuring systems from 32 to 78, and then, at the end of the Eighties, to over 100 without modifying the size (approx. 430 x 150 x 150 [mm]) or weight (approx. 20 kg). These measuring units, models K3600 and MDS-64, can only be extended by combining several basic systems. The problem is that all automobile companies now offer compact, maneuverable town cars with limited trunk space which restricts the usage of existing measuring systems. Apart from that, the trunk space is often filled with actual luggage during many research and development tests in order to simulate real conditions and this space is then no longer available for measuring equipment.

Kayser-Threde recognized this development and started to redesign the on-board measuring equipment two years ago. Intensive discussions with users around the world accompanied these measures. The conclusions drawn from the suggestions from our customers in America, Europe, Japan and Korea soon gave a clear picture of what the automobile industry expects from us:

1. Volume reduction to approx. 50 cm^3 / channel
2. Weight reduction to < 100 g / channel
3. Compact housing variants (approx. 32 channels / unit) that can be fitted to various locations in the vehicle (e.g. behind the seats)
4. Integration of storage batteries for independent operation in the vehicle and reliability during malfunctions
5. Long recording cycles (> 40 s) to allow data recording as soon as the vehicle is started (operation like tape machines)
6. High sampling rates (20 kHz / channel) to satisfy future changes in standards (SAE 211)
7. Implementation of high speed data interfaces to minimize read-out times

8. Retention of all technical features of existing systems: precise amplifier settings from 1 to 10000, analog offset compensation, simultaneous sampling on all channels by using separate ADCs for each channel, programmable bridge voltage supply, high shock resistance (max. 100 g)

9. Full hardware and software compatibility to the existing K3600 and MDS-xx systems.

A completely new data acquisition system with the project name MINIDAU® (Miniaturized Data Acquisition Unit) was developed in intensive cooperation with our partners Audi, BMW and TNO (Holland). Especially the amplifier modules, the power module and the controller module would not have been possible to realize without the use of microsystem technology.

2 Housing and Connector Panel

The demands on size and weight were rigorously implemented. The full system with 32 analog channels, battery and sensor distribution panel has a size of only 270 x 110 x 85 [mm] with 7-pole LEMOSA plugs for each sensor.

3 Amplifier Boards

The MINIDAU® consists of four amplifier boards with 8 channels each. Each amplifier board offers on a size of only 98 cm² eight dedicated A/D-converters (up to 16 Bit resolution), separate programmable bridge supplies for each channel, eight fully programmable gain amplifiers with gain steps from 1 to 10000 in steps of 1, automated transducer offset compensation and a to total of 64 MByte of memory.

With the large memory it is possible to record data for up to 50 s at a sampling rate of 20 kHz per channel. This allows the user to start the recording at a point in time when the test car is still at the launching position. All samples in one system are fully synchronized to fulfill the need of simultaneous sampling an all the channels. The memory is realized as flash EEPROMs. There is no need for a backup battery to retain the data in memory. This increases the data integrity.

4 Controller Module

The Controller Module is the heart of the system. It utilizes a 32 Bit microprocessor that controls the amplifier boards via an I^2C-Bus system.

The size of the memory available in the MINIDAU® demands fast interfaces for data transfer. The user can choose between two interface types: CAN-Bus and Ethernet. Both support high transfer rates and ensure perfect operation even when long umbilical cables are used. Both commands and data are transferred via these interfaces.

5 Power Module

The MINIDAU® is designed to keep the power consumption at a minimum to avoid overheating of the system. Most of the components are based on 3,3 Volt technology. The power module has all the necessary DC/DC converters to generate the operating voltages from the input voltage of 18 to 36 VDC.

6 Backup Battery

One of the main concerns during crash test is the system integrity. Since crash test are so expensive the users have to make sure that the data acquisition does not fail. Therefore, a rechargeable backup battery is built into the system. The battery has an intelligent charge controller that keeps track of all charge cycles of the total life time of the battery. The data is communicated between the system and the battery via the I^2C-Bus.

Microelectronic Technologies for Automotives

Daniel J. Jendritza, Jürgen Engbring and Peter Sommerfeld

Philips GmbH Mikroelektronik Module Werk Krefeld
Kreuzweg 60, D-47 809 Krefeld, Germany, Tel. /Fax: 02151/ 576 - 252 / - 418

1 Introduction

Microelectronic modules for the automotive industry have to adhere to stringent requirements in terms of resistance to various climatic conditions. This results in certain conditions that have to be met by the packaging and interconnection technology as an important interdisciplinary part of the microsystems. To fulfill the ever more stringent quality requirements, the trend is towards modules linked up in networks. The demand for higher reliability, functionality, performance, and miniaturization with a concurrent reduction in costs requires a new type of integration, which can be achieved by the combination of mechanical and electronic components into one unit. At the same time there is an integration of electronics, actuators, sensors, switches and controllers into sub-systems, which are connected to control units for engines, transmission, brakes, etc. These trends have important implications for the packaging and interconnection technologies for microelectronic modules for automotive applications. Thick-film or hybrid circuit consist of screen printed layers of conductive, dielectric, or resistive materials on an alumina substrate. These layers are applied in form of an ink or paste, are subsequently dried and sintered at high temperatures, where they form a strong bond with the alumina substrate. Thus a double-sided substrate with multilayers and integrated resistors is obtained, to which housed components can be soldered and naked dice can be bonded. These dice are covered by epoxy adhesive, or a so-called glob top, to avoid damage to the dice or bond wires. The connections to the hybrid circuit are made by single- or dual-in-line leads (SIL or DIL). The whole module is coated with lacquer for protection.

2 Product Examples

The product GST is a controller module for an automatic transmission of a car. The input consists of a pulse-width modulated signal from the on-board computer. The output is a DC signal which drives a pressure controller via a coil which switches various valves. The hybrid circuit features a voltage standard and a highly stable resistance bridge with matched TCR's and an excellent long-term stability. This is achieved by keeping resistor geometries the same and choosing a compact layout to avoid large temperature changes over the width of the bridge. Furthermore, after passive trimming the products are transferred to high temperature storage to induce and accelerate the inevitable micro-crack formation in the resistors, which causes short-term instability in the resistance value. Only after this treatment are the circuits populated and actively trimmed.

Fig. 1 shows an electronic throttle control. The hybrid thick-film module processes the signal from a magnetoresistive angle sensor, which is attached to the throttle. It then generates the output for the actuator, setting the throttle to the desired position. The hybrid is housed by a metal case, which attaches to the actuator-sensor system. This then forms a highly integrated unit, which is able to optimize the setting of the throttle depending on the current throttle position and the input to the unit.

Fig. 1 Electronic throttle control

Many new technologies were implemented with the production of the product „AltReg" for Ford USA. This is a regulator for the alternator of a car (see Fig. 2). At the heart of the product is a hybrid circuit with many new features.

The power part of the hybrid is represented by a naked die, which dissipates around 5 W. It is mounted on a so-called heatspreader, a plate consisting of copper-invar-copper layers, which by spreading the heat reduces the junction temperatures of the device. The layered structure of the heatspreader is chosen to reduce the mismatch between the thermal expansion coefficients of copper (17 ppm/K) and silicon (2.8 ppm/K) using invar (1.5 ppm/K).
The intelligence of the circuit is provided by so-called Flip-Chips. Those are naked dice with solder bumps electro-plated onto their bond pads, which are placed as surface mounted devices with their active layer at the bottom. he populated hybrid with all the components is reflowed with high temperature solder (350°C). The naked dice are wire bonded to silver bond pads using 200 μm thick aluminium

wire. The leadframe of the housing is soldered to the hybrid by a hydrogen microflame. The circuit is protected by filling the cavity with soft silicone gel and a plastic cover. All parts undergo a final test at 140°C.

Fig. 2 Alternator Regulator (AltReg) versions for Ford Motor Company, USA

3 New Emerging Technologies

There are many new technologies that are under investigation at Philips Microelectronic Modules. Some of these have found their way into new products, which are at various stages of the product life ranging from first samples to volume production. A new product is a dc-dc converter made in collaboration with Philips Magnetic Products. The device is specified with an output power of 12 W and 80% efficiency.It features a screen printed planar transformer on a substrate made from ferrite material. The control circuit is incorporated on the same substrate for higher integration. This required the use of special thick-film pastes that give good adhesion on the ferrite substrate and of the thick-print and multilayer technologies. Fig. 3 shows a testproduct for a Ball-Grid-Array (BGA) multi-chip-module (MCM) used on an automobile trans-mission controller module. The BGA consists of a PCB substrate made from BT material, which has got an improved temperature stability over standard FR4. 5 dice with a total of 250 I/O connections are bonded onto the substrate and electrically connected by either gold or aluminium thin wire bonding. To protect the module, it is incased with a thermoset plastic overmould using transfer moulding. The outer area connections on the reverse side of the substrate are formed by eutectic solder balls, placed into solder paste and reflowed to a semisphere.

Fig. 3.: BGA on a printed circuit board and on a hybrid.

Another emerging technology is called Z-Strate®. Here, copper conductor tracks are deposited onto alumina substrates by a combi-nation of electroless plating, photolitho-graphy, and electro-plating. Through-hole contacts to the flip side of the sample are also made from pure copper. Z-Strate allows very fine copper tracks (down to 50 µm) with the high aspect ratio of 1, i.e. the copper tracks can be made just as high as they are wide. The maximum thickness of the tracks is 150 µm. This gives Z-Strate good properties for high current applications. Z-Strate has even slightly better thermal properties than hybrid thick film because of the thick copper, that can have heat-spreading ability.

4 New Emerging Products and Future Aspects

New developments and new features in the modern motor car increase the number of loads with high power consumption, like the electrically heated windshield and catalytic converter. These new trends in the automotive field seem to suggest that the introduction of a new voltage level for the electrical system of a car is soon to come. This new voltage level will probably lie around 42 V and can therefore drive high power loads much more efficiently than the standard 14 V system, because of the decreased currents.

Active suspension is one of the up and coming fields which is also fueled by new types of actuators. A product which is now being investigated is a control module for an active suspension with electrorheological (ER) fluids. These fluids increase their viscosity dramatically on application of an electric field. Using a suitable construction with high voltage electrodes integrated into a shock absorber, a variable suspension unit can be realized. The control modules for these new suspension modules are best integrated into the unit itsself.

The ever-increasing demand of the automotive industry for smaller, lighter, more efficient, multifunctional, smart and modular sub-systems has to be met by developing and implementing the appropriate technologies for cost-effective production. This results in high automation for an increased process control and a high quality. Advanced interconnection and packaging technology will drive the realization of trends for mechatronic sub-systems in the automotive industry.

5 References

[1] Proceedings of the Forum Vehicle Electrical Systems Architecture organized by SICAN, Hannover 6.3.97 and 19.6.97.

[2] Engbring J., *Hybrid-Technologies at Philips Microelectronic Modules*, Proceedings of the Philips Super Integration Meeting, Eindhoven 14.2.1996.

[3] Sommerfeld, P.K.H., Jendritza, D.J., *Microelectronic Modules for Smart Actuator Applications,* Proceedings of the 21st ICAT International Symposium held at Pennsylvania State University, 11.2.97.

Europractice Competence Center No 1 Microsystems for Automotive Applications and Physical Measurement Systems

W. Riethmüller, B. Wenk, R. Dudde, J. F. Clerc*
Fraunhofer Institut für Silizumtechnologie, D - 25524 Itzehoe
* CEA - LETI, F - 38054 Grenoble

Introduction

Microsystems are today key components for many industrial, consumer and medical products and systems, and their necessesity for the realisation of competetive products in various market segments is growing continously. MST or MEMS is on the way to be a multi-billion ECU world activity. But R&D and production in Microsystem technology (MST) or MEMS, especially the set up of qualified fabrication processes is costly and time consuming.

The European Commission (EC) supports within the Europractice (EP) frame five Manufacturing Clusters (MCs) for Microsystem Technology (MST). MCs are consortia of industrial companies and R&D institutes similar to the sucessfull concept of IC foundries and design houses. They are offering to the outside world (industry as well as universities) their industrial production processes and related services. The MCs are therefore a quick and cheap way for the development of application specific MST protoype devices. Due to the use of industrial production processes the way towards a later mass production of the new device can be organised.

But MST or MEMS is still a young and dynamic technology. Only a few standard processes are available today not sufficient to realise each device requested from the market. Furtheron the support of these fabrication processes by CAD tools is not as good as for ASIC design and the economic aspects are more difficult to evaluate. As consequence since October 1997 a further service is supported by the EC, the so-called Competence Centers (CCs).

Competence Centers are virtual centers consisting of typically 2 leading R&D institutes in the MST field in Europe. The core competences of the 6 different CCs are organised towards products and application areas, rather than around technologies. Each CC will offer the competence in its application field all over Europe. To answer on every customer request the CCs will support each other (networking).

CC1 is a partnership between the Fraunhofer Institute for Silicon Technology (ISiT) and CEA-LETI, with ISiT as the lead partner.

Objective

The main objective of the CCs is to increase the awareness of the European industry on MST, to assist the transfer of R&D results into applications in industry, to guide potential customers on the way to the appropirate manufacturer and to stimulate the development of new devices and fabrication processes. Solutions for European system companies may be based not only on European component suppliers. Using contacts, processes and devices available at US or Japanese companies are also welcome.

Service Offers of CC1

- established routes to and assistance for device prototyping as well as large scale manufacturing at industry
- design service for application specific sensors, actuators and microsystems using well established industrial MST and IC foundries (e.g. AMS CMOS or Bosch surface micromaching process, others are in preparation)
- access to packaging providers specialised on MST products
- experimental feasibility studies for advanced microsystems
- design studies on sensors, actuators and microsystems
- preparation of technical and economical studies, concept evaluation and development
- access gate to all Europractice services
- overview on European MST based sensor manufacturers, distributors and service providers
- Customer in-house and public training courses for engineering staff and management on technical and economical aspects of MST including packaging

Potential Customers of CC1 are:

1. car manufacturers and automotive suppliers already active in MST or interested to use MST in future
2. companies form various market areas interested to use the microsystems know how, technology, production processes or products developed and in use for automotive applications

Customer specific material, studies, designs and training activities must be paid by the customer. All other services are free of charge due to the support of the EC.

Strategy

CC1 will undertake appropirate dissemination and public relations activities, such as attending exhibitions and conferences and establishing web pages to announce its service offers. CC1 will not only act in a reacitve manner on customer requests. It will be proactive contacting potential customers directly and will establish its own view of the market, ie the services required by industry. If a customers request can not be handled by CC1 it will establish trough the EP network contacts to the appropriate CC or MC. CC1 wants to establish itself as a single point of contact to offer the mentioned services in the competence field of CC1.

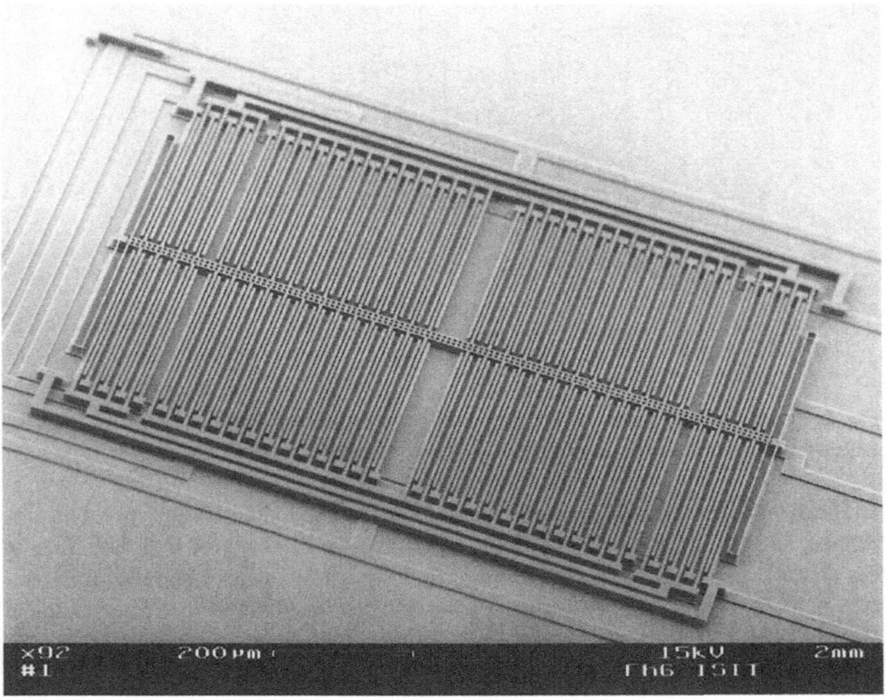

Fig.1. Customer specific 50g accelerometer designed by ISiT for a company, realised with the Bosch surface micromachining process

The support of customers by hardware will be done by using design interfaces to industrial production processes or, in case of advanced devices by using qualified prototype processes at leading R&D institutes like ISiT and LETI. First interfaces to IC processes and to the MST processes of MC1 and MC3 are already existant due to the partnership of ISiT in MC1 and LETI in MC3. Further interfaces to other manufacturers in- and outside of Europractice are in preparation. If requested by customers or if an interest is announced by the supplier such design interfaces will also be generated towards US or Japanes MST manufacturers.

Members of CC1

ISiT, located in Itzehoe and Berlin developed during the past 15 years together with dozens of European and even US and Japanaese companies many basic and complete production processes for sensors and actuators as discrete devices as well as monolithically or hybrid integrated with electronics. Since the 80's serveral car manufacturers and automotive suppliers were using ISiT as R&D partner for MST. ISiT is in parallel deputy partner in CC4 and partner in MC1, the German MC with Bosch as lead partner.

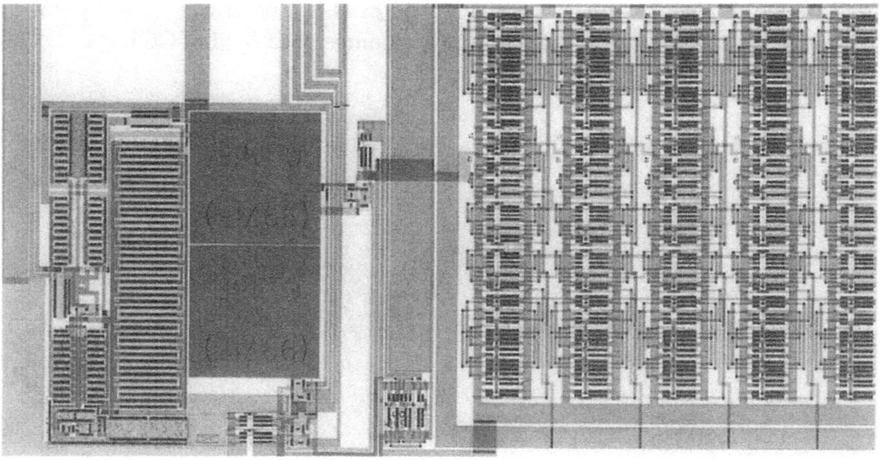

Fig. 2. Layout of the evaluation IC for the 50g accelerometer. Designed by ISiT, realised with a 0.8 μm CMOS process offered by AMS within Europractice.

LETI, a well known leading R&D institute in the MST area located in Grenoble, France, ist part of CEA. LETI is in the MST area well known for their activities in the area of micro- and integrated optics and components for computer pheripherals, e. g. hard disks, but performed also various projects in the area of mechanical sensors. LETI is lead partner of CC3 and is in addition partner in MC2, the French Manufacturing Cluster with Sextant as lead partner.

Fig. 3. 1-D micro scanning unit for a collision avoidance system. Designed and realised by LETI for an industrial customer.

In the area of dissemination and user contacts the two main partners ISiT and LETI are assisted by the Innovation Center IZET in Itzehoe, the VDI/VDE-IT in Teltow (Berlin), well known as manager of the German MST funding programm and by CEA-BEM, the marketing office of the CEA.

Microtechnologies – The Alternative for Automotive Applications Concerning Economic, Technical and Quality Aspects

Johannes Herrnsdorf and Dirk Enderlein

HL-Planartechnik GmbH, Hauert 13, D-44227 Dortmund, Germany

Keywords. Magnetoresistive sensors, cost aspects, temperature sensors, thermopiles, mass airflow sensors

Introduction

Especially in automotive industries information systems grow up. In former times speed, perhaps temperature and concerning only sport cars revolution measurement were of interest for the driver. Nowadays, the driver is the chief of (self working) Management Systems like ABS, ASR, ESP or Air-Bag. Motor Management, Climate control and Navigation systems are the challenges for the next years. The new technological trends in measurement techniques concerning chip and sensor technologies are well suited for automotive applications with their specific requirements as

- low material cost
- high productivity
- high automatic production
- high level of quality control.

Economical and Cost Aspects

Quite similar to automotive production, the cost structure is based on

- high investment and high fixed costs.

A typical production facility needs an investment of about 10 million DM. This means, by thinking of 5 years depreciation time and 200 labour days per year a daily depreciation cost of 20.000 DM. For staff, rent, energy and basic chemicals, you need an amount of 1.700.000 DM for annual fixed costs. This means a technological labour day cost 28.500 DM. We have to produce one car a day or 28.500 sensor components for 1 DM per day.

The High Productivity – Nickel Temperature Sensors

One possibility of measuring temperature is to use platinum or nickel RTDs. The basic physical effect is, that materials change their electrical resistance according to temperature. The innovation step of the last years was characterised by the changing from wire wound components to chip technology. Ni or Pt are deposited on ceramics, glass or silicon substrates and microstructured by lithographic techniques, trimmed by laser and tested to components with an accuracy of 0,5 to 1%.

The actual technology is working on ceramics substrates, but the roughness of the surface allows only a lower number of chips per substrate. Using polished silicon , 12000 components can be produced on one 4" Wafer with pure production costs of about 0,10 DM. Assembly, sales and overhead must be added. Within 72 hours 720.000 components can be manufactured.. This means a yearly production capacity of 16.000.000 units in one shift and 48.000.000 units in 3 shift production. It is only a sales problem.

Stand of the art Ni on ceramics	**Structure width µm**	**Component Area mm2**	**Components per substrate**	**Production cost per chip/DM**
Ni 1000	35	34,29	194	3,56
Ni 500	30	19,05	353	1,95
Ni 100	20	4,18	1676	0,41
Option				
Ni on silicon				
Ni (100-1000)	5	0,706	9000	0,10
Ni (100-1000)	4	0,525	12000	0,08

Combination of Microtechnologies to Microsystems e.g. Thermopile, Mass-Airflow-Sensors and Related Devices

Whereas RTDs are used for contacting temperature measurement, another component for sensing contactless temperature res.. infrared radiation are coming up with increasing request . A series of 100 thermocouples is deposited on a silicon chip (2x2 mm) with microtechnologies. Coating silicon wafers with SiON-layers by CVD-processing and etching silicon from the backside, leads to possible membrane technologies down to typically one micron thickness. If one side of a thermocouple is placed on the membrane and the other on the silicon frame, you receive a voltage which is proportional to the temperature of the membrane, respectively of a black layer deposited on the membrane. This black layer absorbs infrared radiation. So you have a device to measure infrared radiation intensity caused by a surface temperature. An additional nickel RTD on the silicon chip will measure the ambient temperature. Using both signals it is possible to measure contactless surface temperature.

Actual Applications are:
- ear-fever-thermometers
- toasters, microwave ovens
- new concepts in automotive climate control techn.
- anti-fog-regulation on the windshield .

The combination of silicon membrane technology and thin film RTD deposition on substrates, opens the way to small, high speed mass air flow sensors. For this application two heating resistors are deposited side by side on a thin membrane. An additional temperature sensor placed on the substrate is used for ambient temperature measurement. Using this technology it is possible to build an bi-directional mass air flow sensor with reaction time below 5 ms and an accuracy of better 2 %. This will guide to a more effective combustion with lower level of exhaust gases.

Being able to realise combinations of temperature sensors and heaters a very easy modification by depositing metal oxides on the heating area can be a wide area of gas sensing devices. This shows the great variety of products and application basing on a few standard processes.

Bright Variety in Electronic Signal Processing – Magnetoresistive Sensors

Beside the combination of microtechnologies the necessary signal processing opens new dimensions for automotive sensor applications.
Actually magnetoresistive sensors are a coming up product in automotive applications, especially concerning rotation speed measurement in Anti-Breaking-Systems (ABS). The basic physical effect consists of a thin film permalloy deposition on a silicon substrate, changing it's resistance due to external magnetic fields. The permalloy stripes form a Wheatstone bridge, whose output voltage is proportional to the magnetic field. Applying a DC magnetic field in the rear of the sensor, which will be changed by an rotating gear wheel or a polarised magnetic wheel in front of the sensors, rotation speed sensors for ABS and other applications can be constructed easily. The advantage of rotation sensors using MR-sensors are a zero frequency operation and a better signal to noise ratio then known devices. The high sensitivity of the magnetoresistive sensors higher than the magnetic earth field allows compass applications for navigation systems in automotive environment. Each current produces a proportional magnetic field, so magnetoresistive sensors can measure current.
By variation of the chip layout the devices can be fitted to a number of other applications, such as piston detection and angle measurement. Nearly each application needs its own specific electronic signal processing.

Electronic Signal Processing, Hybrid or Monolithic Integration

The combination of signal processing and sensor device on one single chip, the monolithic integration seems to be the highest step of technological evaluation. The practical experience shows economical limits with advantages to hybrid integration.
At stand of the art 3000 MR-components can be produced one one 4" Wafer. The typical chip price is below 1,- DM. Assuming that 3000 ASIC's can be produced for 1 DM per chip on the same wafer size, the system will cost 2.50 DM using hybrid integration. Working with monolithic integration you have no assembly cost for the chip itself but as matter of fact, you can produce only half quantity - 1500 chips per wafer, because you need room for the signal processing part. So the price per chip doubles to 4,-DM system cost.

Hybrid integration		Monolithic integration	
3000 ASIC's produced on 4"-Wafers		1500 ASIC's produced on 4"-Wafers	
Price per chip	1,00 DM	Price per chip	2,00 DM
3000 Sensor devices produced on 4"-Wafers		1500 Sensor devices produced on 4"-Wafers	
Price per chip	1,00 DM	Price per chip	2,00 DM
Assembly costs	0,50 DM	Assembly costs	0
System costs	2,50 DM	System costs	4,00 DM

Batch Fabrication on High Quality Level

Manufacturing sensor devices means working in batch fabrication. Errors in processing reduce the yield of production dramatically, because they are multiplicated to the following procedures. So batch fabrications are typical installed on high automated level with a high degree of process control on each technological production step. The quality philosophy is to make the production sure without any errors in order to have excellent yields first. Within this, it must be approved, that all errors are found.

So batch fabrication comes close to the automotive quality strategy.

Technology and Process Approval Cut Down the Costs for Semiconductor in the Automotive Environment – ES 59001 the Brandnew European Specification for Semiconductors for Automotive Applications

Dipl.–Ing./Eur Ing Arno Bergmann
VDE Testing and Certification Institute
Merianstraße 28 D–63069 Offenbach Germany
Phone: ++ 49–69–8306–535 Fax: ++49–69–8306–636

Introduction

The object of the CENELEC Electronic Components Comitee (CECC) is the System for electronic components of assessed quality and provides European Standards for electronic components which are used all over the world.
With the new specification ES 59001 "Approval scheme for automotive oriented applications within the electronic components industry – Semiconductor stress test qualification" most advanced quality assessment procedures have been established by the Automotive User Group of CECC for all kind of semiconductor devices for the automotive environment. Members of the Automotive User Group are the European Car Manufacturer and their major suppliers. This new specification combines "Technology Approval" the most advanced quality assessment procedure worldwide with the specific automotive oriented requirements.
Technology Approval (TA) is a totally new concept. The objective of TA is to ensure the quality of the finished products by controlling the production processes through the application of modern techniques like process monitoring and SPC rather then selecting the products at the end of all manufacturing processes with final tests into good and bad ones. TA is using the principles of TQM. Management Commitment to Quality, Continuous Improvement Programm, Customer Satisfaction are some of the headings.
As a baseline, the manufacturer's quality system meets the requirements of ISO 9000, enhanced by additional criteria specific to the electronic components industry. An independent organization has confirmed that the company's management has declared and implemented a commitment to quality at the highest level. Process control is applied in all key areas. Superior quality is achieved through continuous improvement programmes and progressive tightening of outgoing quality levels (e.g. ppm, dpm).

Electronic Components will be supplied against CECC specifications or CECC-registered manufacturer's data sheets. CECC certified components can be supplied quickly and conviently as non-approved components. Design and processing improvements can be introduced into component ranges without delay.
Experiences with Technology Approval have shown that it is an universally applicable standard, which is consistent with existing quality assurance systems like QS 9000 and VDA 6.1 and which has sufficient potential to cover most types of electronic components and especially the needs of the car manufacturer.

Technology Approval

What it is and how it works

With Technology Approval (TA) CECC and IECQ has introduced a totally new concept. This procedure recognizes that there has been a change in the quality assurance programs of the companies, moving away from end of line testing to the control of the individual manufacturing processes. "Quality cannot be tested in". The manufacturer has to demonstrate by using Process Monitoring or SPC that all his processes are completely under control instead of selecting out good products from bad ones.
TA is based on the principles of **Total Quality Management** (TQM) with its aim of a strong reduction of time and costs for Qualification of Electronic Components.
Technology Approval is a method of approving a complete technological process (design, process realisation, product manufacture, test and shipment) covering the qualification aspects common to all products as determined by technology under consideration. This method has evolved to meet the needs of users and manufacturers and incorporates many of latest principles and technics in the management of quality ie. TQM. It extents the existing approval concepts by adding the following principles as mandatory aspects of Technology Approval:

1. The foundation of Technology Approval is a formal system for quality management, such as TQM within the organization. This requires that all employees are actively involved in the commitment to quality.
2. It makes use of statistical methods and demonstrates the control of processes rather than relying on end-of-line testing.
3. It arranges for continuous quality improvement and its demonstration.
4. It addresses the overall technology and operation associated with the manufacture rather than the components themselves.
5. It achieves procedural flexibility by basing the quality assurance procedures and documentation on the company's own quality management system and customer requirements.
6. It guarantees rapid approval or extension of approval by making reference to the company's Technology Approval Declaration Document (TADD).

The general requirements of TA are described in EN 100114-6 the technical requirements are described in the relevant **Technology Approval Schedule** (TAS).

One of the general requirements is:

The **Management Commitment**. The manufacturer shall provide a statement of his corporate management's commitment to **Continuous Quality Improvement** and **Customer Satisfaction** (which is the heart of TQM).

If a manufacturer wants to apply for TA he has to act as a "Control Site". The **Control Site** is the location of the manufacturer which has the overall responsibility for the operation within the scope of Technology Approval. The Control Site has to control all adminstrative and technical processes, from the customer order up to the after sales service, independent whether these are inhouse processes or subcontracted to other companies. One of the most critical point is the appropriate control of the interfaces between these processes.

Interfaces are for example:

Customer requirements to design, Design to production, Design to test, Design to maintenance and customer support, Supplier to production, Production to test, Test to delivery, Delivery to customer, Subcontractor to manufacturer.

Therefore a certain **Management Structure** is necessary. The manufacturer shall establish a declared Technology Review Board (TRB) and a Contractor Manufacturer Review Board (CMB) or equivalent organizations.

The **Technology Review Board (TRB)** shall be established to control, stablize, monitor and improve all processes. It is responsible for the overall control of the Technology Approval, and for conducting periodic reviews. The TRB shall have procedures in place for assessing the current status of quality and reliability of components. It is responsible for the development of an overall quality plan and shall consist of representatives of all functions described in that plan such as marketing, sales, design, technology development, manufacture, testing and quality assurance.

An **Contractor Manufacturer Review Board (CMB)** shall be established in the case of several companies are involved in a single Technology Approval to control the interface(s) between the companies. It shall consist of at least one representative from each company and be responsible for the overall control of the interface(s).

The requirements of the **Description of Technology** are defined in the relevant TAS. There are many areas which the component manufacturer has to adress:

Design, Manufacture, Package, Test, Test Vehicles, Process Characterization, Packing and Shipment.

All **Key or Critical Processes** have to be identified and have to undergo Process Control. The manufacturer shall describe in detail his plan for the verification and demonstration of the capability of these processes, using for example: Capability Studies, Capability Indicies (CP, Cpk) and Process *Monitoring*.

For each process **Quality Indicators** (a statistical measure of the relative quality of a process) have to be identified.These Quality Indicators are required for the Quality Improvement Programm, in its first step the present situation should be identified and realistic improvement goals have to be set up. The measurements of improvements shall be based on the Quality Indicators.

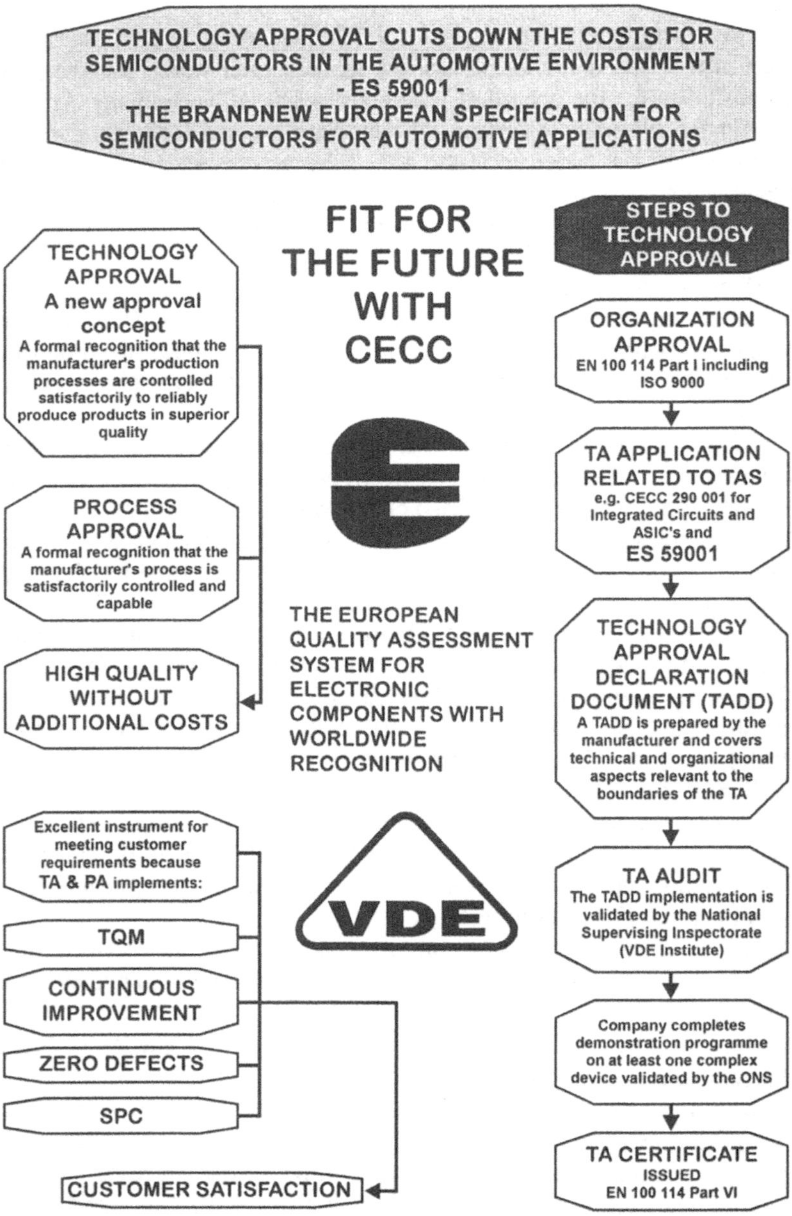

Application of Gas Sensor for Air Damper Control Systems in Automobiles

Nobuaki Murakami and Mitsuharu Kira

Murakami Research Services, Felderhof 186, 40880 Ratingen, Germany
FIS Inc., 2-5-26 Hachizuka, Ikeda, Osaka 563, Japan

1. Introduction

Recently, the demand for the application of an Automatic Air Damper Control System in automobiles has been expanding. In this system; composed of one or two gas sensors and a signal processing software, it is important to detect incoming gasoline and diesel exhaust gases from other cars sensibly and reliably. However, conventionally used tin oxide (SnO_2) based semiconductor gas sensors have high sensitivity to reducing gases (e.g. CO, HC) but are effected by the presence of oxidizing gases (e.g. NO_X). Since the contents of reducing gas in diesel exhaust is relatively lower compared with gasoline exhaust, it is more difficult to detect diesel exhaust from low concentration ranges.

In order to achieve a reliable detection of both types of exhaust gases, it is important to find the most effective combination of effective gas sensor types and signal processing methods. For this purpose, we have made some experiments evaluating the gas sensing performance of commercially available gas sensors (16 models of CO, NO_X, VOCs, Air Quality sensors, etc.). For evaluation, we have used different test methods and various different types of test gases and exhaust gases. Through these experiments, the following various related factors have been reviewed. 1) sensitivity characteristics of gas sensors, 2) an efficient test method to find a suitable combination of sensor performance and signal processing software, 3) the relationship between the sensitivity characteristics of gas sensors and the actual performance on the road.

2. Experimental

2.1. Test System

Figs 1a and 1b show the schematic diagram of the test system.

- Static method (Fig 1a): different types of test gases were injected into the test vessel. This method is effective in evaluating the basic sensitivity characteristics of gas sensors.
- Dynamic method (Fig1b: A): environmental air containing various exhaust gases were supplied through a gas sampling system.
- Pulse gas injection method (Fig 1b: B): different types of exhaust gases (gasoline, diesel exhaust gases from various cars) were sampled in a gas bag. The sampled

gases were injected into the test system through an air pump with a constant flow of clean air. This method is effective to evaluate the dynamic characteristics of sensors, in terms of the combination of sensitivity and response speed.

The same test system was also used in the actual performance test on the car. Using this system, the performance of tested sensors can be compared under the same conditions.

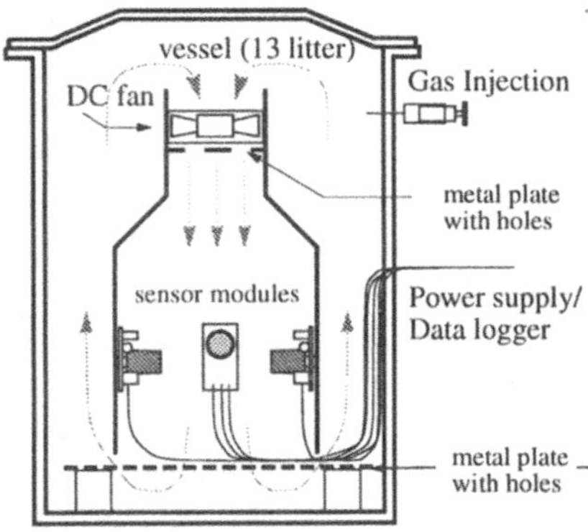

Fig1: Static method

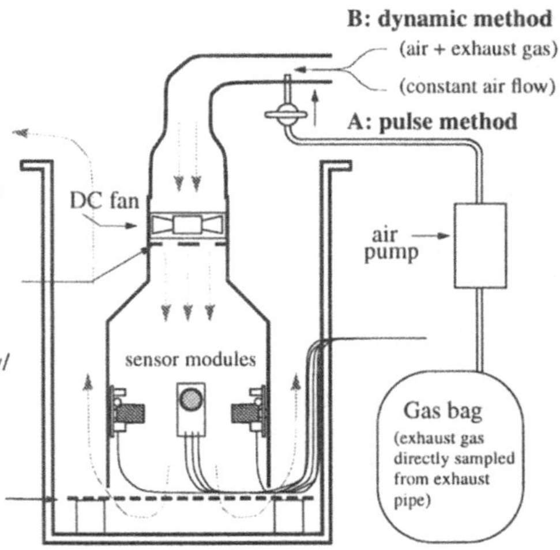

Fig 2: Dynamic method and pulse method

2.2. Sample

The following parameters of gas sensor design were evaluated in this study.

1) sensing material: tin-dioxide (SnO_2) and tungsten oxide (WO_3)

2) sensing element: Tube type, Plate type and Bead type.

A conventional type of gas sensor was used for reference (TGS822: SnO_2 with tube type design). The output voltage across the load resistance from each sensor was recorded in a computer through a data logger. The operating conditions of each sensor are based on the standard specifications.

3. Results

Fig 2a indicates an example of sensor signal changes when different amounts of gasoline exhaust were injected into the test chamber (pulse injection method: with 1 sec, 3 secs, 5 secs, 10 secs and 15 secs of pulse duration). Fig 2b shows an example of response to diesel exhaust gas. The sensitivity of each sensor is compared in terms of sensor resistance change ratio (R/Ro) between the base resistance level in clean air (Ro) and in exhaust gas (R). Y-axis is expressed in the reciprocal number of R/Ro and higher peaks indicate higher sensitivity to reducing gases. These two diagrams represent the different response to gasoline and diesel exhaust gases: much higher sensitivity is obtained in gasoline exhaust detection than in diesel exhaust detection.

In general, the composition of exhaust gas can vary widely and dynamically depending on the types of engine, different stages in engine operation (e.g. idling or hot start, load factors, ambient temperature, etc.). Therefore, it is important to evaluate the senor performance in such variable conditions. Through further comprehensive tests, the following results were obtained.

- The tested sensors have different degrees of cross sensitivity characteristics between reducing and oxidizing gases. As a result, the direction and degree of sensor resistance changes depend on the gas composition. It is found that one of tested diesel exhaust gas sensors using SnO_2 shows a higher sensitivity to oxidizing gas, but is still not selective (B: SnO_2 - plate type - D). Other SnO_2 gas sensor have a good selectivity to oxidizing gas, but response speed is relatively slower than other sensors (A: SnO_2 - bead type - D).
- It is noted that some of tested n-type SnO_2 sensors respond to the reducing gases in the co-presence of oxidizing gases much more sensitively than the conventional sensor (e.g. A: SnO_2 - bead type - G, A: SnO_2 - plate type - G+D, B: SnO_2 - plate type - G, D: SnO_2 - plate dual type- G). This fact is shown in the direction and the degree of sensor resistance changes in Fig 2b.
- In order to improve the performance of this system for reliable exhaust gas detection, the following approaches are suggested.
  1) Improvement in the sensitivity/selectivity to oxidizing gas and response speed: for the detection method of using a combination of gasoline and diesel exhaust sensors.
  2) Application of gas sensors which have higher sensitivity to reducing gases with smaller influence from oxidizing gases combined with a sufficient signal processing method to improve the S/N (signal-noise) ratio: for the detection method of using a single sensor to detect both gasoline and diesel exhaust.

4. Conclusion

Through these experiments, some key factors to improve the performance of an Air Damper Control System has been studied. In the actual application, gas sensor signal can be effected by many influential factors (e.g. air flow rate, temperature, background air quality changes, etc.).

There are two possible approaches of using two sensors and one sensor. The effectiveness of each method depends on the balance between the over all exhaust gas detection performance, reliability and cost. Development of an effective single sensor for both gasoline and diesel exhaust will connect to the further expansion of this system in automobiles (low cost). Further study to find an effective method to cancel such noise factors is an important key as well as the improvement of gas sensor characteristics. In this technology, a combination of an effective sensor type and signal processing method is the target.

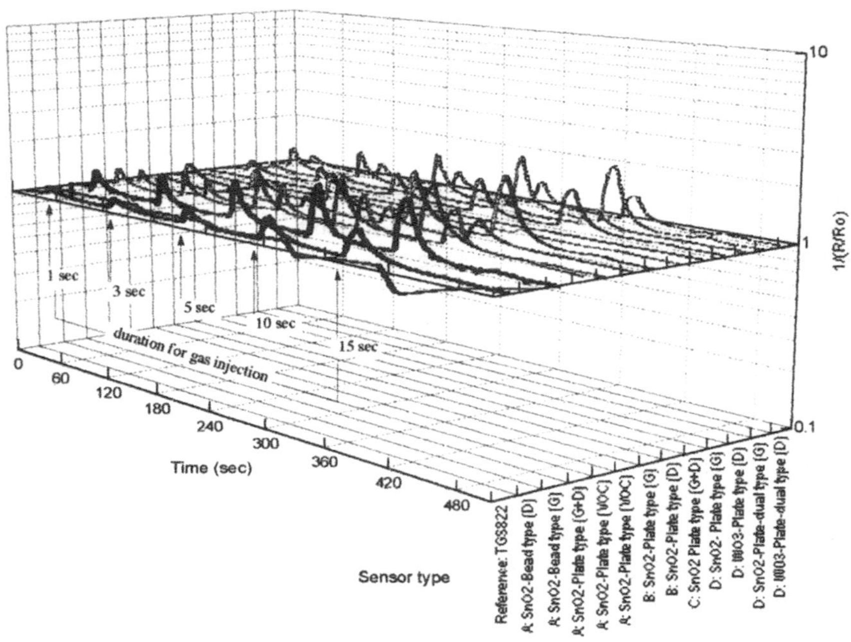

Fig 2a. Response of gas sensors to a gasoline exhaust (OPEL: Astra 1.6l, 1994)

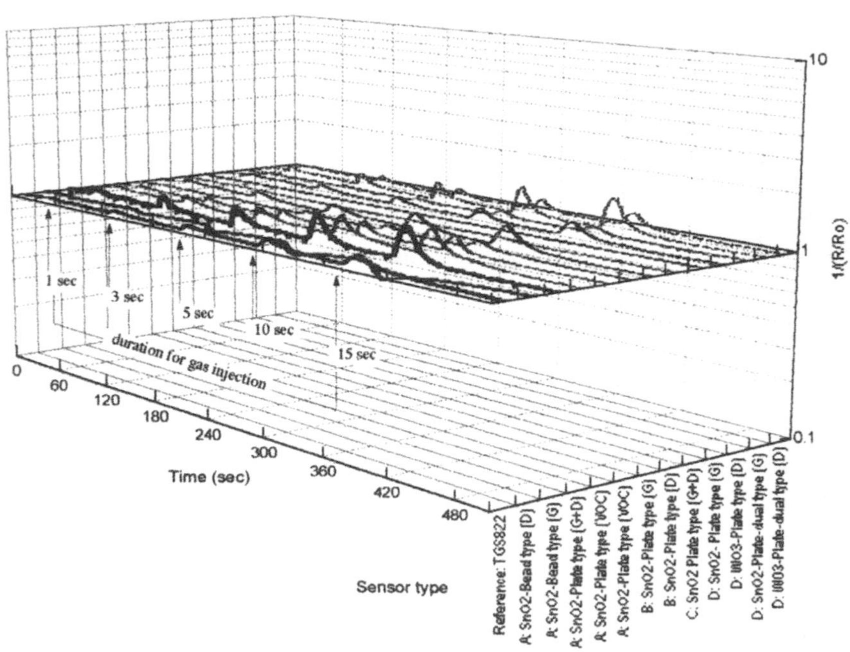

Fig 2b. Response of gas sensors to a diesel exhaust (CITROEN: CXTRD2, 1988)

Protectionsystems for Car Passengers by Surveillance of the Airquality

Christian Voit

Unitronic GmbH, Mündelheimer Weg 9, 40472 Düsseldorf

Abstract. The increasing air-pollution through traffic makes it neccessary to develop new, reliable airconditioning systems for the in-car air-cleaning. Therefore it's not only important to regulate the air insight a car, but also to control the air-supply from the outside. A conventional type of SnO_2 gassensor can detect the unburned components of gasoline, but doesn't respond to exhaust from diesel-engines, which are usually most disturbing for humans. The evaluation of many different materials has shown, that WO_3-gassensors are most effective for the detection of diesel-exhaust. Systems using this sensor are currently tested in various studies. The signal-characteristics are in good relationship with the estimated concentrations of diesel-exhaust and can be calculated with various algorhythms. Systems using a combination of SnO_2 and WO_3 sensors can reliably detect pollutions and supply a signal for closing the damper-flap of the air-inlet.

1 Introduction

The demand of car-passengers regard more and more cleaner air and better comfort, therefore it's important to secure a good air-quality inside a car. On the other hand, there is a public interest for clean air in buildings and offices as well as inside a vehicle. In a car, it's not only important to regulate the internal air-quality, but also to control the air-supply from outside the vehicle. In the past, the driver himself has controlled the damper-flap manually. An automated air-control system would allow him a better concentration on the traffic as well as it also ensures allways sufficient air-supply. With an automatic air-control system, the opening and closing of the flap is controlled by a computer, which enables or eventually disables the incoming air according through measurements from one or more gassensors. The gases, that have to be detected, are the exhaust-gases from gasoline- and diesel-cars.

A conventional type of SnO_2 gassensors can detect gasoline exhaust precisely, but is nearly not able to detect diesel exhaust. In fact, we feel bad, when we drive behind a diesel car, that produces black smoke and contaminants. In various studies, materials were tested, that are sensitive to diesel exhaust.

2 Materials and Their Characteristics

The various gas-sensitive metaloxides are mainly produced by adding ammonia (NH4OH) to a solution of metal chloride. The fall-out is dryed and sintered at an air-temperature of approx. 500°C for one hour. By adding terpentene, the oxides are transduced to a paste. This paste is applied to a substrate of alumina-ceramics (plate approx. 3x1,5mm), which is equipped with electrodes of gold and a small heating-element. Finally, the sensors are heated up to 700°C for 10 minutes.

Operating temperature of a finished sensor is around 350°C. For testing, the sensors are placed in a gas test-chamber and circulated by fresh, active-charcoal filtered air. 15 Minutes later, the sensor resistance in air (R_{air}) is measured. After adding a certain volume of gas into the testchamber, the sensor resistance (R_{gas}) is measured again after four minutes. The sensitivity can then be calculated as follows: R_{gas}/R_{air}. The characteristic of a conventional SnO_2-gassensor to gasoline and to diesel exhaust is shown in diagram 1.

The sensorresistance changes drasticly in the presence of gasoline exhaust. In other words: the sensor can detect gasoline exhaust very good. But the sensitivity to diesel exhaust is quite bad, so that this sensor alone is not suitable for an air-quality control system.

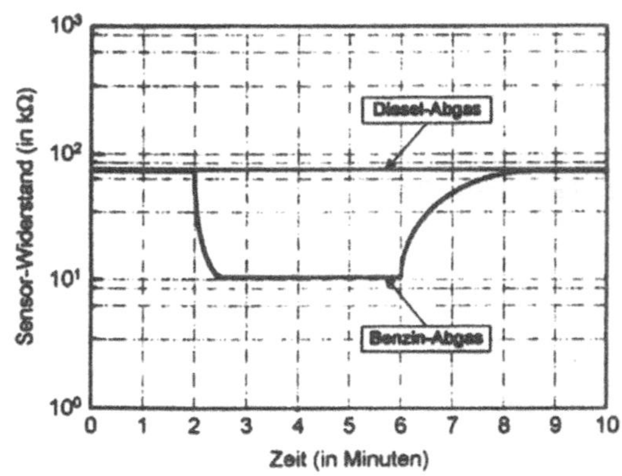

Diagram 1. Typical reaction of a SnO_2-Sensors to diesel- and gasoline-exhaust at a gasconcentration of 2500 ppm

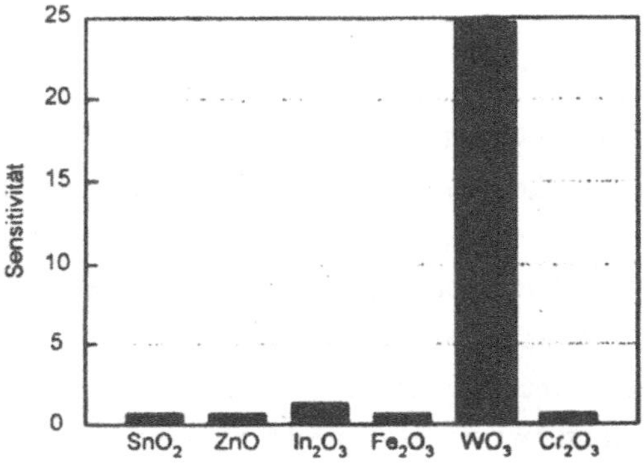

Diagram 2. Sensitivity of various sensormaterials to diesel exhaust (2500 ppm)

Diagram 2 shows the sensitivity of various sensing materials like SnO_2, ZnO, In_2O_3, FeO_2, WO_3 and Cr_2O_3 to diesel exhaust. All materials except WO_3 don't show suitable characteristics for the diesel detection, but WO_3 has a very exposed sensitivity. The reaction of a WO_3-sensor to gasoline- and diesel-exhaust is shown in diagram 3. The sensor resistance increases not much in the presence of gasoline exhaust, but in the presence of diesel exhaust, the resistance goes up a lot. The conclusion is, that WO_3-sensors are capable to detect diesel exhaust sufficiently. The reaction time is approx. 90 seconds.

NOx and flammable gases like H_2, CO and hydrocarbons are normally parts of exhaust gases. It can be shown, that the sensivity of a SnO_2 sensor to NO_2 is smaller compared with a WO_3-sensor, but very large for H_2, CO and EtOH. On the other hand, the resistance of a WO_3-sensor rises about 20 times in the presence of only 1 ppm of NO_2, meanwhile the reaction to other gases (H_2, CO, EtOH) is very small. That means, WO_3-sensors have a high sensivity and selectivity for the detection of NO_2. These results were found in several studies. It's supposed, that SnO_2-sensors react weak to diesel exhaust, because the sensitivity to flammable contents of the exhaust is blocked through the presence of NO_2, from which the concentration in diesel exhaust is normally much higher than in gasoline exhaust.

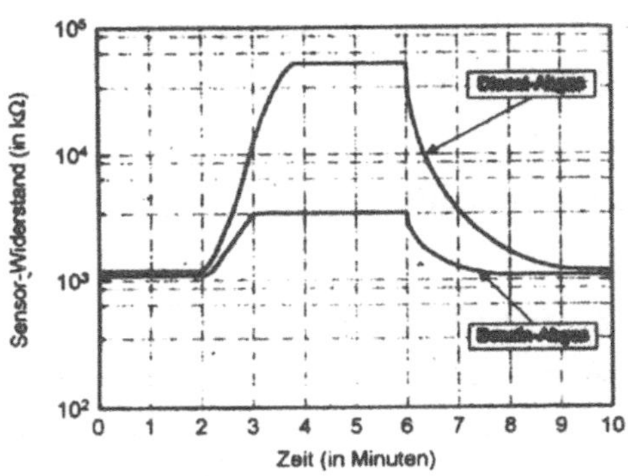

Diagram 3. Typical response of a WO_3-Sensors to 2500 ppm diesel exhaust

3 Practical Tests

For a faster response, the WO_3- and the SnO_2-sensors are mounted in front of the engine. The output signals (V_{out}) and the differentiated (dV_{out}/dt) signals are processed. Changes of the sensorsignal from the WO_3-sensor in the presence of diesel exhaust are relatively slow, so that additional signalprocessing is neccessary, to get a controlsignal for the air-control system. On the other hand, the characteristic of the signal to diesel exhaust is different from the reaction to gasoline exhaust, and there is the additional signal from the SnO_2-sensor. With those additional informations, it's possible to control the system with gassensors. To achieve an input-signal for the damper flap control-unit, the sensor-signals are amplified and then processed in a μ-controller. The output signal is correlated with good accuracy to the presence of air-contaminants and black diesel-smoke. The system is capable to control the damper flap fast and precisely.

The combination of a SnO_2-sensor for the detection of gasoline exhaust and a WO_3- sensor for the detection of diesel exhaust was tested in practice, where the damper flap was closed and opened automatically by a control-signal. To verify the function of the control-system, this control-signal was compared with an output signal of a SnO_2-sensor TGS 822 (Figaro Engineering, Inc.), which was mounted inside the passenger cabin of the vehicle. When the damper flap was continously open, the TGS 822 signal was related to the control-signal from the sensor-unit in the engine-room. As soon, as the automatic damper-control was activated, the TGS 822 signal in the cabin stayed very low, because the contaminated air from outside couldn't get into the passenger cabin. The result was clean air inside of the car through the use of SnO_2- and WO_3- sensors.

4 Conclusion

The practical tests have confirmed, that SnO_2- and WO_3- sensors can detect gasoline- and diesel-exhaust reliably. Further it could be shown, that a system, that uses both sensors (SnO_2 and WO_3) simultaneously, is well suitable to detect the environmental contamination caused by the summary of gasoline- and diesel-exhaust.

Emerging Assembly Technology for Automotive Applications

Katrin Heinricht, Joachim Kloeser, Kai Kutzner, Liane Lauter,
Rolf Aschenbrenner and Herbert Reichl

Fraunhofer Institute FhG/IZM-Berlin
Dept. Chip Interconnection Technologies and Advanced Packages

Gustav-Meyer-Allee 25; D-13355 Berlin, Germany
Phone: ++ 49 30 464 03 155, Fax: ++ 49 30 464 03 161
E-Mail: kloeser@izm.fhg.de

Abstract:

Emerging assembly technology include both flip chip and chip scale packaging. Both of them provide excellent capabilities to fulfill the needs of today's and tomorrow's requirements in automotive applications. These products must withstand high temperature, vibrations, wear and abuse.

A key issue for the introduction of flip chip technology in the markets is the availability of bumped devises and the implementation of low cost bumping processes since the established methods need expensive equipment for metal sputtering and photolithography. At the moment chemical bumping processes based on electroless nickel plating are used by different companies as a low-cost alternative. According to the economical bumping technique the industry needs low cost and high density substrates for use with flip chip devices. For this rigid or flexible laminates have the most promising potential.

Furthermore, the increasing interest in cost effective flip chip technologies leads to the development of various flexible methods for the deposition of solders and adhesives for use on chip or substrate. In addition many suppliers improved the properties of solder materials, solder pastes, solder balls, solder wires and adhesives.

In this work the studies are focused to processes for stencil printing of solder paste. The conventional printing methods and materials are not suitable for applications with fine pitch structures. To achieve reproducible and homogeneous solder deposits the process techniques for fine pitch printing require an improvement of the physical properties of solder paste, of the stencil materials and stencil processing technologies as well as of the printing equipment. Using solder pastes with very small particle sizes, a nitrogen atmosphere and a well controlled temperature profile of the reflow furnace is required. The basic process steps required for the development of a cost effective and flexible flip chip technology are described in this poster (see figure 1).

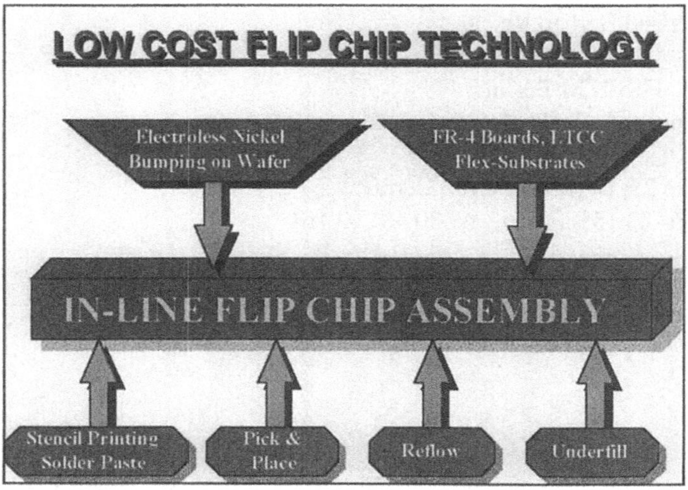

Fig. 1. Basic processes and process steps for low cost flip chip assembly

Besides the technological aspects the key point for the introduction of the flip chip technology in a wide field of applications the equipment for the production is of essential importance. These equipment have to fulfill the demands of manufacturing concerning smaller devices, finer pitches and face down assembly including optical alignment of the chip bumps to the corresponding substrate pads. The machines must be compatible and easy to integrate to already existing production lines for SMT.

The Fraunhofer Institute (FhG/IZM-Berlin) together with several industrial partners has set up a demonstration center for the assembly of flip chips (FC) and chip size packages (CSP). It consists of a complete production line, and additional equipment for quality control and process development (see figure 2). In the center of interest is the implementation of cost effective, high reliability and environmentally friendly processes. To achieve these goals upscaling existing flip chip technologies from laboratory examples to industrial production is necessary. At the same time the technologies will be optimized to guarantee high quality standard and good yield at high volume production. In order to demonstrate the high performance of cost effective flip chip technologies the process flows of different flip chip assembly techniques using solder will be compared and described in detail. It is important to note that flip chip and CSPs can be used in conjunction with standard surface mount technology (SMT) devices. The development of these processes was performed by simultaneous engineering. Finally the yield and the costs will be estimated and the reliability results of a selected flip chip process will be presented.

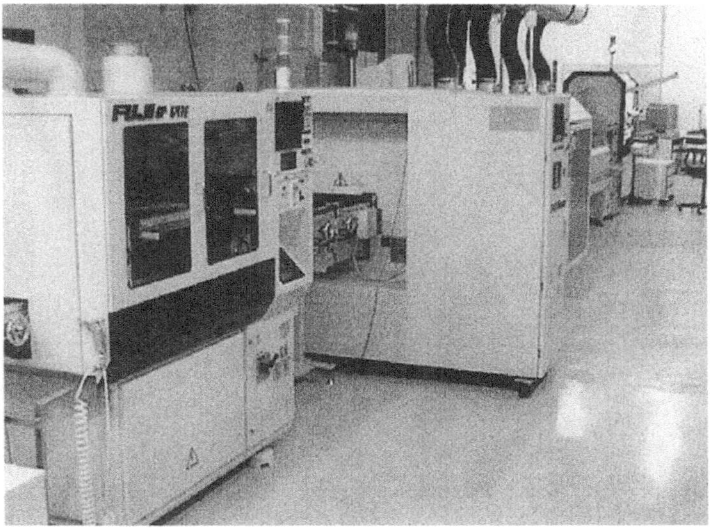

Fig. 2. The flip chip and chip size package demonstration and production line at FhG/IZM – Berlin.

Surveillance of the Air-Quality in the Passenger Cabin of Cars

Christian Voit

Unitronic GmbH, Mündelheimer Weg 9, 40472 Düsseldorf

Abstract. Beside measuring the environmental air-quality for the autodamper-control of a car, the inside air-quality in the passenger cabin can be measured as well by using gassensors. With this information, appropriate filters and air-supply from the outside, a good inside air-quality can be ensured.

1. Introduction

Air-quality sensors are commercialized yet for the control of the damper-flap in a car. New developments are heading to clean the inside air with suitable filters and air-conditioners. Background is, that with the increasing number of air-conditioners a sufficient air supply is not allways ensured. For example on a hot, sunny sommerday, a lot of drivers drive with closed windows and inlet flaps to increase the cooling effect of the air-condition. If more passengers are in the car, who eventually smoke, it can easily happen, that the concentration of contaminants in the passenger cabin can reach comfort- and health-affective levels. If the route goes through an area with high environmental pollution, the opening of the damper-flap could cause an inlet of unhealthy and unconvenient exhaust gases in the cabin.

2. State of the Art

In Japan, air-cleaning units for the passenger cabin are allready available. On demand, they can be switched on to filter out cigarrette smoke or unconvenient smell. In those air-cleaners, gassensors are used to detect exceedings over a maximum pollution-level, and to switch on the filter, if necessary. The supply with fresh air from outside can either be controlled manually by the driver, or automatically by a damper control system with gassensors, which ensure, that no polluted air from outside can reach the passengers.

3. Integrated Air-Control Systems

Developments are heading clearly to integrated air-quality surveillance systems, where the outside air-quality is measured as well as the inside air-quality. According to those measurements, fresh air from outside is automatically leaded to the passenger cabin on demand, or, if there is bad air inside as well as outside, the filter will be switched on until the outside air-quality is good again. Such an integrated system has several opportunities:

- Slow changes of the air-quality are normally not recognized, because the human nose is much more sensitive to changes than to constant levels. The concentration of the driver is nevertheless affected through continously high levels of CO_2, poor O_2-supply or cigarrette-smoke. Gassensors are able to detect higher levels and can ensure sufficient air-supply.

- CO_2 has no smell, but it causes an unrecognizable, slowly appearing tiredness. Gassensors can measure the CO_2-level in the passenger-cabin reliable.

- The convenience increases, because smells can be reduced.

- The lifetime of active-charcoal filters increases, if they are only switched on when it's necessary.

4. Sensors for Gasdetection

Important for the air-quailty in a closed room are mainly the levels of CO_2, O_2 and pollutions caused by cigarrette-smoke or smell emitting sources.

4.1. CO_2-Sensors

Most suitable for the detection of CO_2 is it's peculiarity to absorb infrared light of a specific wavelengh. The principal is shown in fig. 1. Affordable sensors are available as standard-products on the market for some time now with the possibility, to build modified solutions for the integrati-

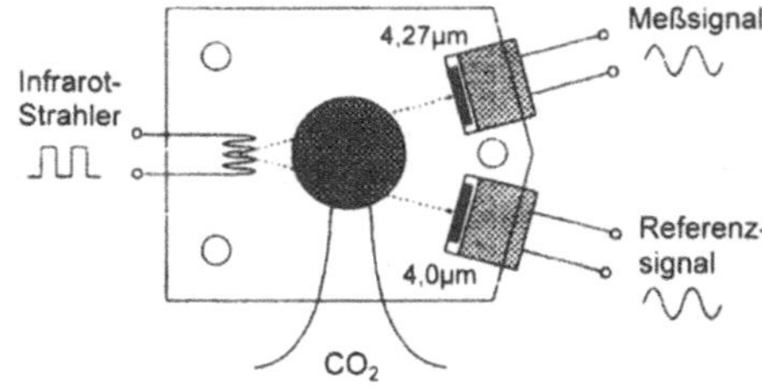

Fig. 1. Principal of an infrared-optical CO_2-sensor

Fig. 2. Housings of infrared-optical gassensors

on in customspecific designs for cars (fig. 2). Infrared-optical gassensors are well known for their good selectivity, reliability and long lifetime, because it's a totally physical measuring principal. The power-consumption is low, and the sensors are immediately useable after power-on. The infrared-optical principal is as well as for CO_2 suitable for CO and hydrocarbons, but only for higher concentration-ranges, and not for those concentrations, that are important for air-quality systems.

4.2. O_2-sensors

Measuring O_2 is in the low-cost range at the moment only possible with electrochemical gassensors. Fig. 3 shows a possible construction of such an O_2-sensor. Allthough there is a chemical reaction in the sensor, a lifetime of up to 5 years can be achieved with modern sensor-cells. Using an electrochemical gassensor is very simple, because the output signal is a voltage, that's linear proportional to the oxygen-concentration. Because of the small formfactor (fig. 4), it's easy to integrate the devices in a car. The only problem at this time is the limited operating temperature, because the sensor contains electrolytes based on water, that may freeze at low temperatures.

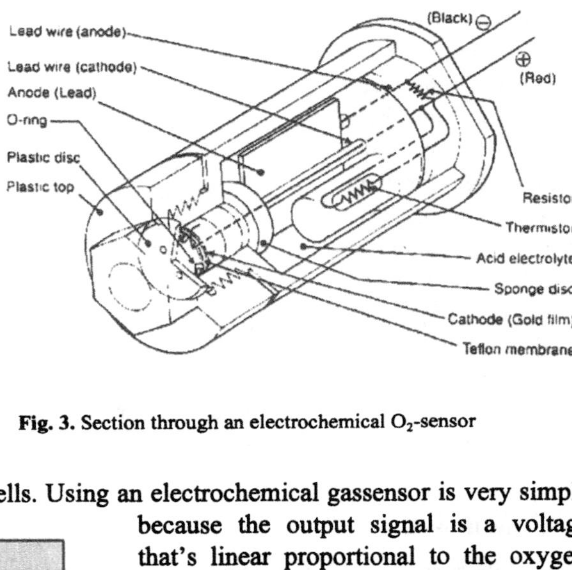

Fig. 3. Section through an electrochemical O_2-sensor

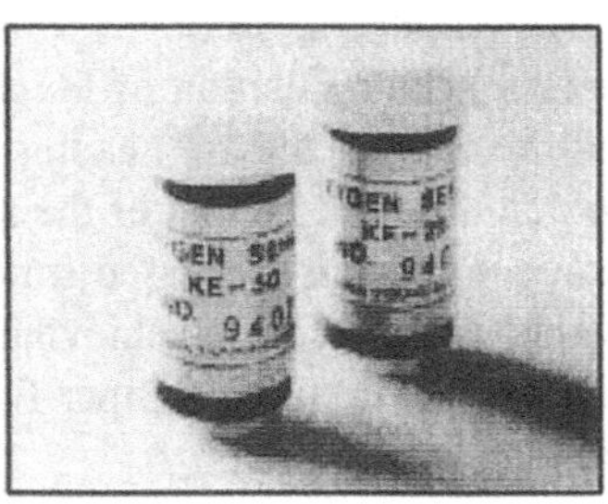

Fig. 4. Housings of electrochemical O_2-sensors

4.3. Semiconductor gassensors

Gassensores based on SnO_2 use the effect, that the surface conductivity of the material changes in presence of reducing or oxidizing gases, when the sensing material is heated up to approx. 350°C (fig. 5). They are currently commercially used for the detection of gasoline- or diesel-exhaust to control the damper-flap. Semiconductor gassensors are well known for their high reliability and long lifetime, and for their ability to detect very small concentrations of air pollutants. Therefore it's less important, to select a single component out of a complex mixture. It's more important to measure the summary of all contaminations, for which those sensors are ideally suitable.

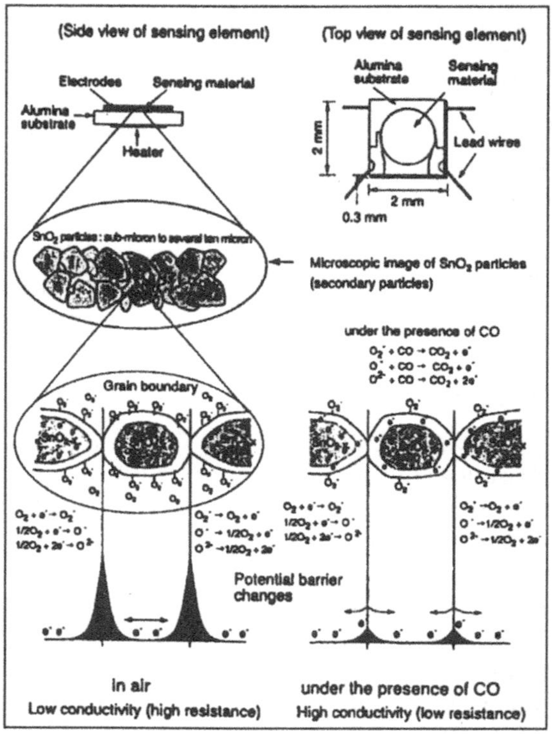

Fig 5. Construction and principal of semiconductor gassensors

5. Conclusion

Meanwhile today spreaded systems with gassensors for the regulation of the air-quality in a car are allready commercialized, future developments are heading to the target of an integration of existing systems, so that different factors of the air-quality can be detected separatly, and in a common control-unit processed simultaneously. With it's output signals, several different actions for re-achieving a convenient air-quality can be initiated, like opening or closing of the damper-flap, switching of smell-filters, de-humidifier or other climate-conditioners.

List of Contact Addresses

Oliver Berberig
ZEXEL Corporation
Ihmepassage 4
D-30449 Hannover
Phone: +49/ 511/ 924 51 17

Arno Bergmann
VDE Prüf- und Zertifizierungsinstitut
Merianstr. 28
D-63069 Offenbach
Phone: +49/ 69/ 83 06 - 6 32

Georg Bischopink
Robert Bosch GmbH
Tübinger Str. 123
D-72703 Reutlingen
Phone: +49/ 7121/ 35 - 20 64

Markus Böhm
Universität-Gesamthochschule Siegen
Hölderlinstr. 3
D-57068 Siegen
Phone: +49/ 271/ 740 32 94

Gianfranco Burzio
Centro Ricerche FIAT
Strada Torino, 50
I-10043 Orbassano (TO)
Phone: +39/ 11/ 902 30 66

Jean-Frédéric Clerc
CEA - LETI
17, Rue des Martyrs
F-38054 Grenoble Cedex 9
Phone: +33/ 4/ 76 88 - 94 47

Wolfram Geiger
Institut für Mikro- und
Informationstechnik
Wilhelm-Schickard-Str. 10
D-78052 Villingen-Schwenningen
Phone: +49/ 7721/ 943 - 226

Wolfgang Geßner
VDI/VDE Technologiezentrum
Informationstechnik GmbH
Rheinstraße 10B
D 14513 Teltow
Phone +49/ 3328/ 435 - 173

Wolfgang Golderer
Robert Bosch GmbH
Postfach 1342
D-72703 Reutlingen
Phone: +49/ 711/ 811 - 18 32

Thierry Goniva
International Electronics &
Engineering
2b, route de Trèves
L-2632 Luxembourg
Phone: +352/ 42 47 37 - 224

Gerhard Haas
Kayser-Threde GmbH
Perchtinger Str. 3
D-81379 München
Phone: +49/ 89/ 724 95 - 346

Johannes Herrnsdorf
HL-Planartechnik
Hauert 13
D-44227 Dortmund
Phone: +49/ 231/ 97 40 - 0

Thomas Huth-Fehre
Institut of Chemical and
Biochemical Sensor Research
Mendelstr. 7
D-48149 Münster
Phone: +49/ 251/ 980 19 68

Robert Ingenbleek
ZF Friedrichshafen AG
Allmansweilerstr. 25
D-88046 Friedrichshafen
Phone: +49/ 7541/ 77 - 76 24

Daniel Jendritza
Philips GmbH
Kreuzweg 60
D-47809 Krefeld
Phone: +49/ 2151/ 576 - 252

Jean-Michel Karam
TIMA-CMP
46, av. Félix Viallet
F-38031 Grenoble Cedex
Phone: +33/ 4/ 76 57 46 20

Joachim Kloeser
Fraunhofer-IZM
Gustav-Meyer-Allee 25
D-13355 Berlin
Phone: +49/ 30/ 464 03 - 155

Gerhard Krötz
Daimler Benz Aerospace AG
Postfach 80 04 65
D-81663 München
Phone: +49/ 89/ 607 255 44

Anton Leidl
Fraunhofer-IFT
Hansastr. 27 d
D-80686 München
Phone: +49/ 89/ 547 59 - 227

Henrik Lind
AB Volvo
Dept. 06950, CTP
S-41288 Göteborg
Phone: +46/ 31/ 772 40 78

Takeshi Matsui
Denso Corporation
1 - 1 Showa-cho
Kariya-shi
J-448 Aichi-Ken
Phone: +81/ 566/ 25 69 85

Guy Meynants
IMEC
Kapeldreef 75
B-3001 Leuven
Phone: +32/ 16/ 28 14 92

Nobuaki Murakami
Murakami Research Services
Felderhof 186
D-40880 Ratingen
Phone: +49/ 2102/ 44 47 11

Alfred Pohl
Technische Universität Wien
Gußhausstr. 27-29
A-1040 Wien
Phone: +43/ 1/ 588 01 - 37 16

Uwe Regensburger
Daimler Benz AG
D-70546 Stuttgart
Phone: +49/ 711/ 17 - 418 84

David B. Rich
Delphi Delco Electronics Systems
One Corporate Center M. S Fab A
P.O. Box 9005
Kokomo, Indiana 46901
USA
Phone: +1/ 765/ 451 - 77 16

Detlef Egebert Ricken
VDI/VDE Technologiezentrum
Informationstechnik GmbH
Rheinstraße 10B
D 14513 Teltow
Phone +49/ 3328/ 435 - 242

Dieter Schödlbauer
ruf electronics GmbH
Bahnhofstr. 26-28
D-85635 Höhenkirchen
Phone: +49/ 8102/ 781 - 441

Ulrich Seger
IMS Chips
Allmandring 30a
D-70569 Stuttgart
Phone: +49/ 711/ 685 - 31 00

Peter Seitz
Centre Suisse D'Electronique et
de Microtechnique SA
Badenerstr. 569

Hideyuki Tamura
Nissan Motor Co., Ltd.
6-1, Daikoku-cho. Tsurumi-ku
J-230 Yokohama-City
Phone: +81/ 45/ 505 - 84 36

Christian Voit
Unitronic GmbH
Mündelheimer Weg 9
D-40472 Düsseldorf
Phone: +49/ 211/ 95 11 - 171

Beatrice Wenk
Fraunhofer- ISiT
Fraunhoferstr. 1
D-25524 Itzehoe
Phone: +49/ 4821/ 17 - 45 08

Michael Wollitzer
Daimler Benz AG
Wilhelm-Runge-Str. 11
D-89081 Ulm
Phone: +49/ 731/ 505 - 20 48

Innovation Relay Centres Supporting the Conference

France

Chambre Régionale de Commerce
et d`Industrie de Bourgogne, ARIST
Dijon

Spain

Sociedad para la Promocion y
Reconversion Industrial, SPRI
Bilbao

Sweden

IVF - The Swedish Institute of Production
Research Engineering
Göteborg

United Kingdom

South West
University Gate
Bristol

Welsh Development Agency, WDA
Cardiff